Chemistry and Technology of Surfactants

Chemistry and Technology of Surfactants

Edited by

Richard J. Farn
Consultant and former Director
of the British Association
for Chemical Specialities

Editorial Offices:
Blackwell Publishing Ltd, 9600 Garsington Road, Oxford OX4 2DQ, UK
Tel: +44 (0) 1865 776868
Blackwell Publishing Professional, 2121 State Avenue, Ames, Iowa 50014-8300, USA
Tel: +1 515 292 0140
Blackwell Publishing Asia Pty, 550 Swanston Street, Carlton, Victoria 3053, Australia
Tel: +61 (0)3 8359 1011

First published 2006 by Blackwell Publishing Ltd

ISBN-13: 978-14051-2696-0
ISBN-10: 1-4051-2696-5

Library of Congress Cataloging-in-Publication Data

Chemistry and technology of surfactants / edited by Richard J. Farn.
p. cm.
Includes bibliographical references and index.
ISBN-13: 978-1-4051-2696-0 (acid-free paper)
ISBN-10: 1-4051-2696-5 (acid-free paper) 1. Surface chemistry. 2. Surface active agents.
I. Farn, Richard J.
QD506.C446 2006
541′.33–dc22 2005017738

A catalogue record for this title is available from the British Library

Set in 10/12pt Minion & Optima
by TechBooks, New Delhi, India
Printed and bound in India
by Replika Press Pvt, Ltd, Kundli

The publisher's policy is to use permanent paper from mills that operate a sustainable forestry policy, and which has been manufactured from pulp processed using acid-free and elementary chlorine-free practices. Furthermore, the publisher ensures that the text paper and cover board used have met acceptable environmental accreditation standards.

For further information on Blackwell Publishing, visit our Web site:
www.blackwellpublishing.com

Contents

Contributors xi

Preface xiii

Glossary xv

1 **What Are Surfactants?** 1
- 1.1 History and applications of surfactants *David R. Karsa* 1
  - 1.1.1 Introduction 1
  - 1.1.2 Properties and other criteria influencing surfactant choice 3
  - 1.1.3 Surfactant applications 5
  - 1.1.4 Conclusion 7
- Appendix: Application guide 8
- 1.2 Surfactant market overview: importance in different industries *Joel Houston* 14
  - 1.2.1 Introduction 14
  - 1.2.2 Consumer 14
  - 1.2.3 Industrial 21

2 **The Basic Theory** *Hatice Gecol* 24
- 2.1 Molecular structure of surfactants 24
- 2.2 Surface activity 26
  - 2.2.1 Surface tension 26
  - 2.2.2 Interfacial tension 28
  - 2.2.3 Surface and interfacial tension reduction 28
  - 2.2.4 Test methods for surface and interfacial tension measurements 31
- 2.3 Self-assembled surfactant aggregates 32
  - 2.3.1 Micelles and critical micelle concentration 33
  - 2.3.2 Aggregate structures and shapes 35
- 2.4 Adsorption of surfactants at surfaces 38
  - 2.4.1 Adsorption at liquid–gas and liquid–liquid interfaces 38
  - 2.4.2 Adsorption at liquid–solid interface 39

Acknowledgement 43
References 43

3 **Applied Theory of Surfactants** *Peter Schmiedel and Wolfgang von Rybinski* 46
3.1 Introduction 46
3.2 Detergency 47
3.2.1 Fundamental processes 47
3.2.2 Basic formulae of detergents and cleansers 48
3.2.3 Adsorption at the solid–liquid interface 48
3.2.4 Surface tension and wetting 54
3.2.5 Interplay of surfactants with other detergent ingredients 60
3.3 Phase behaviour of surfactants 62
3.3.1 Introduction 62
3.3.2 Surfactant phases 62
3.3.3 Impact of the phase behaviour on detergency 66
3.4 Emulsions 69
3.4.1 Introduction 69
3.4.2 Emulsion types 70
3.4.3 Breakdown of emulsions 74
3.5 Foaming and defoaming 76
3.5.1 Introduction 76
3.5.2 Stabilising effects in foams 77
3.5.3 Correlation of foamability with interfacial parameters 78
3.5.4 Foam control 81
3.6 Rheology of surfactant solutions 82
3.6.1 Introduction 82
3.6.2 Rheological terms 83
3.6.3 Rheological behaviour of monomeric solutions and non-interacting micelles 83
3.6.4 Entanglement networks of rod-like micelles 84
3.6.5 The rheological behaviour of bilayer phases 86
References 88

4 **Anionic Surfactants** *John Hibbs* 91
4.1 Sulphonates 92
4.1.1 Alkylbenzene sulphonates 93
4.1.2 α-Olefin sulphonates 102
4.1.3 Paraffin sulphonates 104
4.1.4 Sulphonated methyl esters 106
4.1.5 Sulphonated fatty acids 108
4.1.6 Sulphosuccinates 110
4.2 Sulphates 113
4.2.1 Alkyl sulphates 113
4.2.2 Alkyl ether sulphates 118
4.3 Phosphate esters 122

4.4 Carboxylates 124
4.4.1 Soap 124
4.4.2 Ether carboxylates 126
4.4.3 Acyl sarcosinates 127
4.4.4 Alkyl phthalamates 128
4.4.5 Isethionates 129
4.4.6 Taurates 130
References 132

5 **Non-ionic Surfactants** *Paul Hepworth* 133
5.1 Introduction 133
5.2 General alkoxylation reactions 133
5.3 Alkyl phenol ethoxylates 135
5.4 Fatty alcohol ethoxylates 136
5.5 Polyoxethylene esters of fatty acids 139
5.6 Methyl ester ethoxylates 140
5.7 Polyalkylene oxide block co-polymers 141
5.8 Amine ethoxylates 142
5.9 Fatty alkanolamides 143
5.10 Amine oxides 144
5.11 Esters of polyhydric alcohols and fatty acids 145
5.12 Glycol esters 146
5.13 Glycerol esters 146
5.14 Polyglycerol esters 146
5.15 Anhydrohexitol esters 147
5.16 Polyoxyalkylene polyol esters 148
5.17 Alkyl poly glucosides 149
5.18 Gemini surfactants 150
References 151

6 **Other Types of Surfactants** 153
6.1 Cationics *J. Fred Gadberry* 153
6.1.1 Introduction and background 153
6.1.2 Manufacturing processes 153
6.1.3 Applications of cationic surfactants 156
6.1.4 Industrial applications of cationic surfactants 165
References 166
6.2 Amphoteric surfactants *Richard Otterson* 170
6.2.1 Introduction 170
6.2.2 Aminopropionates and Iminodipropionates 170
6.2.3 Imidazoline-based amphoteric surfactants 172
6.2.4 Betaine surfactants 180
6.2.5 Other amphoteric surfactants 184
6.2.6 Summary 185
References 185

6.3 Silicone surfactants *Randal M. Hill* 186
6.3.1 Introduction 186
6.3.2 Structures 187
6.3.3 Synthesis 189
6.3.4 Hydrolytic stability 191
6.3.5 Surface activity 191
6.3.6 Wetting 192
6.3.7 Phase behavior 194
6.3.8 Ternary systems 195
6.3.9 Applications 196
References 199
6.4 Polymerizable surfactants *Guido Bognolo* 204
6.4.1 Introduction 204
6.4.2 Reactive surfactants 204
6.4.3 Emulsion polymerization 221
Acknowledgements 224
References 224
6.5 Fluorinated surfactants *Richard R. Thomas* 227
6.5.1 Introduction 227
6.5.2 Uses 227
6.5.3 Applied theory 228
6.5.4 Environmental considerations 231
6.5.5 Latest developments 231
References 235

7 **Relevant European Legislation** 236
7.1 Biodegradability *Paul J Slater* 236
7.1.1 Biodegradation of surfactants 236
7.1.2 Sewage treatment plants 237
7.1.3 Measurement of biodegradability 238
7.1.4 Legislation 239
7.1.5 Detergents Regulation 243
References 246
7.2 Classification and labelling of surfactants *Richard J Farn* 248
Acknowledgement 248
References 249
7.3 The European Commission's New Chemicals Strategy (REACH) *Philip E. Clark* 250
7.3.1 Introduction 250
7.3.2 History of chemicals legislation 250
7.3.3 The principles behind REACH 251
7.3.4 REACH 251
7.3.5 The impact on the surfactant industry 257

7.3.6 Testing cost 258
7.3.7 Conclusion 258
References 259
7.4 The Biocidal Products Directive *Mike Bernstein* 260
7.4.1 Introduction 260
7.4.2 The Directive 260
7.4.3 Some definitions 260
7.4.4 Requirements and operation 261
7.4.5 Costs 262
7.4.6 Transitional measures 263
7.4.7 Data protection and 'free-riding' 265
7.4.8 Impact 265
7.4.9 Final comment 267
References 268

8 **Relevant Legislation – Australia, Japan and USA** 269
8.1 Relevant legislation – Australia *John Issa* 269
8.1.1 Introduction 269
8.1.2 National Industrial Chemicals Notification and Assessment Scheme 269
8.1.3 Food Standards Australia New Zealand 276
8.1.4 National Drugs and Poisons Scheduling Committee 277
8.1.5 Therapeutic Goods Administration 278
8.1.6 Hazardous substances 278
8.1.7 Dangerous goods 280
8.1.8 Eco labelling in Australia 282
References 283
8.2 Japanese legislation relating to the manufacture and use of surfactants *Yasuyuki Hattori* 284
8.2.1 Chemical substances control law and industrial safety and health law 284
8.2.2 Pollutant release and transfer register system 288
References 293
8.3 Relevant US legislation *Arno Driedger* 294
8.3.1 General 294
8.3.2 TSCA 294
8.3.3 FDCA 295
8.3.4 FIFRA 297
8.3.5 Other pertinent regulations 297
References 298

9 **Surfactant Manufacturers** *Richard J Farn* 300

Index 311

Contributors

M. Bernstein ChemLaw UK, MWB Business Exchange, 494 Midsummer Boulevard, Central Milton Keynes, MK9 2EA, UK

G. Bognolo WSA Associates, Schuttershof 2 B-3070, Everberg, Belgium

P. E. Clark Lakeland Laboratories Limited, Peel Lane, Astley Green, Tyldesley, Manchester M29 7FE, UK

A. Driedger Arno Driedger Consulting, 3131 Wilson Street, Conklin, MI 49403, USA

R. J. Farn Phoenix House, Arkholme, Carnforth, Lancashire LA6 1AX, UK

J. F. Gadberry Surfactants America Research, Akzo Nobel Chemicals – Surface Chemistry, Dobbs Ferry, New York, USA

H. Gecol Chemical Engineering/MS170, University of Nevada Reno, Reno, NV 89557, USA

Y. Hattori Technical Regulatory Affairs Centre, Product Quality Management Division, Kao Corporation, 2-3-1 Bunka Sumida-ku, Tokyo 131-8051, Japan

P. Hepworth Merrington Cottage, Oulston, York YO61 3RA, UK

J. Hibbs Mclntyre Limited, Holywell Green, Halifax West Yourkshire, HX4 9DL. UK

J. Houston R M Hill Dow Corning Corporation, ATVB Materials Science, 2200 W Salzburg Road, Midland, MI 18686, USA

J. Issa Cintox Pty Ltd, 121 Carlton Crescent, PO Box 168, Summer Hill, NSW 2130, Australia

D. R. Karsa TensioMetrics Ltd., 10 Barnfield Road East, Stockport SK3 8TT, UK

R. Otterson McIntyre Group, 24601 Governors Highway, University Park, IL 60466, USA

P. Schmiedel Dept VTR- Physical Chemistry, Henkel KGaA, Henkelstrasse 67, 40191 Dusseldorf, Germany

P. J. Slater Shield Consulting, 17 Oregon Walk, Wokingham, Berkshire RG40 4PG, UK

R. R. Thomas OMNOVA Solutions Inc., 2990 Gilchrist Road, Akron, OH 44305-4418, USA

W. von Rybinski Dept VTR – Physical Chemistry, Henkel KGaA, Henkelstrasse 67, 40191 Dusseldorf, Germany

Preface

This book is designed to give practical help to those involved with the use of surface active agents or surfactants as they are more generally known. It is intended particularly for new graduate and post graduate chemists and chemical engineers at the beginning of their industrial careers and for those who, in later life, become involved with surfactants for the first time. It aims to give practical help to the formulator by providing a straightforward and application led survey of the manufacture, chemistry and uses of surfactants.

Surfactants are not new: the oldest surfactant is soap which dates back well over 2000 years although the modern surfactant industry has developed essentially since the Second World War, utilising the expansion of the petrochemical industry as one of its main sources of raw materials. Chapter 1 covers the development of the industry and elaborates on the importance of surfactants in modern day living and the very many areas where they find application.

Surfactants are generally classified by ionic types which relate to their chemical structure and are described as anionic, non-ionic, cationic and amphoteric. Following descriptions of the theory behind surfactants, each category is considered with a brief summary of methods of manufacture but with the main emphasis on properties and applications.

In choosing surfactants for a given application, it is no longer sufficient to discover a single product or, more generally, a blend which will do the job: one must now take into account relevant legislation which can restrict the use of some materials. In particular, the European Detergents Regulation is now in force requiring, inter alia, ultimate biodegradation or mineralisation of surfactants for certain applications. Other regulations such as the Dangerous Substances and Preparations Directives, the Biocidal Products Directive and the proposed REACH (Registration, Evaluation and Authorisation of Chemicals) legislation are relevant and restrictive throughout the European Community. Other, sometimes similar, legislation is in force throughout different parts of the world and all this is covered towards the end of the book.

Finally, commercial availability at the right price is of major importance and a list of the main manufacturers of the different types of surfactants is included.

Acknowledgements

The Editor would like to thank the authors of each chapter or section for their time and effort in contributing to this book which provides a state-of-the-art review of the surfactant industry. Thanks are also due to their employers, be they companies or universities, for their support and permission to publish.

Richard J. Farn MBE

Glossary

ABA	Branched Alkyl Benzene Sulphonate
ACCC	Australian Competition & Consumer Commission
AD	Alkylolamide
ADG	Australian Dangerous Goods
ADR	International Carriage of Dangerous Goods by Road
AE	Alcohol Ethoxylate
AEEA	Aminoethylethanolamine
AELA	Australian Environmental Labelling Association
AES	Alcohol Ether Sulphate
AFFF	Advance Fire Fighting Foams
AFM	Atomic Force Microscopy
AGES	Alkyl Glyceryl Ether Sulphonates
AGO	Australian Greenhouse Office
AHA	Alpha Hydroxy Acid
AIBN	Azobisisobutyronitrile
AICS	Australian Inventory of Chemical Substances
AISE	Association Internationale de la Savonnerie, de la Detergence et des Produits D'entretien (European Association for Cleaning & Maintenance Products)
AO	Amine Oxide
AP	Alkyl Phosphate
AOS	Alpha Olefin Sulphonate
APE	Alkyl Phenol Ethoxylate
APG	Alkyl Polyglucoside
APODS	Alkyl Diphenyl Oxide Disulphonate
APVMA	Australian Pesticides & Veterinary Medicines Authority
AQIS	Australian Quarantine Inspection Service
AS	Alkyl (alcohol) Sulphate
ASTM	ASTM Standards
BAB	Branched Alkyl Benzene
BGA	Bundesverband des Deutschen Gross und Aussenhandles BV (The Federal Association of German Large & Foreign Trade)

BHA	Butylated Hydroxy Anisole
BHT	Butylated Hydroxy Toluene
BiAS	Bismuth Active Substance
BO	Butylene Oxide
BOD	Biological Oxygen Demand
BPD	Biocidal Products Directive
CAC	Critical Aggregation Concentration
CAHA	Colin A Houston & Associates Inc.
CAS	Chemical Abstracts Service
CAS-RN	CAS Registered Number
CBI	Confidential Business Information
CDEA	Coco Diethanolamide
CDMABr	Tetradecyl Dimethyl Amine Bromide
CDMAO	Tetradecyl Dimethyl Amine Oxide
CED	Cohesive Energy Density
CEFIC	European Chemical Industry Council
CESIO	Committee Europeen des Agents de Surface et Leurs Intermediaires (European Committee of Surfactants and their Intermediates)
CHPS	Chlorhydroxy Propane Sulphonate
CIR	Cosmetic Ingredients Review
CMC	Critical Micelle Concentration
CMEA	Coco Monoethanolamide
CPL	Classification, Packaging & Labelling
CRM	Carcinogenic,Mutagenic & Toxic to Reproduction
CSA	Chlorsulphonic Acid
CTAB	Cetyl Trimethyl Ammonium Bromide
CTFA	Cosmetic, Toiletry & Fragrance Association
CWC	Critical Wetting Concentration
CWM	Coal/Water Mixtures
DAT	Dialkyl Tetralin
DEA	Diethanolamine
DID	Detergent Ingredients Database
DMAPA	Dimethylaminopropylamine
DOC	Dissolved Organic Carbon
DOSS	Dioctyl Sulphosuccinate
DOT	Department of Transport
DOTARS	Department of Transport & Regional Services
DP	Degree of Polymerisation
EC	European Community
ECB	European Chemicals Bureau
EEC	European Economic Community
EINECS	European Inventory of Existing Commercial Chemicals
ELINCS	European List of Notified Chemical Substances
EN	European Normalisation
EO	Ethylene Oxide
EOR	Enhanced Oil Recovery

EP	Emulsion Polymerisation
EPA	Environmental Protection Agency
EU	European Union
FDA	Food & Drug Administration
FDCA	Food, Drug & Cosmetics Act
FIFRA	Federal Insecticide, Fungicide & Rodenticide Act
FSANZ	Food Standards Australia New Zealand
GA	Glucosamine
GC	Gas Chromatography
GHS	Global Harmonisation System
GIC	General Industry Charge
GLP	Good Laboratory Practice
GRAS	Generally Recognised as Safe
H2P	High 2-phenyl
HALS	Hindered Amine Light Stabiliser
HDD	Heavy Duty Detergent
HDL	Heavy Duty Liquid Detergent
HMTA	Hazardous Materials Transportation Act
HPLC	High Performance Liquid Chromatography
HSE	Health & Safety Executive
I & I	Institutional & Industrial
IATA	International Air Transport Association
ICAO	International Civil Aviation Organisation
IMDG	International Maritime Dangerous Goods
INCI	International Nomenclature for Cosmetic Ingredients
ISO	International Organisation for Standardisation
IUCLID	International Uniform Chemical Database
IUPAC	International Union of Pure & Applied Chemistry
L2P	Low 2-phenyl
LAB	Linear Alkyl Benzene
LABS (also LAS)	Linear Alkyl Benzene Sulphonate
LAS (also LABS)	Linear Alkyl Benzene Sulphonate
LDL	Light Duty Liquid Detergent
LRCC	Low Regulatory Concern Chemicals
M14	Sodium Tetradecyl Maleate
MAN	Mutual Acceptance of Notification
MATC	Maximum Allowable Toxic Concentration
MBAS	Methylene Blue Active Substance
MCA	Monochlor Acetic Acid
MES	Methyl Ester Sulphonate
MLABS	Modified LABS
Mn	Number Average Molecular Weight
MSDS	Material Safety Data Sheet
MW	Molecular Weight
NCI	National Chemical Inventories
NDPC	National Drugs & Poisons Scheduling Committee

NDPSC	National Drugs & Poisons Schedule Committee
NICNAS	National Industrial Chemicals Notification & Assessment Scheme
NOHSC	National Occupational Health & Safety Commission
NP	Nonyl Phenol
NRA	National Registration Authority for Agricultural & Veterinary Chemicals
NTA	Nitrilo Triacetic Acid
NTP	National Technology Programme
O/W	Oil in Water Emulsion
OECD	Organisation for Economic Co-operation & Development
OH&S	Occupational Health & Safety
OSHA	Occupational Safety & Health Act
OSPAR	Oslo & Paris Commission
OTC	Over the Counter
PARCOM	Paris Commission
PBT	Persistent, Bioaccumulative & Toxic
PCB	Polychlorinated Biphenyls
PE	Phosphate Ester
PEC	Predicted Environmental Concentration
PEG	Polyethylene Glycol
PEO	Polyethylene Oxide
PFOA	Perfluorooctanoic Acid
PFOS	Perfluorooctyl Sulphate
PLC	Polymers of Low Concern
PMN	Pre Manufacture Notification
PNEC	Predicted No-effect Concentration
PO	Propylene Oxide
POE	Polyoxyethylene
PRTR	Pollutant Release & Transfer Register
PT	Product Type
PTB	Persistent, Bioaccumulative & Toxic
PTC	Phase Transfer Catalyst
PUF	Polyurethane Foam
QAC	Quaternary Ammonium Compound
RAFT	Reversible Addition Fragmentation Chain Transfer
RBS	Rutherford Backscattering Spectroscopy
REACH	Registration, Evaluation & Authorisation of Chemicals
SAS	Paraffin Sulphonate
SCI	Sodium Cocyl Isethionate
SFA	Sulphonated Fatty Acid
SHOP	Shell Higher Olefins Process
SIEF	Substance Information Exchange Forum
SIME	Strategic Information Management Environment
SLS	Sodium Lauryl Sulphate
SME	Sulphonated Methyl Ester
SME	Small & Medium Enterprise
SPE	Silicone Polyethers

SPF	Sun Protection Factor
STPP	Sodium Tripoly Phosphate
SUSDP	Scheduling of Drugs & Poisons
Te	Tonne
TGA	Therapeutic Goods Administration
TGD	Technical Guidance Document
TGO	Therapeutic Goods Order
THF	Tetrahydro Furan
TQM	Total Quality Management
TSCA	Toxic Substances Control Act
TTAB	Trimethyl Ammonium Bromide
UCST	Upper Critical Solution Temperature
UN	United Nations
UNCED	UN Conference on Environment & Development
UOM	Unsulph(on)ated Organic Matter
USP	US Pharmacopoeia
VOC	Volatile Organic Compound
vPvB	Very Persistent, Very Bioaccumulative
W/O	Water in Oil Emulsion
W/O/W	Water in Oil in Water Emulsion
WHO	World Health Organisation

Chapter 1
What Are Surfactants?

1.1 History and Applications of Surfactants

David R. Karsa

1.1.1 Introduction

Surfactants (or 'surface active agents') are organic compounds with at least one lyophilic ('solvent-loving') group and one lyophobic ('solvent-fearing') group in the molecule. If the solvent in which the surfactant is to be used is water or an aqueous solution, then the respective terms 'hydrophilic' and 'hydrophobic' are used. In the simplest terms, a surfactant contains at least one non-polar group and one polar (or ionic) group and is represented in a somewhat stylised form shown in Figure 1.1.

Two phenomena result from these opposing forces within the same molecule: adsorption and aggregation.

For example, in aqueous media, surfactant molecules will migrate to air/water and solid/water interfaces and orientate in such a fashion as to minimise, as much as possible, the contact between their hydrophobic groups and the water. This process is referred to as 'adsorption' and results in a change in the properties at the interface.

Likewise, an alternative way of limiting the contact between the hydrophobic groups and the water is for the surfactant molecules to aggregate in the bulk solution with the hydrophilic 'head groups' orientated towards the aqueous phase. These aggregates of surfactant molecules vary in shape depending on concentration and range in shape from spherical to cylindrical to lamellar (sheets/layers). The aggregation process is called '*micellisation*' and the aggregates are known as '*micelles*'. Micelles begin to form at a distinct and frequently very low concentration known as the '*critical micelle concentration*' or '*CMC*'. Figure 1.2 illustrates the various types of micelle described above.

In simple terms, in aqueous media, micelles result in hydrophobic domains within the solution whereby the surfactant may solubilise or emulsify particular solutes. Hence, surfactants will modify solution properties both within the bulk of the solution and at interfaces.

The hydrophilic portion of a surfactant may carry a negative or positive charge, both positive and negative charges or no charge at all. These are classified respectively as anionic, cationic, amphoteric (or 'zwitterionic') or non-ionic surfactant.

Typical hydrophilic groups are illustrated in Table 1.1.

In the case of the non-ionic polyoxyethylene moiety, n can vary from 1 to >100, affording a broad spectrum of hydrophilicity and, as a consequence, surface active properties. The

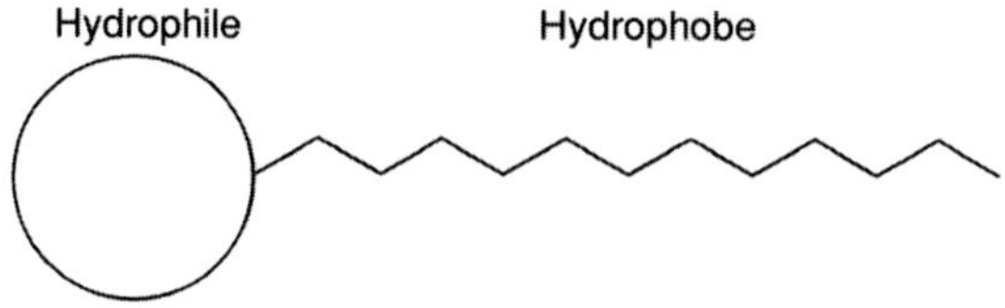

Figure 1.1 Simplified surfactant structure.

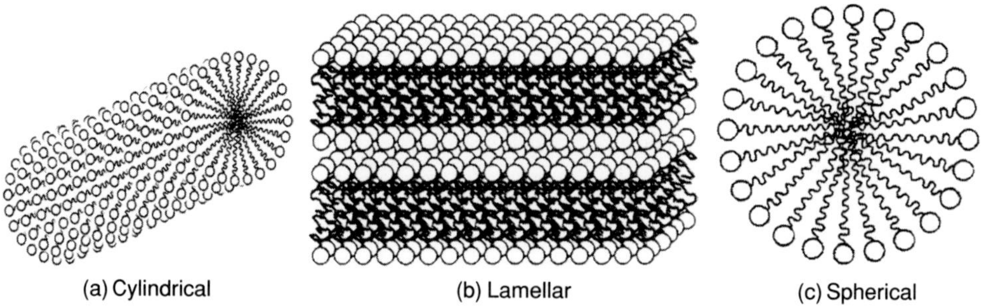

Figure 1.2 Typical micelle configurations.

Table 1.1 Typical hydrophilic groups

Ionic type	Example	Structure
Anionic	Sulphate	$-OSO_2O^-$
	Sulphonate	$-SO_2O^-$
	Ether sulphate	$-(OCH_2CH_2)_nOSO_2O^-$
	Ether phosphate	$-(CH_2CH_2O)_nP(O)O^-$
	Ether carboxylate	$-(CH_2CH_2O)_nCO_2^-$
	Carboxylate	$-C(O)O^-$
Cationic	Primary ammonium	$-N^+H_3$
	Secondary ammonium	$-N^+(R)H_2$
	Tertiary ammonium	$-N^+(R)_2H$
	Quaternary ammonium	$-N^+(R)_3$
Amphoteric	Amine oxide	$-N^+(R)_3O^-$
	Betaine	$-N^+(R)_3(CH_2)_nC(O)O^-$
	Aminocarboxylates	$-N^+H(R)_2(CH_2)_nC(O)O^-$
Non-ionic	Polyoxyethylene (an 'ethoxylate')	$-(OCH_2CH_2)_nOH$
	Acetylenic	$-CH(OH)C{\equiv}CH(OH)-$
	Monoethanolamine	$-NHCH_2CH_2OH$
	Diethanolamine	$-N(CH_2CH_2OH)_2$
	Polyglucoside	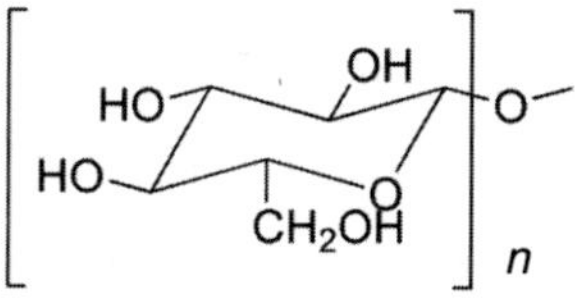

Table 1.2 Typical hydrophobic groups

Group	Example	Structure
Alkylbenzene	Linear dodecyl-benzene	$CH_3(CH_2)_5CH(C_6H_4)(CH_2)_4CH_3$[a]
Linear alkyl[b] (saturated)	n-dodecyl	$CH_3(CH_2)_{10}CH_2-$
Branched alkyl[b] (saturated)	2-ethyl hexyl	$CH_3(CH_2)_3CH-CH_2-(CH_2CH_3)$
Linear alkyl[b] (unsaturated)	Oleyl	(*cis*–)$CH_3(CH_2)_7{=}CH(CH_2)CH_2-$
Alkylphenyl (branched)	Nonylphenyl	$C_9H_{19(\text{branched isomers})}C_6H_4-$
Polyoxypropylene		$-[OCH_2CH(CH_3)]_n-$
Polysiloxane		$(CH_3)_3Si[OSi(CH_3)]_nOSi(CH_3)_3$ |

[a] Alkylbenzene has a linear alkyl chain with, in the case of dodecyl, the phenyl group distributed between the second and sixth positions on the aliphatic chain. The C_6 isomer is illustrated above.
[b] Alkyl groups, whether linear, branched/saturated or unsaturated, are usually within the C_8 to C_{18} chain length range.

so-called polyglycosides have only a low 'degree of polymerisation' (m) and therefore do not have a broad spectrum of properties.

Likewise there are many types of hydrophobe to choose from and Table 1.2 illustrates some of the common commercially available ones.

Even from the few examples given in Tables 1.1 and 1.2, there can be clearly very many combinations of hydrophobe and hydrophile, which afford a spectrum of surface active properties.

1.1.2 Properties and other criteria influencing surfactant choice

The principle 'surface active properties' exhibited by surfactants are

- Wetting
- Foaming/defoaming
- Emulsification/demulsification (both macro- and micro-emulsions)
- Dispersion/aggregation of solids
- Solubility and solubilisation (hydrotropic properties)
- Adsorption
- Micellisation
- Detergency (which is a complex combination of several of these properties)
- Synergistic interactions with other surfactants

Many surfactants possess a combination of these properties.

In addition, depending on the chemical composition of a particular surfactant, some products may possess important ancillary properties including

- Corrosion inhibition
- Substantivity to fibres and surfaces

Table 1.3 Additional criteria to be met in specific applications

Application	Criteria
Domestic, institutional and industrial cleaning products	Surfactants must be biodegradable
Toiletry and personal care products	Surfactants must be biodegradable Low skin and eye irritation Low oral toxicity
Crop protection formulations used in agriculture	Compliance with EPA regulations (not mandatory/customer requirement) Low phyto toxicity Low aquatic toxicity
Oil field chemicals (off shore) and oil spill chemicals	Must meet marine aquatic toxicity requirements in force in that location
Food grade emulsifiers	Must meet vigorous food additive standards for toxicity, etc.
Emulsion polymers for coatings, inks and adhesives	Emulsifier must comply with FDA or BGA regulations for some applications, e.g. direct/indirect food contact

- Biocidal properties
- Lubricity
- Stability in highly acidic or alkaline media
- Viscosity modification

Thus, by defining the properties required to meet a specific application need, the choice of surfactant is narrowed down. It is further reduced by other criteria dictated by the end use, often underpinned by regulations or directives.

Table 1.3 illustrates a few application areas where other parameters have to be met.

Likewise other regional criteria may also come into play, e.g.

- Use of a surfactant may be banned in defined applications in one region of the world, such as the EU and not elsewhere.
- The surfactant must be listed on the regional inventory of approved chemicals, e.g. EINECS in Europe, TSCA in the United States, etc., to be used in that region.
- In cleaning applications and in uses resulting in discharges of effluent to the environment, the surfactant will be required to meet biodegradability criteria but test methods and biodegradability 'pass levels' vary worldwide.
- Customers may insist on compliance with specific regulations, usually governing permitted use levels (based on toxicological data), even though there is no legal requirement as such (e.g. compliance with FDA, EPA, BGA regulations).
- Specific by-products in some surfactants may give rise to concern on toxicological grounds and permitted by-product levels and/or use of levels of that surfactant may be restricted in particular formulations, e.g. nitrosamine levels in diethanolamides when used in personal care formulations.

1.1.3 Surfactant applications

The oldest surfactant is soap, which may be traced back to the ancient Egyptians and beyond. Synthetic surfactants had been produced in the first half of the 20th century but it was only after World War II, with the development of the modern petrochemical industry, that alternative feedstocks to oleochemicals became readily available. Hence chloroparaffins and/or alphaolefins and benzene were used to produce alkylbenzene (or 'alkylate'), processes were developed to produce a range of synthetic fatty alcohols and alkylene oxide chemistry resulted in ethylene oxide and propylene oxide building blocks becoming readily available.

Figure 1.3 illustrates the use of fatty alcohols as a surfactant feedstock. Derived from either oleochemical or petrochemical sources, they may be needed to produce several families of both non-ionic and anionic surfactants.

Coconut oil and tallow have been traditional oleochemical raw materials for many years. However, the significant increase in the production of palm oil over recent decades has had a marked influence on the availability of such feedstocks.

Methyl fatty esters, derived from oils and fats or fatty acids, are another key raw material for surfactant production and this is illustrated in Figure 1.4.

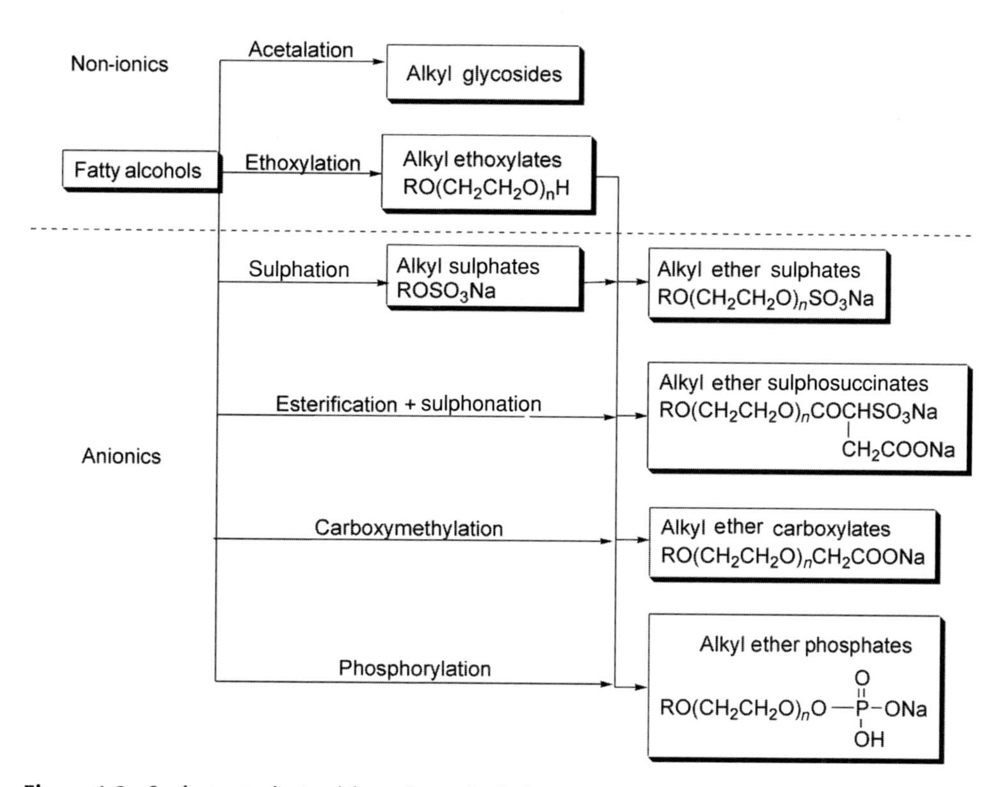

Figure 1.3 Surfactants derived from fatty alcohols.

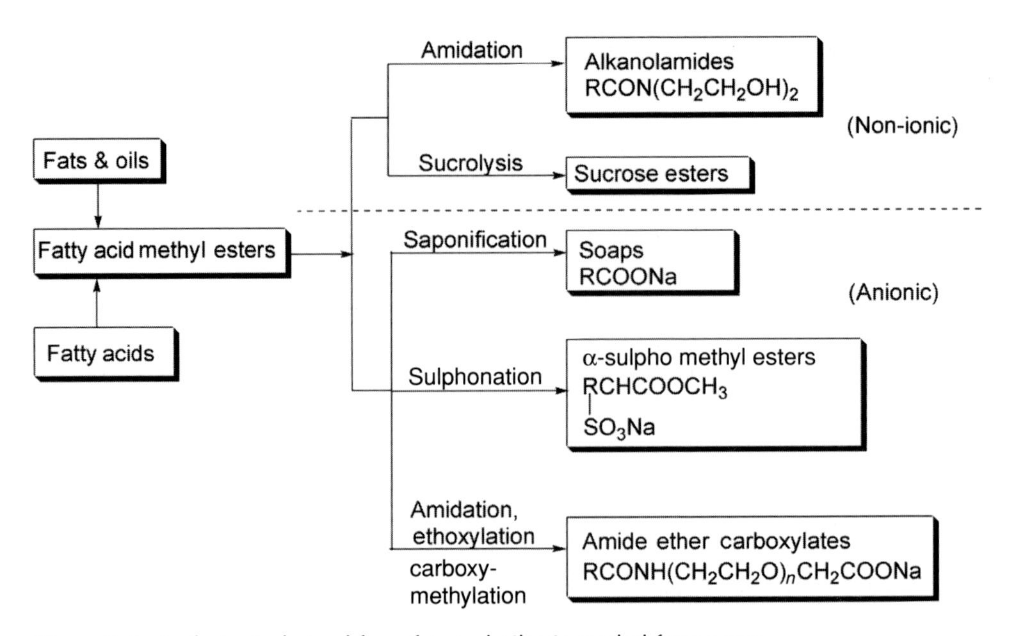

Figure 1.4 Surfactants derived from fats and oils via methyl fatty esters.

On a global basis, the 11 million tonnes or so of surfactants produced each year (excluding soap) utilises approximately equal volumes of oleochemical and petrochemical feedstocks. However, it is interesting to note that, in the case of fatty alcohols, the balance has changed over the last 20 years (Table 1.4).

In the mid-1990s, synthetic surfactant production finally overtook soap production, both of which were running at approximately 9 million tonnes per annum.

Approximately 60% of all surfactants are used in detergents and cleaning products, ranging from household detergents and cleaners to personal care and toiletry products and a range of specialised hygiene products used in institutional and industrial applications. The other 40% finds application in a broad spectrum of agrochemical and industrial uses where 'detergency' is not required. The appendix attempts to illustrate many of these applications.

If Western Europe is taken as an example, applications may be illustrated by Table 1.5.

Table 1.4 Fatty alcohols; natural versus synthetic

Year	K tonnes	% synthetic	% natural
1980	720	60	40
1990	1300	53	47
2000	1680	40	60

Table 1.5 Surfactant usage W. Europe 1990/2000 (K tonnes)

Application	1990	2000
Household detergents	1033	1300
Toiletries/personal care	95	110
Industrial and institutional cleaning	160	210
Textiles	133	140
Pulp and paper	233	250
Construction	120	140
Others (including polymers, paints and coating, leather, oil field chemicals, agro-chemical formulations, etc.)	152	190
	1926	2.340

In 2003, the household and industrial cleaning market in Europe had a value of nearly 30 billion euro the distribution of which, in value terms (%), may be described as

- Household laundry products 41%
- Industrial and institutional cleaners 18%
- Hard surface household cleaners 11%
- Dishwash household products 11%
- Domestic maintenance products 10%
- Soaps 5%
- Domestic bleach products 4%

The industrial and institutional sector (value worth 5.7 billion euro) may be further broken down into

- Kitchen and catering 30%
- General surfaces 24%
- Industrial hygiene 17%
- Laundry 15%
- Others 14%

The source is AISE internal data/AISE collaboration with A.C. Nielsen.

1.1.4 Conclusion

The previous sections are an attempt to illustrate the diversity of surfactants, their many properties and the factors influencing their selection for a specific application. Practically everything that has an impact on our everyday life has a connection with surfactants, whether from a detergent and hygiene aspect or their use as process aids in the production of the objects around us.

Their selection for a specific use is not only governed by their intrinsic surface active properties but must also be considered in terms of toxicological, environmental, regulatory and application-specific requirements which may dictate their suitability for purpose.

Appendix

Application guide

The following is far from being a comprehensive list but is designed to illustrate the variety of surfactants and their versatility in a wide range of applications.

Agrochemical formulations		
Emulsifiable concentrates of crop protection chemicals in solvents	Calcium and amine salts of linear or branched alkylaryl sulphonates	Anionic emulsifiers for solvent based concentrates
	Alkylphenol ethoxylates Fatty alcohol ethoxylates Castor oil ethoxylates Ethylene oxide–propylene oxide co-polymers Polyarylphenol ethoxylates Alkylphenol ethoxylate – formaldehyde condensates	Non-ionic emulsifiers
	Several of above products	Anionic/non-ionic blended emulsifiers
Aqueous flowables (or 'suspension concentrates')	Phosphate esters Sulphonated fatty acids Ethylene oxide–propylene oxide co-polymers Various anionic/non-ionic blends	Suspending and dispersing agents
Water dispersible granules	Naphthalene sulphonic acid – formaldehyde condensates and derivatives	Dispersing agents
	Alkylphenol ethoxylate – formaldehyde condensates Sulphosuccinates	Wetting agents
Wettable powders	Sodium di-alkyl naphthalene Sulphonates Fatty alcohol ethoxylates	Wetting agents
	Naphthalene sulphonic acid – formaldehyde condensates	Dispersants
Adjuvants/activity enhancers	Fatty amine ethoxylates and other non-ionic ethoxylates	Wetting, penetration aids
Tank additives	Quaternary ammonium compounds Betaines	Dispersants/emulsifiers

Civil engineering		
Bitumen additives to give wet adhesion to road aggregates	Quaternary ammonium compounds Imidazoline derivatives Amine ethoxylates	Wetting agents
Foaming of urea – formaldehyde resins for cavity wall insulation	Alkylbenzene sulphonic acids and salts	Foaming agents
Lightweight concrete and cement and production of gypsum plasterboard	Alcohol ether sulphates Betaines Alkylphenol ether sulphates	Foaming agents

Cosmetics and toiletries		
Shampoos, bubble baths and shower gels	Alcohol and alcohol ether sulphates Monoester sulphosuccinates	Detergency/foaming
	Alkanolamides	Foam booster
	Betaines and amido-betaines	Foam booster
Cosmetic and pharmaceutical creams and lotions	Polyglycol esters Long chain fatty alcohol ethoxylates Sorbitan esters and ethoxylates	Emulsifiers
Perfume solubilisers/ emulsifiers for essential oils	Fatty alcohol ethoxylates Polyglycol esters	Solubilisers/emulsifiers

Detergents		
Powder and liquid detergent bases and concentrates	Alkyl benzene sulphonates Methyl esters of sulphonated fatty acids Alpha-olefin sulphonates	Detergency/some foaming
	Alkyl ether sulphates	Secondary surfactants
	Fatty alcohol ethoxylates Alkanolamides	Detergency/foam refinement
Dishwashing liquids	Alkyl benzene sulphonates Alcohol ether sulphates Fatty alcohol ethoxylates	Detergency/foaming
Machine dishwashing Powders, tablets, rinse aids, bottle washing, 'cleaning-in-place"	Fatty alcohol ethoxylates End-blocked fatty alcohol ethoxylates Amine-derived ethylene oxide – propylene oxide co-polymers	Low foam wetters in soil conditions/defoamers/ good rinsibility
Hard surface and other industrial cleaners	Fatty alcohol ethoxylates Alkyl and alkyl ether sulphates Phosphate esters Cocoimino dipropionates	Detergency/solubilisation
Sanitisers	Quaternary ammonium compounds	Biocidal activity

Food industry (other than industrial hygiene)		
Food grade emulsifiers	Wide range of surfactant esters	Emulsifiers
Production of sugar from sugar beet	PEG esters Monoesters of fatty acids Polypropylene glycols	Defoamers

Household products		
Carpet shampoos and upholstery cleaners	Diester sulphosuccinates Fatty alcohol sulphates	Detergents/wetting agents and foamers
Rinse aids	Fatty alcohol EO/PO co-polymers Fatty alcohol EO/BO co-polymers	Low foam wetting agents affording good rinse properties
Toilet blocks	Fatty alcohol long chain ethoxylates Alkyl benzene sulphonate powder	Solid, high foamers
Acid toilet cleaners	Fatty amine ethoxylates Fatty alcohol EO/PO co-polymers	Wetting agents with acid stability
Hard surface cleaners	Alkyl benzene sulphonates Alkanolamides Fatty alcohol ethoxylates Potassium oleic acid sulphonate Shorter chain alcohol ether sulphates	Detergency high/low foaming

Miscellaneous industrial applications		
Bottle washing in dairies and breweries	Fatty alcohol alkoxylates End-blocked fatty alcohol ethoxylates Phosphate esters	Low foam wetting agents with good rinse off properties Hydrotrope
Dry cleaning	Dialkyl sulphosuccinates Phosphate esters	Wetting agents/ emulsifiers used in solvent conditions
Dairy and brewery sanitisers ('iodophors')	Fatty alcohol ethoxylates and alkoxylates Phosphate esters	Complex with iodine to provide iodine in non-staining form
Emulsifiers for (a) Aliphatic solvents		Emulsifiers
	Fatty alcohol ethoxylates Alkylbenzene sulphonates (calcium salts) Polyglycol ethers	
(b) Aromatic solvents	Fatty alcohol ethoxylates Alkylphenol ethoxylates	

(c) Chlorinated solvents/ many mineral and vegetable oils	Polyglycol ethers Alkanolamides Alkyl ether sulphates Alkylbenzene sulphonates (amine salts)	
(d) Waxes	Alkylphenol ethoxylates Alkanolamides Alkylbenzene sulphonates (amine salts)	
Dispersants for organic and inorganic pigments and minerals		
(a) Organic pigments	Polyaryl and alkylaryl ethoxylates Phosphate esters EO/PO co-polymers	Dispersants
(b) Talc	EO/PO co-polymers Alkylphenol ether sulphates	
(c) Gypsum	Phosphate esters	
(d) Coal–water mixtures (CWM) fuels	Poly substituted ethoxy lated aromatics and their phosphated and sulphated derivatives	

Leather	
Fatty alcohol ethoxylates Sulphated oils Sulphated fatty acids and esters	Wetting agents
Fatty alcohol ethoxylates Alkyl sulphates Alkyl benzene sulphonates	Soaking and de-greasing agents
Poly substituted phenol ethoxylates Fatty alcohol ethoxylates	Dispersants for pigment pastes

Metal and engineering		
Pickling and plating bath	Phosphate esters Fatty acid sulphonates Alkanolamides Complex amphoterics	Wetting agents
Cutting and drilling oils	Alkanolamides Phosphate esters	Wetters and emulsifiers

(*Continued*)

Metal and engineering (*Continued*)		
Solvent degreasers	Fatty alcohol ethoxylates EO/PO co-polymers Amine ethoxylates Polyglycol esters	Emulsifiers
De-watering fluid	Amine ethoxylates Alkanolamides	De-watering
Hydraulic fluids	EO/PO co-polymers	Emulsifiers

Paints, inks, coatings, adhesives		
Flow promoters and viscosity modifiers	Amine ethoxylates Fatty alcohol ethoxylates EO/PO co-polymers	Viscosity modifiers
Resin emulsifiers	Alkylphenol ether sulphates Fatty alcohol ether sulphates Fatty alcohol ethoxylates	Emulsifiers
Pigment and filters	Poly substituted phenol ethoxylates Phosphate esters Short chain amine EO/PO co-polymers Lignin sulphonates	Wetters and dispersants

Paper and pulp		
Pulp production	Polyglycol esters	Low foam wetters
De-inking of wastepaper	Fatty alcohol ethoxylates Sodium/potassium soaps	Wetting/flotation
De-foamers for pulp and paper	EO/PO co-polymers and esters Propoxylates	Components
Wax emulsions for chipboard manufacture	Fatty alcohol ethoxylates Amine ethoxylates	Emulsifiers

Petroleum and oil		
Oil spill chemicals	Polyglycol esters Di-alkyl sulphosuccinates	Dispersants/emulsifiers
Demulsifiers for crude oil recovery	Alkylphenol ethoxylate – formaldehyde condensates	Demulsifiers

Plastics, rubber and resins		
Emulsion polymer Production	Fatty alcohol and alkylphenol ethoxylates	
(a) Primary emulsifiers	Fatty alcohol and nonylphenol ether sulphates	
	Fatty alcohol and nonylphenol ether phosphates	Emulsifiers
	Mono and di-ester phosphates	
	Mono and di-ester sulphosuccinates	
	Oleic acid sulphonates	
	Alkylbenzene sulphonates	
	Alcohol sulphates	
	Lignin sulphonates	
(b) Secondary emulsifiers	Fatty alcohol ethoxylates	Emulsifiers
	EO/PO co-polymers	
Plastisol viscosity modifiers	Fatty alcohol ethoxylates	Viscosity modification
	Sodium dialkyl sulphosuccinates	
Carpet and textile backing latex	Sodium octadecyl sulphosuccinamate	
	Alcohol ether sulphates	Foaming agents
	Alcohol sulphates	
	Ammonium stearate	
	Coco imido dipropionate	
	Fatty alcohol ethoxylates	
	Alkyl sulphates	Wetting agents
	Sodium dialkyl sulpho succinates	
PVC antistatic agents	Quaternary ammonium compounds	Antistatic properties

Textiles and fibres		
Detergents, scouring and wetting agents for wool	Fatty alcohol ethoxylates	Detergency and wetting
	EO/PO co-polymers	
	Alkylbenzene sulphonates	
	Phosphate esters	
	Amphoterics	
Antistatic agents	Polyglycol ethers	
	Alkanolamides	Antistatic properties
	Amine ethoxylates	
	Phosphate esters	
	Quaternary ammonium compounds	
Fibre lubricants	Polyglycol esters	Lubricity
	EO/PO co-polymers	
	Phosphate esters	
Dye levelling and dispersing	Solutions of alcohol, alkylphenol and amine ethoxylates	Wetting/dispersion
Self-emulsifiable oils	Fatty alcohol ethoxylates	Emulsification
	Polyglycol esters	

1.2 Surfactant Market Overview: Importance in Different Industries

Joel Houston

1.2.1 Introduction

Surfactants are a group of chemicals that touch our everyday lives in countless ways. They are present in our food, our drinks, the products that we use to clean ourselves, cars that we drive and clothes that we wear. Surfactants affect us all and enable us to clean, prepare and process countless articles around us. The quality of our lives and our health is related to the availability and safe use of surfactants.

Since the advent of surfactants (other than soap) in the twentieth century, we have come to take surfactants for granted. The use of surfactants has matured but preferences in their applications have evolved. We have come to understand their optimal use in different application areas since their functionality can be made to vary depending on the objectives. Table 1.6 indicates the wide range of applications where surfactants are used today. There are two major areas of use: consumer products and industrial uses. Industrial applications include the use of surfactants as cleaning agents and emulsifiers but their use stands out in the very important area of process aids. In the consumer product sector they are typically viewed as household cleaning agents and personal care products. Within the industrial area there is the industrial and institutional (I&I) cleaning product sector, which mirrors many of the consumer applications. The other industrial use areas comprise product modifiers (concrete, drilling) or aids in manufacturing. Table 1.6 lists most of the applications of surfactants, but since these tend to be ubiquitous, one could never compile a complete list.

Figure 1.5 indicates surfactant consumption by major application area. Surfactants are the backbone of the household laundry industry where they play their most important role, in washing anything from laundry to the kitchen sink. This is also the single largest application for them and, although soap continues to play an important role in many areas, it is not included in Figures 1.5 and 1.6. When used in a product, the formulation level of soap, in laundry bars for instance, is used at a much higher level and including this in consumption figures tends to skew the analysis.

Regionally, the market for surfactants (not including soap) can be broken down into four main regions with Asia accounting for the largest market with a 31% share. In 2003, the global surfactant market had a value estimated to have been worth $14.3 billion.

Without surfactants in cleaning formulations, soils would not be removed with washwater, they would not be emulsified.

1.2.2 Consumer

1.2.2.1 Household

Recent figures have placed the value of the global cleaning product market at about 29 billion euro (U. Lehner, CESIO 2004). Surfactants play a critical role in carrying out domestic cleaning, chief of which is carrying soils away with washwater. Other tasks include

Table 1.6 Surfactant application areas

Consumer products		
Household cleaners	Personal care products	
Automatic dish detergents	Bath and shower products	
Fine fabric detergents	Cream rinse/conditioners	
Hard surface cleaners	Hair preparations	
Heavy duty laundry detergents	Shampoo	
Powders and tablets	Shaving creams	
Liquid	Skin creams and lotions	
Laundry aids	Toothpaste	
Fabric softeners	Others	
Pre-wash	Aftershaves	
Bleach	Denture cleaners	
Light duty dish liquids	Deodorants	
Toilet soaps	Hair sprays	
Scourers	Hair dyes	
Specialty cleaners	Lipstick	
Rug cleaners	Mouthwash	
Oven cleaners	Nail polishes	
Toilet bowl cleaners		
Window cleaners		
Metal polishes		
Commercial/industrial Industrial and institutional cleaners industrial processes		
Car wash products	Agricultural chemicals	
Carpet cleaners	Asphalt	
Commercial dishwashing	Cement	
Dairy and food plant cleaners	Corrosion inhibitors	
Dry cleaning	Dispersants	
Electroplating baths	Food and beverage	
Hard surface cleaners	Leather	
I&I hand cleaners	Metalworking fluids	
I&I laundries	Oilfield chemicals	
Metal cleaners	Oil spill control	
Printed circuit board cleaners	Ore flotation	
Transport vehicle cleaners	Paint	
Aircraft	Paper	
Bus	Petroleum additives	
Truck	Plastics and elastomers	
Marine	Polishes	
Railroad	Slurries	
Agents	Textiles	
Processing	Wallboard	
Foam	Miscellaneous	
Remediation	Adhesives	Mould release agents
	Animal feed	Sugar processing
	Fire extinguishing	Polyurethane foam
	Inks	Soil remediation
	Pharmaceutical	Medical

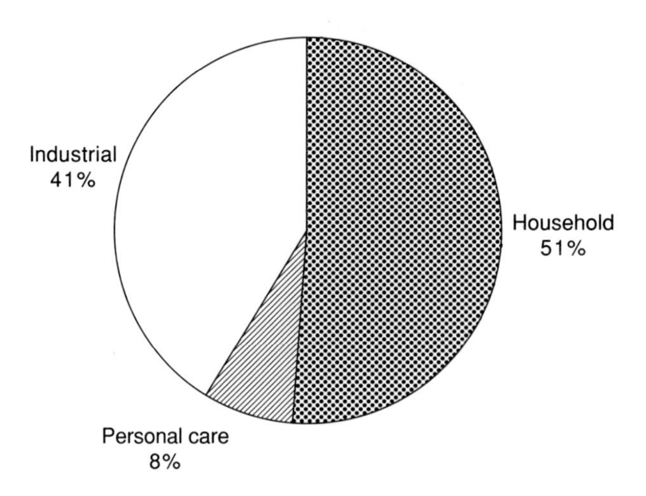

Figure 1.5 World surfactant consumption by major application area, 2000 (total = 10.5 million metric tonnes).

emulsification, solubilisation, thickening and other formulation aids. There is considerable segmentation of products, most of which use surfactants. Figure 1.7 shows consumption of products by segment where surfactants are used in the $14 billion U.S. household market.

The largest outlet for surfactants in household is in the laundry sector, specifically in heavy duty detergents. There is a large amount of commodity surfactant used in household as indicated in Figure 1.8. The U.S. market is unique in that the liquid detergents have become very popular, accounting for over 70% of the heavy duty sector. Europe has experienced strong growth in the use of liquid detergents in recent years and their share has reached the 30% level but penetration in other regions is less dramatic although the trend exists in many areas. The effect of liquid detergents has been to increase the amount of surfactant per washload as detergent builder systems are less effective than with powders. The penetration of liquids drives the high use rate of alcohol ether sulphates in the United States as seen in Figure 1.8.

During the 1970s and early 1980s, many European countries saw rapid expansion of their surfactant markets as laundry soap was displaced by detergents. The use of detergent builders to sequestrate calcium ions in the washwater was very important to the effective functioning

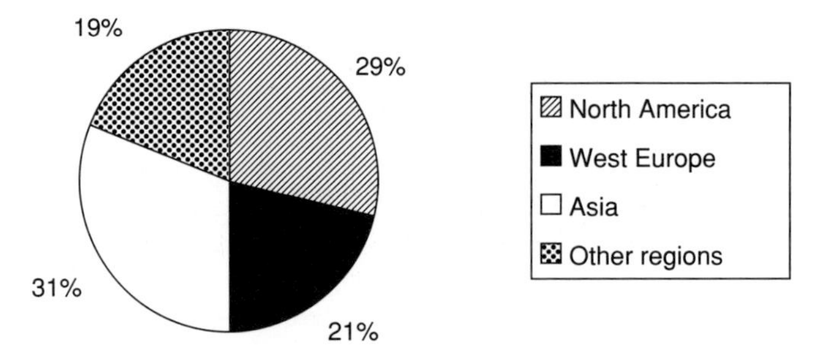

Figure 1.6 Surfactant consumption by region, 2000.

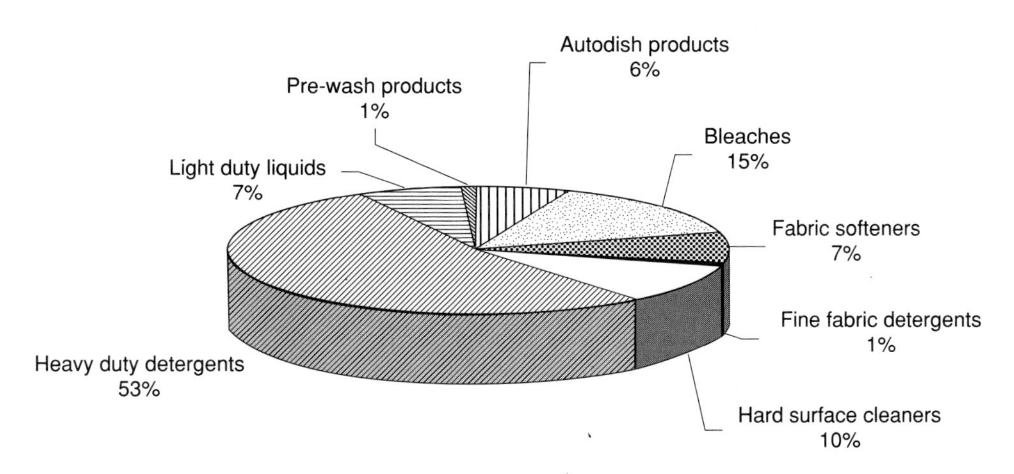

Figure 1.7 Household product consumption, 2000 (total = 2.7 million metric tonnes) in the United States.

of the surfactants in laundry systems. Various phosphates were used and a crisis developed in the 1960s when phosphates in laundry products were singled out as the root cause of eutrophication of lakes and streams. With legislative bans being enacted, the industry sought replacement materials. The only comparable replacement material found was nitrilo triacetic acid (NTA) but it was threatened by a toxicity claim and the material never found widespread use. Formulators were left with the problem of working with inferior builder systems which increased the performance requirements of surfactants. The problem has abated somewhat with the development of increasingly effective detergent polymer materials.

As heavy duty liquids (HDL) have evolved, surfactant selection has reflected the change. In particular, as builder systems for HDL changed, surfactant performance requirements increased considerably. Surfactants needed to be more tolerant to water hardness since builders in most liquids could not achieve the level seen in powders. In the United States, where laundry liquids were becoming increasingly popular for the top loading machines found there, alcohol ether sulphates (AES) and non-ionic use grew rapidly. In the 1990s, in response to the growing interest in surfactants based on renewable materials the use of

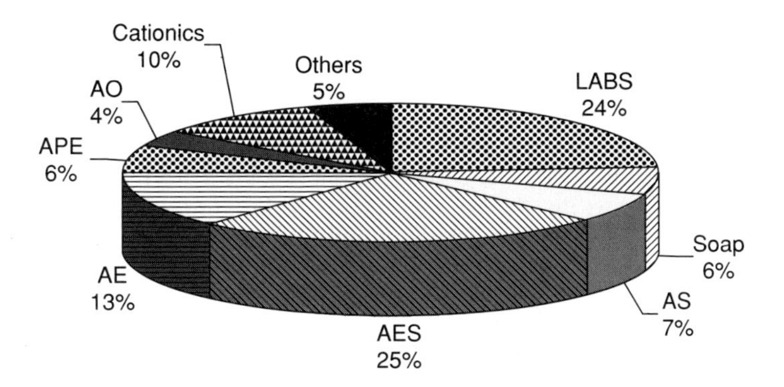

Figure 1.8 Surfactant consumption in household products, 2000 in the United States.

alkyl glucosamides (GA) and alkyl poly glucosides (AGL) took off in laundry applications. Towards the end of the decade, as cost consciousness of consumers took over, the use of these materials tapered off.

The rising energy costs which hit the world in the 1970s pushed consumers to save energy, and washwater temperatures began to decline. This event affected the use of C16–18 alcohol sulphates, which had been a favoured component of detergents for their attractive cost and efficacy in cleaning cotton. But as wash temperatures declined, their use in laundry products disappeared. In some areas, lighter cut oxo alcohol sulphates continued to be in use as the temperature performance profile remained attractive and problems of spray drying also were not severe.

Another impact of rising energy costs was an effort by detergent manufacturers to reduce the energy input in production. The effort evolved and, in 1986, detergents were revolutionised by the introduction of the laundry detergent Attack by Kao Soap in Japan. This product modified the process to rely less on spray drying and more on agglomeration to produce concentrated and highly soluble detergents and this development drew other producers to follow. By working to eliminate or reduce the use of spray drying, producers were challenged to find ways to effectively mix ingredients and form powders that functioned well on storage and in the washer. Various production systems evolved, which agglomerated detergent ingredients in high-speed mixers. Surfactants which had been too temperature sensitive to spray dry well benefited as levels could be increased. Developmental efforts in this direction continue today with recent introduction of powders with greater solubility and faster washing times.

Partly in reaction to greater concentration of detergents, there was a backlash in consumer preferences for the new product forms in some areas. The concentrates were considered too expensive for the size of the package by many consumers who did not appreciate the full benefits in what came to be known as the 'compacts'. There was a reversion to what were considered less costly, lower density products similar to what had been seen in the past. This effect was not universal but it has affected wide areas, especially in Europe. Coinciding with the cost consciousness was a more widespread move towards less costly detergents in a broader geographic region and this was partly brought on by changes in the retail structure and the growing popularity of private label detergents. Private label has not evolved evenly around the world as the perception of these products varies greatly from region to region. But most regions have been hit increasingly since the 1990s with a crisis in product costs brought on by competition from the private label sector and retailers' relentless pursuit of lower prices. Various strategies are being employed to counter it but, generally, it has had the effect of challenging premium quality products and the average level of surfactants seen in the market has declined to some extent. Other formulation shifts have resulted in a stagnation in surfactant demand in this sector. Regions where economic development continues rapidly are still experiencing strong surfactant growth. Generally, products have not evolved greatly in many developing areas and premium formulations make up a small sector in the detergent arena. As a result, demand for surfactant varieties has been low and heavily favours the use of LABS.

In North America, efforts by regulators to drive lower wash water use and temperatures have resulted in energy guidelines for machine manufacturers. As these guidelines developed in the 1990s, so did efforts to find surfactants that would be effective at very low temperatures.

Branched chain surfactants are an important direction for finding surfactants that have greater solubility at low temperatures. An alcohol sulphate based on a branched C17 alcohol was developed by Shell Chemical and patented for laundry use by Procter & Gamble. This material is in use today and offers an important avenue to the future of detergent materials.

Laundering is the world's largest recycling program. Today, the development for surfactants for detergents is focusing on two major avenues. First is the continued development in the production and use of methyl ester sulphonates (MES). These sulphonates are potentially cost advantaged and based on renewable materials such as palm stearin, a by-product of palm olein manufacture. These materials are not completely new, as they were used in the late 1970s and early 1980s in a popular French detergent, 'Le Chat' by Union Generale des Savonneries. Subsequently the major producers have explored their potential and worked to overcome various problems associated with their use. In the late 1980s in Japan, Lion Corp. introduced MES use in detergents there. Development continues and recently in the United States, Huish Detergents has been manufacturing and formulating with them. Many problems have been resolved while others remain only partially improved upon. In future, with developments in products and processes, these products could become mainstream surfactants.

Another material on the horizon is MLABS (modified LABS), a branched chain development for lower wash temperatures. Procter & Gamble patented this material in the late 1990s but it has not found commercial support due to the low margins in the industry and the need for substantial investment to modify the feedstock. Interest in MLABS remains as the modifications have overcome most of the performance deficiencies of linear alkyl benzene sulphonate (LABS).

Dishwashing is the second largest of the household surfactant uses. In hand dishwashing products, there has been a considerable evolution in surfactant selection. First there was a race to achieve greater performance in terms of foaming and soil suspension and, later, the market evolved to milder systems as consumers experienced 'dishpan' hands – or an erythema of the skin. Some of the first dishwashing detergents were more in the line of universal detergents that could also be used on laundry or other hard surfaces. Powders were displaced by liquid product forms. Alkylbenzene sulphonates and alkylphenol ethoxylates were seen in products in the 1940s and foam boosters were added to enhance foaming. Alcohol sulphates and ether sulphates came to dominate formulations at many producers once detergent alcohols became available from petrochemical sources. Some producers developed a preference for secondary alkane sulphonates whilst betaines and alkylpolyglycosides further increased product mildness levels. By the 1990s, Western market use became saturated and more consumers had access to machine dishwashing; hence, the market stagnated. Concentrates became popular and the markets in many countries converted to these. In order to further pursue mildness, producers have been experimenting with various inclusions containing skin care ingredients such as vitamins and oils whilst other aesthetic features such as colour and perfumes have increased in importance. Performance has also been enhanced by the formulation of enzymes and oxygen-based cleaning system additives in some products.

In Western Europe, the evolution of surfactant selection is shown from 1976 to 2000 in Figure 1.9. The use of surfactants followed a classical development. Rapid growth in the early period was partly the result of conversion from powder to liquid products in many countries. There was also a notable development in surfactant preferences as formulations became milder. The rise of betaines is an indication of this development, replacing alkanolamides.

	1976	1980	1984	1990	1995	2000
LABS	68	98	125	131	68	84
AES/AS	33	39	52	86	142	162
SAS	27	51	41	35	38	18
AE	6	7	7	12	13	23
AD	9	19	12	7		
AGL				1	13	6
AO				5	1	4
Betaine				10	9	24
Other/unsp.				2	9	16
	143	214	237	289	293	337

Figure 1.9 Surfactant evolution in light duty dish detergents, 1976–2000 (000 metric tonnes) in Western Europe.

Alcohol ether sulphates (AES) and alcohol sulphates (AS) are shown combined in Figure 1.9 since it is rarely possible to differentiate when products are formulated to contain both materials. Cost competition and producer preferences show their effect on the often erratic consumption of the surfactants over time.

Personal care. Personal hygiene is the first line of defence against the carriers of illness and germs and cleaning reduces the number of these pathogens to be found on our bodies and clothes. Furthermore, personal appearance is key in terms of psychological well-being. It can influence how we feel about ourselves and how we interact with others whilst the 19th century increase in hygienic practices has extended the life span of the average person today. Figure 1.10 indicates that surfactant use in personal care today is relatively evenly divided across the different regions of the world.

A century ago, the personal care market consisted of just toilet soaps and tooth powder but, today, the consumer is offered a plethora of choices, with a product for every conceivable task. To illustrate this point, the *major* categories of personal care products available in the U.S. market are shown in Figure 1.11.

In personal care, washing involving the body, hair and teeth is the fundamental application of surfactants and there are also important roles played by surfactants as emulsifiers in skin care products. In the late 1970s the hand soap market was revolutionised by the introduction of liquid soaps based on 'synthetic' surfactants as opposed to the bar soaps based on natural

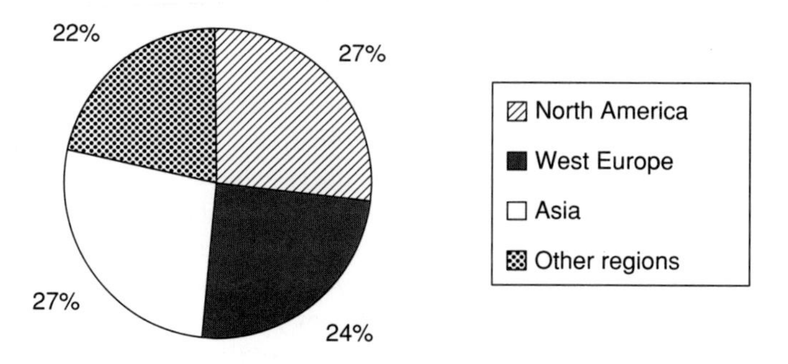

Figure 1.10 Surfactant consumption in personal care by region, 2000.

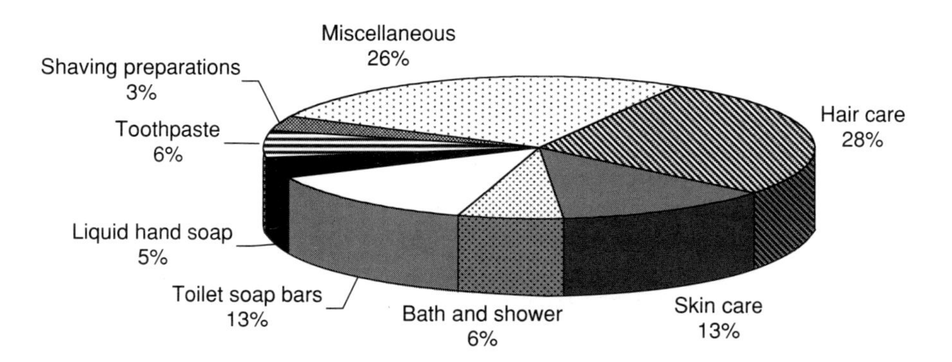

Figure 1.11 Personal care product consumption, 2000 (total = 2.2 million metric tonnes) in the United States.

fats and oils. Early formulations favoured the use of alpha olefin sulphonates and the formulations looked very much like very mild hand dishwashing products. Recent consumption of surfactants in personal care in the United States is shown in Figure 1.12. Although toilet soap is included in categories of Figure 1.11, the consumption of soap as a surfactant is not included in Figure 1.12.

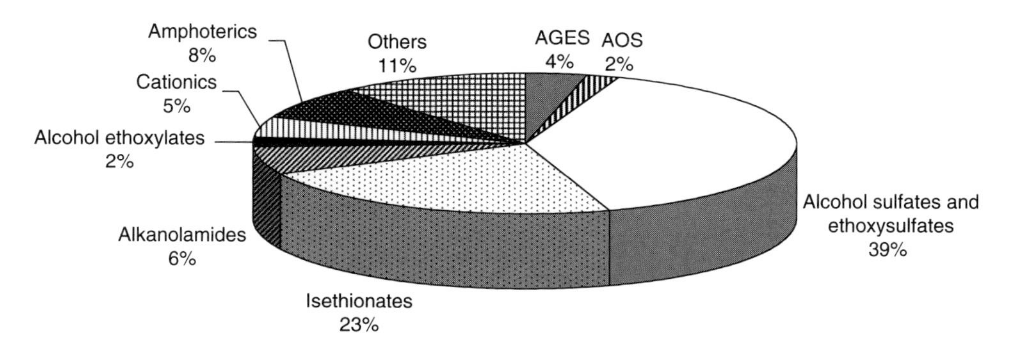

Figure 1.12 U.S. surfactants consumed in personal care products, 2000 (total = 0.2 million metric tonnes).

1.2.3 Industrial

There are a number of industrial process areas where surfactants are used and are important as the means by which the process succeeds. Table 1.6 has presented a fairly comprehensive list of industrial applications for surfactants. Surfactant consumption in the industrial area amounted to more than 4.3 million metric tonnes in 2000 and is shown regionally in Figure 1.13. Regionally, North America is the most important market consuming the largest volume of materials.

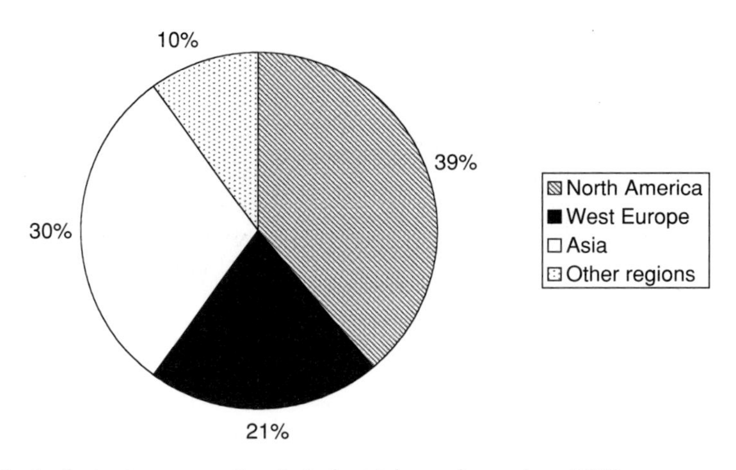

Figure 1.13 Surfactant consumption in industrial uses by region, 2000.

The number of surfactant types consumed in industrial applications is much greater than that used in consumer products and includes a large number of specialty materials. While consumer products employ a defined set of surfactants, one could say that industrial applications employ many of the surfactant types available and, within these types, there is a high degree of specificity towards the identities that function for a given application. For instance, the selection of the hydrophobe or degree of ethoxylation is an example of the characterisations which can critically affect performance. As a result, industrial applications are less apt to see switching of the surfactants in use due to offerings from other suppliers.

The industrial arena is highly dependent today on surfactant functionality for smooth and efficient running. Surfactants can play critical roles here which are often overlooked, especially by the public at large and even by those working in different industries. Typically, surfactants are part of additive packages that only a few of the chemists and the engineers in an industry understand and appreciate but, if one tries to find entrepreneurs trying to gain success in making something new, one will soon find that surfactants have played a role in developing their business.

The markets for surfactants in the industrial field are diverse and few producers try to service large numbers of them. Specialisations have evolved over the years with the development of proprietary approaches to problems. Today much of the technology is mature but new applications do evolve, albeit slowly. More dramatically affecting industrial use of surfactants are government regulations and efforts to evolve technologies into being more efficient. This often results in either surfactant performance being compromised or new approaches being taken. Some of the areas where surfactants are used today are briefly discussed below.

Petroleum and petroleum additives are the largest consumers of surfactants in the industrial field. Many of the materials here are seen only in this vital industry: i.e. alkyl phenates, polybutene succinimides and salycylates, which are used in petroleum products to improve their effectiveness. Surfactants are also regularly used to break the water and oil mixture that is produced from the ground. Without the rapid dewatering step carried out at the well site, crude oil would not be economical to transport and process in the fashion that is common today. There are many other important applications in the establishment and maintenance of oil and gas wells.

Emulsification is a key role that surfactants play in agricultural chemicals. They also benefit formulations by assisting in bringing the active ingredients to adhere to surfaces. Plant surfaces tend to have waxy compounds, which help them shed water. Surfactants assist in preventing the biologically active chemicals from being shed and lost. The major surfactants used in ag chem today include alkylphenol ethoxylates, alkylbenzene sulphonates, alcohol ethoxylates, phosphate esters, ethoxylated amines, lignosulphonates, naphthalene sulphonates and block polymers.

The adhesion of various coatings can depend on the properties of surfactants. Pigment wetting and dispersancy in the formulation of a product is initially of importance but the wetting out of the coated surface is also often dependent on the properties of surfactants. Even highway marking tapes that indicate temporary lanes during construction phases are areas where surfactants play a critical role in the effectiveness of the products.

In metal working, surfactants are used to emulsify lubricants for metal processing. To consider the losses from inadequate processing and finishing of metals would reduce the efficiency of manufacturing and under today's expectations of total quality management, the result would be unacceptable and so costly that it would grind to a halt. The cost of the loss of protection and replacement would add an unacceptable burden which would cripple the world economy.

Emulsion polymerisation used for the production of various elastomers is carried out in a number of processes. Generally surfactants are used in EP to foam the material consistently to give it certain properties. Without surfactants many products would cease to exist. Can one imagine a world without rubber products, or one reduced to hand-made, weak-performing materials?

There are both process aids and cleaning application uses of surfactants in the textiles field. Surfactants are used to help produce and process fibres into cloth which, once produced, needs to be cleaned prior to the dyeing steps. Surfactants also find use as auxiliaries to the dyeing process to help lay down consistent levels of dyes on fibre surfaces.

There is a range of uses in industrial and institutional cleaners, which, in many cases, mirror those in household applications. While maintenance activities are important, surfactants find use in many industries in cleaning the equipment or products which are made. Electronics, food and metal product manufacturing are some examples of this kind of application. Restaurants, hospitals, hotels, as well as factories and offices regularly employ surfactants to maintain the health and safety of people utilising these facilities. A distinct industry has evolved to supply products in these applications where leaders include Ecolab and Johnson-Diversey but many, many more producers are also involved. Surfactant producers supply these companies with bulk surfactants and blends which are compounded into products for discrete applications.

The roles for the applications for surfactants are nearly limitless. These 'humble molecules' are at work around us constantly, enabling the 21st century lifestyle that much of the developed world has come to take for granted and what so many of the less fortunate aspire to. Without surfactants this world would come to a screeching halt, plunged back into a 19th century system of drudgery forced on by the necessity to replace today's chemical energy with human labour.

Chapter 2
The Basic Theory

Hatice Gecol

The words *surf*ace *act*ive *ag*e*nts* are combined to form surfactant. Surfactant is characterised by its tendency to adsorb at surfaces and interfaces. It is an organic compound and widely used in agriculture, pharmaceutical, biotechnology, nanotechnology, cosmetic, detergent, printing, recording, microelectronics, petroleum, mining and other industries. It exists in both natural and synthetic forms. Surfactants such as phospholipids are the main components of the cell membranes and sustain life by organising the order of chemical reactions. For a compound to be a surfactant, it should possess three characteristics: the molecular structure should be composed of polar and non-polar groups, it should exhibit surface activity and it should form self-assembled aggregates (micelles, vesicles, liquid crystalline, etc.) in liquids.

2.1 Molecular structure of surfactants

A surfactant molecule consists of two structures: polar (hydrophilic, lipophobic or oleophobic) head groups and non-polar (hydrophobic, lipophilic or oleophilic) tail groups. The hydrophilic group makes the surfactant soluble in polar solvents such as water. The hydrophobic group makes the surfactant soluble in non-polar solvents and oil. The relative sizes and shapes of the hydrophobic and hydrophilic parts of the surfactant molecule determine many of its properties. Surfactant molecules can have one hydrophilic head and one hydrophobic tail; one hydrophilic head and two hydrophobic; or one hydrophobic tail terminated at both ends by hydrophilic groups (bolaform surfactants or α, ω surfactants); hydrophilic heads of two surfactants are combined with a rigid spacer, which is a linear or ring organic structure (gemini surfactants), and a number of hydrophilic (more than two) hydrophobic groups, with both groups linked in the same molecule by covalent bonds (polymeric surfactants) [1, 2]. Some examples of these structures are shown in Table 2.1.

Hydrophilic molecules are composed of ions (such as sulphonate, sulphate, carboxylate, phosphate and quaternary ammonium), polar groups (such as primary amines, amine oxides, sulphoxides and phosphine oxide) and non-polar groups with electronegative atoms (such as oxygen atom in ethers, aldehydes, amides, esters and ketones and nitrogen atoms in amides, nitroalkanes and amines). These molecules associate with the hydrogen bonding network in water.

Table 2.1 Schematics of molecular surfactant structures and sample surfactants

Schematic of surfactant structure	Sample surfactants
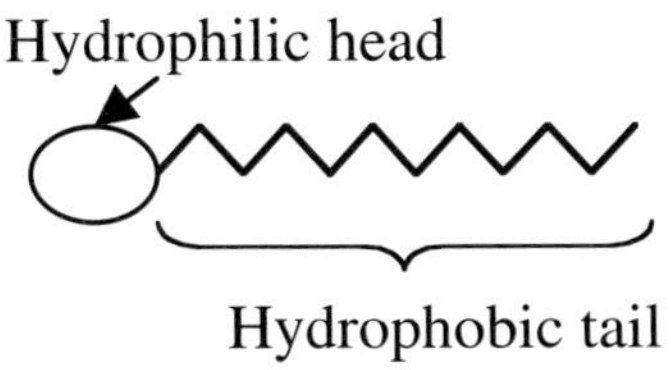 Hydrophilic head Hydrophobic tail	Soap (sodium salt of fatty acids) Alkyltrimethylammonium salts Polyoxyethylene alkyl ether Alkyldimethylamine oxide
	Alkylbenzene sulphonate Phosphatidyl choline (phospholipids) Alkyl secondary amines
	Bolaform quaternary
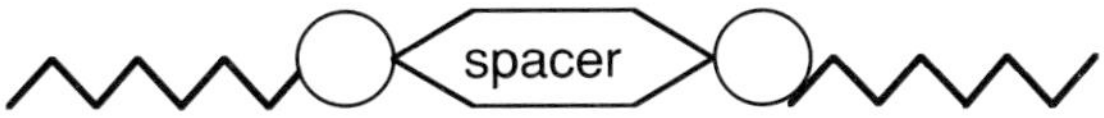 spacer	Gemini phosphate esters
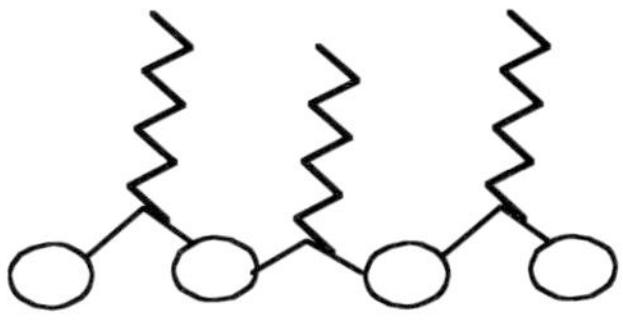	Polymeric alkyl phenol ethoxylates Silicone polymeric surfactants Polyester surfactants

Depending on the hydrophilic groups, surfactants are classified as anionic, cationic, non-ionic or amphoteric. Anionic surfactants dissociate in water into a negatively charged ion and a positively charged ion and the hydrophilic head is negatively charged (anion). Anionic surfactants are the most common and inexpensive surfactant. They are sold as alkali metal salts or ammonium salts and are mainly used in detergent formulations and personal care products. Cationic surfactants also dissociate in water into a negatively charged ion and a positively charged ion and the hydrophilic head is positively charged (cation). Due to the positive charge of the head group, cationic surfactants strongly adsorb onto negatively charged surfaces such as fabric, hair and cell membrane of bacteria. Therefore, they are used as fabric softeners, hair conditioners and antibacterial agents. Non-ionic surfactants, on the other hand, do not dissociate in water and the hydrophilic head has a neutral charge. Non-ionic surfactants are commonly used in the formulation of emulsifier, dispersant and low-temperature detergents. Depending on pH, the hydrophilic head of amphoteric surfactants in water has a positive, negative or both positive and negative charges. They are cations in acidic solutions, anions in alkaline solutions and zwitterions (both ionic groups show equal ionisation and behave uncharged) in an intermediate pH range. They are commonly used in toiletries, baby shampoos, daily cleaners and detergents [1–5]. For further description of surfactant hydrophilic groups and their detail applications, the reader is referred to [1, 2, 5–9].

The major cost of the surfactant comes from the hydrophobic group because the hydrophobic group except for high ethylene oxide non-ionics is the largest part of the surfactant molecule [1]. The hydrophobic group in the surfactant structure is made up of hydrocarbon chains, fluorocarbon chains, combination of fluorocarbon and hydrocarbon chains or silicone chains [1–4]. The majority of the commercially available surfactants (99%) have hydrocarbon chains and are synthesised from natural animal fats, natural vegetable oils or petrochemicals. Hydrocarbons synthesised from natural sources exclusively contain even number of hydrocarbon chains because their structures are built up from ethylene [1, 10]. On the other hand, hydrocarbons derived from petrochemicals contain mixtures of odd and even carbon chains because they are synthesised by cracking higher hydrocarbons. The hydrocarbon chains can be linear or branched and include polycyclic, saturated, unsaturated or polyoxypropylene structures. The linear structure is desirable due to its biodegradability [1, 2]. Fluorocarbon and silicone chain surfactants in water and non-aqueous systems reduce the surface tension lower than the hydrocarbon chain surfactants. Both fluorocarbon and silicone chain surfactants have better thermal and chemical stability than hydrocarbons and provide excellent wetting for low-energy surfaces. Due to their costs, these surfactants are used in limited applications [1].

2.2 Surface activity

For a compound to be qualified as a surfactant, it should also exhibit surface activity. It means that when the compound is added to a liquid at low concentration, it should be able to adsorb on the surface or interface of the system and reduce the surface or interfacial excess free energy. The surface is a boundary between air and liquid and the interface is a boundary between two immiscible phases (liquid–liquid, liquid–solid and solid–solid). Surface activity is achieved when the number of carbon atoms in the hydrophobic tail is higher than 8 [3]. Surfactant activities are at a maximum if the carbon atoms are between 10 and 18 at which level a surfactant has good but limited solubility in water. If the carbon number is less than 8 or more than 18, surfactant properties become minimal. Below 8, a surfactant is very soluble and above 18, it is insoluble. Thus, the solubility and practical surfactant properties are somewhat related [1].

In order to understand how surfactant reduces surface and interfacial tension, one must first need to understand the concept of surface and interfacial tension.

2.2.1 Surface tension

The attractive forces between molecules in the bulk liquid are uniform in all directions (zero net force). However, the molecules at the liquid surface cannot form uniform interaction because the molecules on the gas side are widely spaced and the molecular interactions are mainly between surface molecules and the subsurface liquid molecules (non-zero net force). As a result, the molecules at the liquid surface have greater free potential energies than the molecules in the bulk liquid. This excess free energy per unit area that exists in the surface molecules is defined as surface tension (γ). Surface tension is a thermodynamic property and can be measured under constant temperature and pressure and its value represents

Table 2.2 Surface tension of some liquids and solids and interfacial tension of some immiscible liquids

Substance	Temperature (°C)	Surface tension (dyne cm^{-1})
Liquid		
Diethyl ether	20	17.01
n-octane	20	21.8
Ethyl alcohol	20	22.3
Methyl alcohol	20	22.6
Chloroform	20	27.14
Benzene	20	28.88
Benzene	25	28.22
Phenol	20	40.9
Glycerol	20	63.4
Water	20	72.8
Water	25	72
Mercury	20	476
Aluminum	700	900
Copper	1140	1120
Solid		
Copper	1050	1430–1670
Iron	1400	1670
Liquid–liquid interface		
Diethylether–water	20	10.7
n-octane–water	20	50.8
Benzene–water	20	35
Benzene–water	25	34.71
Chloroform–water		

the amount of minimum work required per unit area to create a greater surface area. In measuring surface tension, one is measuring the free energy per unit area of the surface between liquid and the air (erg cm^{-2} or J m^{-2}). Surface tension is also quantified as the force acting normal to the interface per unit length of the surface at equilibrium (dyne cm^{-1} or mN m^{-1}). Due to this force, liquid surface has a propensity to occupy minimum surface area. Therefore, a liquid drop in a gas phase and bubbles in a liquid phase adopt a spherical shape. The surface tension of some liquid and solids are shown in Table 2.2.

As seen in Table 2.2, surface tension of the substances decreases with increasing temperature because increasing temperature reduces the cohesive energy between molecules. At the critical temperature, surface tension becomes zero. For example, the critical temperature for chloroform is 280°C [11].

The surface tension of water at 20°C (72.8 dyne cm^{-1}) is higher than the surface tension of chloroform (27.14 dyne cm^{-1}) but lower than the surface tension of mercury (476 dyne cm^{-1}). This indicates that the attractive forces between the water molecules are stronger than the attractive forces between the chloroform molecules but weaker than the attractive forces between the mercury molecules.

2.2.2 Interfacial tension

Interfacial tension is the tension that is present at the interface of two immiscible phases and it has the same units as surface tension. The value of interfacial tension generally lies between the surface tension of two immiscible liquids as seen in Table 2.2, although it could also be lower than the surface tension of both liquids (water–diethyl ether). The interfacial tension between phases A and B, γ_{AB}, is expressed by:

$$\gamma_{AB} = \gamma_A + \gamma_B - 2\psi_{AB} \tag{2.1}$$

where γ_A, γ_B and ψ_{AB} are surface tension of A, surface tension of B, and interaction energy between A and B per unit area, respectively.

The value of γ_{AB} also shows how similar the molecules at the interface are. The interfacial tension (γ_{AB}) will be small if the molecules of the two phases are similar (large ψ_{AB}). The greater the similarity, the larger the ψ_{AB} and smaller the γ_{AB}. For example, the interfacial tension between water and ethanol (short chain alcohol) is almost zero because the OH group of ethanol orients itself towards the water phase and interacts with water molecules via hydrogen bonding ($2\psi_{AB} = \gamma_A + \gamma_B$). As a result, the interface disappears ($\gamma_{AB} = 0$) and the two phases form a single phase. If one phase (phase B) is gas, the interface forms at the surface of the condensed phase (phase A) and the interfacial tension is equivalent to the surface tension of the condensed phase ($\gamma_{AB} = \gamma_A$). It means that molecules in the gas phase are widely spaced, so the tension produced by molecular interaction in the gas phase and gas phase–condensed molecules phase is negligible [2–4].

2.2.3 Surface and interfacial tension reduction

Expansion of the interface by unit area can be achieved by the movement of enough molecules from bulk to the interface. However, the potential energy difference between the interface molecules and bulk molecules hinders this move. A minimum amount of work is required to overcome this potential energy difference between the molecules. The interface free energy per unit area or interfacial tension is a measure of this minimum work. When surfactant is added in such a system, surfactant molecules move towards the interface and the hydrophobic tail of the molecule either lies flat on the surface (few surfactant molecules at the interface) or aligns itself to the less polar liquid (sufficient number of surfactant molecules at the interface) while the hydrophilic head orientates itself towards the polar phase. The surfactant molecules destroy the cohesive forces between polar and non-polar molecules and replace the polar and non-polar molecules at the interface. The molecular interaction at the interface occurs between the hydrophilic head of the surfactant and the polar phase molecules and between the hydrophobic tail of surfactant and the non-polar phase molecules. This phenomenon lowers the tension across the interface because the newly developed interactions are stronger than the interaction between the non-polar and polar molecules. If one of the phases is gas or air, tension reduction at the interface is named as surface tension reduction since gas or air molecules are mainly non-polar. Surfactant at low concentration has a tendency to adsorb at the surface or interface and significantly reduce the amount of work required to expand those interfaces. The stronger the tendency, the better the surfactant and the denser the surfactant packing at the interface, the larger the reduction in surface tension.

The surface activity of surfactant is one of the most commonly measured properties and can be quantified by the Gibbs adsorption equation:

$$d\gamma = -\sum_i \Gamma_i d\mu_i \tag{2.2}$$

where $d\gamma$ is the change in surface or interfacial tension of the solution (erg cm^{-2} = dyne cm^{-1} or mJ m^{-2} = mN m^{-1}), Γ_i is the surface excess concentration of solute per unit area of surface or interface (mol cm^{-2} or mmol m^{-2}) and $d\mu_i$ is the change in chemical potential of the solute in the solution. At equilibrium between the interfacial and bulk phase concentrations, $d\mu_i = RTd\ln a_i$ where R is the gas constant (8.314×10^{-7} erg mol^{-1} K^{-1} or 8.314 J mol^{-1} K^{-1}), T is the absolute temperature and a_i is the activity of solute in the solution. Then, the Gibbs adsorption equation becomes:

$$\begin{aligned} d\gamma &= -\sum_i \Gamma_i d\mu_i = -RT\sum_i \Gamma_i d\ln a_i = -RT\sum_i \Gamma_i d\ln(x_i f_i) \\ &= -RT\sum_i \Gamma_i d(\ln x_i + \ln f_i) \end{aligned} \tag{2.3}$$

where x_i is the mole fraction of solute in the bulk phase and f_i is the activity coefficient of solute. For dilute solution containing only one type of non-ionic surfactant (10^{-2} M or less) and containing no other solutes, the activity coefficient of surfactant can be considered as constant and the mole fraction can be replaced by its molar concentration, C. Thus:

$$d\gamma = -RT\Gamma \quad d\ln C \tag{2.4}$$

If the surface or interfacial tension is reduced with the addition of a solute (surfactant), Γ is positive (concentration of the solute at the solution surface is higher than that in the bulk liquid). If the surface tension is elevated with the addition of a solute (such as K_2CO_3), Γ is negative (concentration of the solute at the solution surface is lower than that in the bulk liquid).

For dilute solution (10^{-2} M or less) containing one ionic surfactant that completely dissociates (A^+B^-), the Gibbs adsorption equation is:

$$d\gamma = -RT(\Gamma_A d\ln a_A + \Gamma_B d\ln a_B) \tag{2.5}$$

$\Gamma_A = \Gamma_B$ due to electroneutrality and $a_A = a_B$. Then:

$$d\gamma = -2RT\Gamma \; d\ln C \tag{2.6}$$

For the mixture of non-ionic and ionic surfactants in water with no electrolyte, the coefficient decreases from 2 to 1 with a decrease in the ionic surfactant concentration at the interface [12]. For the ionic surfactant solution in the presence of electrolyte such as NaCl, KCl, NaBr and KBr [13–15] the Gibbs adsorption equation is:

$$d\gamma = -yRT\Gamma \; d\ln C \tag{2.7}$$

where

$$y = 1 + \frac{C}{C + C_{NaCl}}$$

For the ionic surfactant solution in the presence of electrolyte containing non-surfactant counterion, the surface activity can be quantified with eqn 2.4. The more complicated Gibbs

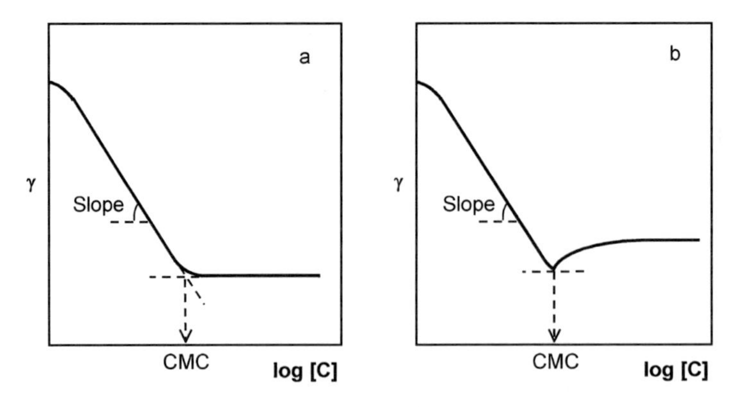

Figure 2.1 Surface tension as a function of bulk surfactant concentration in the aqueous phase (a) for pure surfactant (b) for surfactant containing impurities.

adsorption equations for multi-valent ion systems or the systems containing surfactant concentration more than 10^{-2} M are discussed in [2, 14, 15].

The surface excess concentration (Γ), which is the surface concentration of surfactant, can be determined by the representative Gibbs adsorption equation. The Γ can be obtained from the slope of a plot shown in Figure 2.1 (γ versus $\log[C]$ at constant temperature).

The CMC in the curve represents the critical micelle concentration (CMC) at which the surfactant molecules start forming aggregates known as micelles. Below CMC, surfactant molecules are in monomeric form and the surface or interfacial tension reduces dramatically with the increase of surfactant concentration in the bulk. The slope of the curve below the CMC is constant and reaches its maximum value since the surface or interface is saturated with surfactant monomers [2, 16]. For the dilute solution of non-ionic surfactant:

$$\Gamma = -\frac{1}{RT}\,(\text{slope}/2.303) \tag{2.8}$$

For the dilute solutions of ionic surfactant:

$$\Gamma = -\frac{1}{2RT}\,(\text{slope}/2.303) \tag{2.9}$$

For the ionic surfactant solution in the presence of electrolyte:

$$\Gamma = -\frac{1}{yRT}\,(\text{slope}/2.303) \tag{2.10}$$

The area occupied per surfactant molecule (A) at the surface or interface can be obtained from Γ by using:

$$A = \frac{10^{\alpha}}{N\Gamma} \tag{2.11}$$

where α is 16 for A in square angstrom and Γ in mol cm^{-2}. N is Avogadro's number.

The degree of surfactant concentration at an interface depends on the surfactant structure and the characteristics of the two phases of the interface. Hence, there is no single surfactant that is suitable for all applications. The choice is dependent on the application. Surfactant provides significant surface activity if it has good but limited solubility in the system where it is used. If the solubility of the surfactant is high in the solvent, the surface or interfacial tension does not decrease significantly [2]. A more detailed discussion for the efficiency and effectiveness of a surfactant in reducing surface tension and interfacial tension can be found in Chapter 5 of [2].

2.2.4 Test methods for surface and interfacial tension measurements

A number of methods are available for the measurement of surface and interfacial tension of liquid systems. Surface tension of liquids is determined by static and dynamic surface tension methods. Static surface tension characterises the surface tension of the liquid in equilibrium and the commonly used measurement methods are Du Noüy ring, Wilhelmy plate, spinning drop and pendant drop. Dynamic surface tension determines the surface tension as a function of time and the bubble pressure method is the most common method used for its determination.

For the Du Noüy ring method, a precision-machined platinum/iridium ring (wire diameter being 0.3 mm and the circumference of the ring being 2, 4 or 6 cm), which is suspended from a force measuring balance, is lowered into the liquid placed in a glass container and gradually withdrawn (or the container of liquid is raised and then lowered). As the ring is withdrawn, surface tension causes the liquid to adhere to the underside of the ring. The weight of the ring increases due to the added weight of the adherent liquid and the maximum vertical force increase is a measure of the surface tension. A detailed description of the test procedure can be found in the ISO Standard 304 [17] and ASTM D1331-89 (2001) [18]. This method is not direct and the result should be adjusted by using a correction factor, which accounts for the dimensions of the ring (the perimeter, ring wire thickness and the effect of the opposite inner sides of the ring on the measurement). Furthermore, the ring should be carefully handled and stored to avoid dimensional deformation.

For the Wilhelmy plate method, a thin plate with a perimeter of about 4 cm is lowered to the surface of a liquid and the downward force directed on the plate is measured. Surface tension is the force divided by the perimeter of the plate. For this method to be valid, the liquid should completely wet the plate before the measurement, which means that the contact angle between the plate and the liquid is zero. Furthermore, the position of the plate should be correct, which means that the lower end of the plate is exactly on the same level as the surface of the liquid. Otherwise the buoyancy effect must be calculated separately.

The pendant drop technique measures the shape of a liquid drop suspended from the tip of a capillary needle. The drop is optically observed and the surface tension is calculated from the shape of the drop. This method is not as precise as the force measurement method because it depends on the eye of the operator or the sophistication of detection hardware and analysis software.

The spinning drop method is used to measure low surface tension (μN m^{-1}). In this method a drop of the liquid sample is injected into a thin tube containing another immiscible liquid with higher density. When the tube is spun along its long axis with high speed, the drop is forced to the centre by centrifugal forces and its shape elongates. The interfacial surface tension is calculated from the angular speed of the tube and the shape of the drop.

Dynamic surface tension is the time trajectory of surface tension before equilibrium is reached. Dynamic surface tension tracks the changes during surface formation when surfactants are added. The bubble pressure method is the one most commonly used for the determination of dynamic surface tension. The details of this method are described in ASTM D3825-90 (2000) [19]. In this method a capillary tube is immersed in a sample liquid and a constant flow of gas is maintained through the tube forming bubbles in the sample liquids. The surface tension of the sample is calculated from the pressure difference inside and outside the bubble and the radius of the bubble.

The interfacial tension methods are described in ISO 6889 [20], ISO 9101 [21] and ASTM D1331-89 (2001) [18]. The method described in ISO 6889 is a simple method and applicable for the systems if the interfacial values are between 4 and 50 dyne cm^{-1}, the immiscible liquids are water and organic liquids and the systems contain non-ionic or anionic surfactants but not cationic surfactants. The repeatability is within about 2 dyne cm^{-1}. On the other hand, the drop volume method as described in ISO 9101 can be used for viscous liquids and liquids containing all types of surfactants. This method can measure the interfacial tension as low as 1 dyne cm^{-1} with 0.5 dyne cm^{-1} accuracy. If the interfacial tension is lower than 1 dyne cm^{-1}, the spinning drop will be the suitable method.

2.3 Self-assembled surfactant aggregates

A surfactant at low concentration in aqueous solution exists as monomers (free or unassociated surfactant molecules). These monomers pack together at the interface, form monolayer and contribute to surface and interfacial tension lowering. Although this phenomenon is highly dynamic (surfactant molecules arrive and leave the interface on a very rapid timescale), molecules at the interface interact with the neighbouring molecules very strongly which enables measurement of the rheological properties of the monolayer.

As the surfactant concentration increases, the available area at the surface for surfactant adsorption diminishes and surfactant monomers start accumulating in the solution. However, the hydrophobic tail of the surfactant molecules has extremely small solubility in water and the hydrophilic head has extremely small solubility in non-polar solvents. Hence, the hydrophobic effect will drive surfactant monomers to form self-assembled aggregates above certain aggregate concentration. These aggregates are micelles, vesicles, liquid crystals and reverse micelles and exist in equilibrium with the surfactant monomers. All of these structures are dynamic in nature and surfactant molecules constantly join and leave the microstructure on a timescale of microseconds. As a result, these microstructures have a limited lifetime. For example, the lifetime of spherically shaped micelle is about milliseconds [22]. Furthermore, the difference in energy between various microstructures is small so that the physical forces of the interaction become dominant. As a result, surfactant molecules can be transformed between several types of aggregates by small changes in temperature, concentration, pH or electrolyte strength. Also, the properties of the solution show sharp changes around the critical aggregation concentration. As shown in Figure 2.2, formation of self-assembled aggregates is evidenced by an increase in turbidity and organic dye solubility, a decrease in electrical conductivity (ionic surfactants only) and stability in surface tension, interfacial tension and osmotic pressure around the critical aggregation concentration.

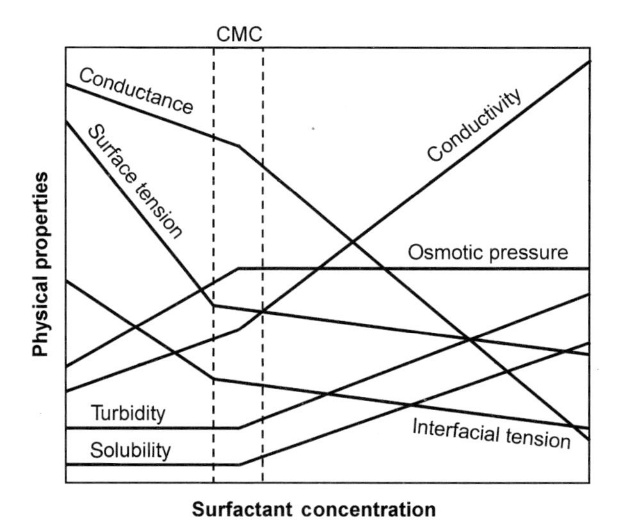

Figure 2.2 Physical properties of aqueous surfactant solution as a function of surfactant concentration.

2.3.1 Micelles and critical micelle concentration

The properties of surfactant at low concentration in water are similar to those of simple electrolytes except that the surface tension decreases sharply with increase in concentration. At a certain concentration, surfactant monomers assemble to form a closed aggregate (micelle) in which the hydrophobic tails are shielded from water while the hydrophilic heads face water. The critical aggregation concentration is called the critical micelle concentration (CMC) when micelles form in an aqueous medium. The CMC is a property of the surfactant. It indicates the point at which monolayer adsorption is complete and the surface active properties are at an optimum. Above the CMC, the concentrations of monomers are nearly constant. Hence, there are no significant changes in the surfactant properties of the solution since the monomers are the cause of the surface activity. Micelles have no surface activity and any increase in the surfactant concentration does not affect the number of monomers in the solution but affects the structure of micelles.

The typical CMC values at room temperature are 10^{-3}–10^{-2} M for anionic surfactants, 10^{-3}–10^{-1} M for amphoteric and cationic surfactants and 10^{-5}–10^{-4} M for non-ionic surfactants. The CMC of several surfactants in aqueous media can be found in [2, 23].

Surfactant structure, temperature, the presence of electrolyte, existence of organic compounds and the presence of a second liquid have an effect on the CMC. The following factors contribute to CMC decrease [1, 2, 24–30]:

(a) an increase in the number of carbon atoms in the hydrophobic tails
(b) the existence of polyoxypropylene group
(c) fluorocarbon structure
(d) an increased degree of binding of the counterions
(e) the addition of electrolyte to ionic surfactants
(f) the existence of polar organic compounds (such as alcohols and amides)
(g) the addition of xylose and fructose

The following factors contribute to CMC increase [1, 2, 24–30]:

(a) branch hydrophobic structure
(b) double bonds between carbon atoms
(c) polar groups (O or OH) in hydrophobic tail
(d) strongly ionised polar groups (sulphates and quaternaries)
(e) hydrophilic groups placed in the surfactant molecule centre
(f) increase in the number of hydrophilic head
(g) trifluoromethyl groups
(h) an increase in the effective size of hydrophilic head
(i) an increase in the pH of weak acids (such as soap)
(j) a decrease in pH from isoelectric region and increase in pH from isoelectric region for amphoteric surfactants (low CMC at the isoelectric region and high CMC outside the isoelectric region)
(k) addition of urea, formamide, and guanidinium salts, dioxane, ethylene glycol and water soluble esters

The CMC decreases with temperature to a minimum and then increases with further increase in temperature. The minimum appears to be around 25°C for ionic surfactants and 50°C for non-ionic surfactants [31, 32].

Several empirical correlations are available for the estimation of CMC values. For straight and saturated single tail ionic surfactants, the CMC can be calculated from [33]:

$$\log \mathrm{CMC} = A - Bn \tag{2.12}$$

where n is the number of carbon atoms in the hydrophobic tail, and A and B are temperature dependent constants for a given type of surfactant. The value of B is around $0.3(= \log 2)$ for the ionic surfactants because the CMC of the ionic surfactants is halved for each carbon atom added to the hydrophobic tail. B value is about $0.5(= 0.5 \log 10)$ for the non-ionic and amphoteric surfactants because the CMC will decrease by a factor of 10 for each of the two methylene groups added to the hydrophobic tail. The values of A and B for some surfactants can be found in [2] and [34].

The effect of electrolyte concentration on the CMC of ionic surfactant is given by [35]:

$$\log \mathrm{CMC} = a - b \log C \tag{2.13}$$

where a and b are constants for a given ionic hydrophilic head at a certain temperature and C is the total counter ion concentration in equivalent per litre. The effect of electrolyte concentration on the CMC of non-ionic and amphoteric surfactants is given by [26, 36]:

$$\log \mathrm{CMC} = x - yC_e \quad C_e < 1 \tag{2.14}$$

where x and y are constants for a given surfactant, electrolyte and temperature, and C_e is the concentration of electrolyte in moles per litre. Further discussion of the theoretical CMC equations can be found in [2, 37–39].

In non-polar solvents, hydrophilic head groups interact due to dipole–dipole attractions and produce aggregates called reverse micelles. With this structure, head groups of surfactant molecules orientate towards the interior and the hydrophobic tails orientate towards the non-polar solvents. In the absence of additives such as water, the aggregation numbers of reverse micelles are small (mostly less than 10). On the other hand, in polar solvents such as glycol, glycerol and formamide, surfactant aggregates are thought to be similar to the aggregates in water since these polar solvents have multiple hydrogen bonding capacity. In general, the CMC of ionic and non-ionic surfactants is higher in nonaqueous solvent than in water [40].

The CMC is a useful tool for the selection of surfactants for specific applications or properties. For example, surfactants with a low CMC are less of an irritant than those with high CMC. The CMC can be determined by measuring the changes in physical properties such as electrical conductivity, turbidity, surface tension, interfacial tension, solubilisation and auto diffusion. Detail evaluation of different methods for the determination of CMC can be found in [23]. Amongst these methods, the surface tension method is most commonly used in practice and ISO Standard 4311 [41] describes this method which is applicable to all types of surfactants in both commercial and pure forms. It requires the strict control of test temperature for precise and reproducible values. According to procedure, 16 surface tension values are measured over the range of surfactant concentrations; among these, six values should be in the region close to CMC. Each value is repeated three times and measurements are made within 3 h of solution preparation. The average of each set of three values is plotted as surface tension versus the log of the surfactant concentration. For a pure surfactant, the break point at the CMC is sharp and well defined as shown in Figure 2.1a. The minimum in the plot as shown in Figure 2.1b indicates the existence of impurities in the surfactant. The concentration at the minimum surface tension gives the CMC value.

Most formulators use more than one surfactant to improve the properties of products. In addition, commercial surfactants are mixtures because they are made from mixed chain length feedstock and they are mixtures of isomers and by-products depending on their synthesis. Purifying the surfactant to a great extent is not economically feasible. Furthermore, a mixture of surfactants was found to perform better than single surfactants in many applications such as emulsion formation, detergents and enhanced oil recovery. The CMC of the mixture is either the intermediate value between the CMC values of each surfactant, less than any of the surfactant CMC (positive synergism) or larger than any of the surfactant CMC (negative synergism). The CMC_M of the mixture, if the mixture contains two surfactants and mixed micelles, is an ideal mixture (activity coefficients of free surfactant monomers for each surfactant type in the mixture are equal to unity):

$$\frac{1}{\text{CMC}_M} = \frac{x_1}{\text{CMC}_1} + \frac{(1 - x_1)}{\text{CMC}_2} \tag{2.15}$$

where x_1 is the mole fraction of surfactant 1 in solution on a surfactant base, and CMC_1 and CMC_2 are the critical micelle concentrations of pure surfactants 1 and 2 respectively. Details and the equations for the nonideal surfactant mixtures can be found in [42].

2.3.2 Aggregate structures and shapes

A theory for the aggregate structure was developed based on the area occupied by the hydrophilic and hydrophobic groups of surfactant [43, 44]. For a stable formation of a surfactant aggregate structure in an aqueous system, the internal part of the aggregate should contain the hydrophobic part of the surfactant molecule while the surface of the aggregate should be made up of the hydrophilic heads. The polar head groups in water, if ionic, will repel each other because of same charge repulsion. The larger the charge, the greater the repulsion and the lower the tendency to form aggregates. The hydrophilic heads have also strong affinity for water and they space out to allow water to solvate the head groups. On the other hand, hydrophobic tails attract one another due to hydrophobic effect. When the surfactant concentration is high enough, the surfactant molecules pack together due to the interaction of the two opposing forces between the surfactant molecules. The shape and the size of the aggregate can be determined by using the surfactant packing parameter which is the ratio of the hydrophobic group area (ν/l_c) to the hydrophilic head area (a_o). The ν and l_c are the volume and length of the hydrophobic tail in the surfactant aggregate:

$$\nu = 27.4 + 26.9n \quad (2.16)$$

$$l_c \leq 1.5 + 1.265n \quad (2.17)$$

where n is the total or one less than the total number of carbon atoms of the hydrophobic tail in the surfactant aggregate, ν is in cubic Angstrom (Å^3) and l_c is in Å. For saturated straight chain, l_c is 80% of the fully extended chain [43]. The structures of surfactant aggregates as a function of surfactant packing parameter and shape are shown in Figure 2.3 [44].

Spherical micelles are formed where the value of surfactant packing parameter is less than 1/3 (single chain surfactants with large head group areas such as anionic surfactants). The spherical aggregates are extremely small and their radius is approximately equal to the maximum stretched out length of the surfactant molecule.

Cylindrical micelles are formed where the surfactant packing parameter is between 1/3 and 1/2 (single chain surfactants with small head group areas such as non-ionic surfactants and ionic surfactants in high salt concentration). Any change in solution properties which causes a reduction in the effective size of hydrophilic head groups will change the aggregate size and shape from spherical to cylindrical form. For example, the addition of electrolyte reduces the effective hydrophilic area of ionic surfactants because the increased counterions reduce the repulsion between ionic polar head groups. Addition of co-surfactant with a smaller head group size also contributes to mixed micelle formation of cylindrical shape. Increasing the temperature reduces the ethoxylated non-ionic head groups. Furthermore, changing the pH changes the degree of protonation of amphoteric surfactants and affects the head size.

Vesicles, liposomes and flexible bilayers are formed where the surfactant packing parameter is between 1/2 and 1 (double chain surfactants with large head group areas such as phospholipids, surfactants with bulky or branched tail groups and the mixture of anionic and cationic surfactants with single chain at nearly equimolar concentration). These types of surfactants cannot pack themselves into a close micelle and they form bilayers (lamellar structure). As the packing parameter approaches unity, the lamella becomes flat and planar (double chain anionic surfactants in high salt concentration). Only the flexible lamellar bilayer bends around and joins in a sphere (vesicle). This structure keeps aqueous solution both inside and outside of the sphere. Liposomes are concentric spheres of vesicles (layers of an onion arrangement): they are more than a micrometer in size and formed by gentle shaking of surfactant in water. The internal bilayer structures of the liposomes are optically active. Hence, they can easily be identified with a polarising light microscope. Vesicles are formed from liposomes by ultrasonication, ultrafiltration or microfluidisation. They are nanometre in size and can only be detected by electron microscopy. Vesicles are used as drug delivery agents, model components for cell membranes and cationic softeners in detergency.

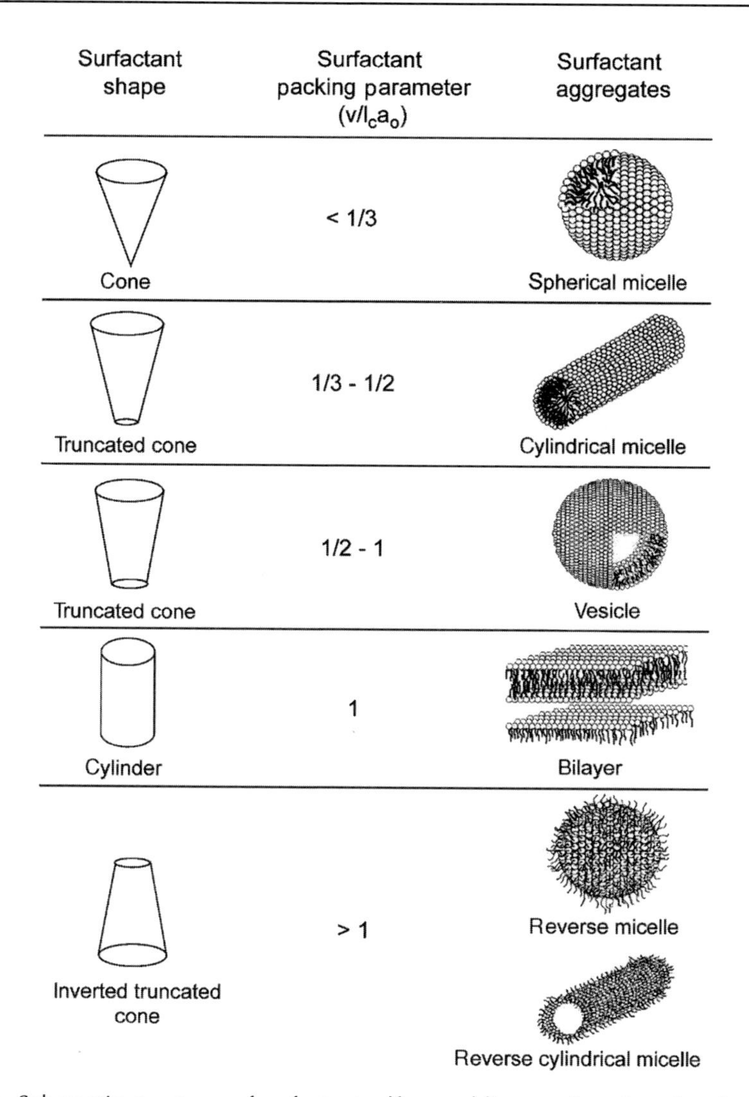

Figure 2.3 Schematic structures of surfactant self-assemblies as a function of surfactant packing parameters and shape.

Inverted or reverse micelles are formed where the surfactant packing parameter is greater than 1 (surfactants with small head groups or large tail groups such as double tailed anionic surfactants). These structures are formed in non-polar solvents. In these structures, head groups are clustered together and tails are extended towards the solvent. They have the capacity to take water into their cores and hence they form water-in-oil microemulsions. Hydrophilic materials can also be solubilised into the reverse micellar core (engine oil additives, hydraulic oils and cutting oils). Inverse micelles are often used for the separation of biological molecules such as proteins.

The surfactant phase diagrams for several surfactants have been developed in order to understand the phase structure of surfactants in solution at high concentration. With these phase diagrams, the effects of concentration and temperature on the surfactant aggregate structures and viscosity can be determined. Hence, they provide significant information for many industrial applications. Detail discussion of the phase behaviour of concentrated surfactant systems can be found in Chapter 3 of [45].

2.4 Adsorption of surfactants at surfaces

The adsorption mechanisms of surfactant at interfaces have been extensively studied in order to understand their performance in many processes such as dispersion, coating, emulsification, foaming and detergency. These interfaces are liquid–gas (foaming), liquid–liquid (emulsification) and liquid–solid (dispersion, coating and detergency).

2.4.1 Adsorption at liquid–gas and liquid–liquid interfaces

As discussed in Section 2.2, surfactant has a tendency to adsorb at interfaces since the polar head group has a strong preference for remaining in water while the hydrocarbon tail prefers to avoid water. The surfactant concentration affects the adsorption of surfactants at interfaces. Surfactant molecules lie flat on the surface at very low concentration. Surfactant molecules on the surface increase with increasing surfactant concentration in the bulk and surfactant tails start to orient towards gas or non-polar liquid since there is not enough space for the surfactant molecules to lie flat on the surface. Surfactant molecules adsorb at the interface and form monolayer until the surface is occupied at which point surfactant molecules start forming self-assembled structures in the liquid (Section 2.3).

Adsorption can be measured by direct or indirect methods. Direct methods include surface microtome method [46], foam generation method [47] and radio-labelled surfactant adsorption method [48]. These direct methods have several disadvantages. Hence, the amount of surfactant adsorbed per unit area of interface (Γ) at surface saturation is mostly determined by indirect methods namely surface and interfacial tension measurements along with the application of Gibbs adsorption equations (see Section 2.2.3 and Figure 2.1). Surfactant structure, presence of electrolyte, nature of non-polar liquid and temperature significantly affect the Γ value. The Γ values and the area occupied per surfactant molecule at water–air and water–hydrocarbon interfaces for several anionic, cationic, non-ionic and amphoteric surfactants can be found in Chapter 2 of [2].

Adsorption isotherms are used to relate the bulk surfactant concentration (C) to Γ. Surfactant solutions are generally represented by the Langmuir adsorption isotherm:

$$\Gamma = \Gamma_{\max} \frac{C}{C + a} \tag{2.18}$$

where $\Gamma_{\max}$ is the maximum surfactant adsorption at infinite dilution in mol cm^{-2} and a is a constant in mol cm^{-3}. a is expressed as:

$$a = \frac{\Gamma_{\max}}{\delta} \exp\left(\frac{\Delta G^0}{RT}\right) \tag{2.19}$$

where δ is the thickness of the adsorption layer, ΔG^0 is the free energy of adsorption at infinite dilution, R is ideal gas constant and T is the absolute temperature [2, 11, 49, 50]. The linear form of eqn 2.18 is:

$$\frac{C}{\Gamma} = \frac{C}{\Gamma_{max}} + \frac{a}{\Gamma_{max}} \tag{2.20}$$

The plot of C/Γ versus C is a straight line if the surfactant adsorption is of the Langmuir type. Furthermore, the values of Γ_{max} and a can be determined from the slope $(1/\Gamma_{max})$ and intercept (a/Γ_{max}) of this plot.

The Szyszkowski equation [51] relates surface tension (γ) to C:

$$\pi = \gamma_0 - \gamma = RT\,\Gamma_{max} In\left(\frac{C}{a} + 1\right) \tag{2.21}$$

where π is the surface pressure of a solution, γ_0 is the surface tension of pure solvent and γ is the surface tension of the surfactant solution. The Frumkin equation [52] derived from eqns 2.18 and 2.21 shows the relationship between γ and Γ_{max}:

$$\pi = \gamma_0 - \gamma = RT\,\Gamma_m \ln\left(1 - \frac{\Gamma}{\Gamma_{max}}\right) \tag{2.22}$$

If the interaction between the adsorbed surfactant molecules and electrostatic charge of surfactant ions is incorporated in eqn 2.18, the Langmuir equation becomes [53]:

$$\Gamma = (\Gamma_{max} - \Gamma)\frac{C}{a \exp\left(-b\frac{\Gamma}{\Gamma_m}\right) \exp\left(\frac{Ze\psi_s}{kT}\right)} \tag{2.23}$$

where b is a constant representing the non-electrostatic interaction between adsorbed surfactant molecules, Z is the valence of surfactant ion (zero for non-ionic surfactant), e is the elementary charge, ψ_s is the surface electric potential and k is Boltzmann's constant.

The efficiency of surfactant adsorption is determined as a function of minimum bulk surfactant concentration, C that produces saturation adsorption (Γ_{max}) at the liquid–gas or liquid–liquid interface. This minimum concentration is defined as pC_{20} which is $(-\log C_{20})$ reducing the surface or interfacial tension by 20 dyne cm^{-1} ($\pi = 20$ dyne cm^{-1}). With C_{20}, Γ lies between 84 and 99.9% of Γ_{max}. The larger the pC_{20} (smaller the C), the more efficient the surfactant is in adsorbing at the interface and reducing the surface tension at liquid–gas or interfacial tension at liquid–liquid interfaces. The pC_{20} values for several surfactants can be found in Chapter 2 of [2].

2.4.2 *Adsorption at liquid–solid interface*

Surfactants adsorb on solid surfaces due to hydrophobic bonding, electrostatic interaction, acid–base interaction, polarisation of π electrons and dispersion forces. Hydrophobic bonding occurs between the hydrophobic surfactant tail and the hydrophobic solid surface (tail down adsorption with monolayer structure) or between the hydrophobic tails of the surfactant adsorbed on the hydrophilic solid surface and the hydrophobic tails of the surfactant from the liquid phase (head down adsorption with bilayer structure) [54, 55].

Electrostatic interactions occur between the ionic head groups of the surfactant and the oppositely charged solid surface (head down adsorption with monolayer structure) [56]. Acid–base interactions occur due to hydrogen bonding or Lewis acid–Lewis base reactions between solid surface and surfactant molecules (head down with monolayer structure) [57]. Polarisation of π electrons occurs between the surfactant head group which has electron-rich aromatic nuclei and the positively charged solid surface (head down with monolayer structure) [58]. Dispersion forces occur due to London–van der Waals forces between the surfactant molecules and the solid surface (hydrophobic tail lies flat on the hydrophobic solid surface while hydrophilic head orients towards polar liquid) [59].

Adsorption of surfactant on solid surfaces is generally described by adsorption isotherms. For this purpose, a simple adsorption experiment can be performed at a constant temperature by dispersing known amounts of solid adsorbent into a constant volume of dilute surfactant solution at which the initial surfactant concentrations are varied and shaking the mixture until equilibrium is reached. The moles of surfactant adsorbed per unit mass of the solid (N_s) for each solution can be determined from:

$$N_s = \frac{(C_0 - C_e)V}{m} \tag{2.24}$$

where C_0 is the initial concentration of surfactant in the liquid phase before adsorption, C_e is the concentration of surfactant in the liquid phase after the equilibrium is reached, V is the volume of liquid phase and m is the mass of the adsorbent. Then, the Langmuir adsorption isotherm can be expressed in linear form:

$$\frac{C}{N_s} = \frac{C}{N_{s,\max}} + \frac{a}{N_{s,\max}} \tag{2.25}$$

where $N_{s,\max}$ is the maximum moles of surfactant adsorbed per gram of adsorbent at equilibrium. The slope and intercept of a plot of C/N_s versus C are $1/N_{s,\max}$ and $a/N_{s,\max}$, respectively. Furthermore, the surface concentration of surfactant on solid surface, Γ (mol/area), can be determined from:

$$\Gamma = \frac{N_s}{A_s} = \frac{(C_0 - C_e)V}{A_s m} \tag{2.26}$$

where A_s is the surface area per unit mass of the solid adsorbent. Then, eqn 2.20 is used for plotting and determining the Langmuir isotherm parameters.

The characteristics of surfactant adsorption isotherm on solid surface are generally analysed by the plot of log N_s versus log C_e based on eqn 2.24 or the plot of log Γ versus log C_e based on eqn 2.25. These plots show four region isotherms as shown in Figure 2.4.

Region 1 represents the adsorption at low surfactant concentration and in this region, linear adsorption isotherm exhibits a slope of 1 which can be explained by Henry's adsorption isotherm. In this region, surfactant molecules adsorbed on the solid surface as seen in Figure 2.5 do not interact with each other and the zeta potential of the solid surface stays unchanged. Non-ionic surfactant adsorbs on solid surface by hydrogen bonding (hydrophilic solid surface) or hydrophobic bonding (hydrophobic solid surface) [54, 55, 60]. Ionic surfactants adsorb on solid surface by electrostatic interaction (hydrophilic solid surface) or hydrophobic bonding (hydrophobic solid surface) [56, 60]. In region 2, adsorption isotherm shows an increase with a slope greater than 1. In this region (Figures 2.4 and 2.5),

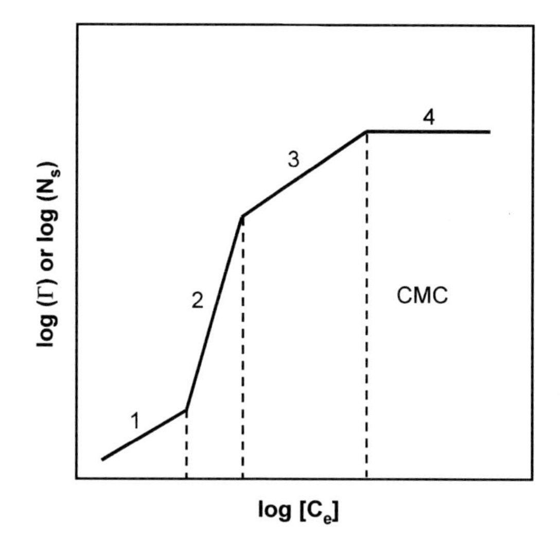

Figure 2.4 Four region isotherms of surfactant adsorption.

surfactant molecules move toward the solid surface and form hemimicelle on the hydrophilic solid surface (head down adsorption) or monolayer on the hydrophobic solid surface (tail down adsorption). As the bulk concentration increases, surfactant molecules interact with the previously adsorbed surfactant molecules by hydrophobic bonding and create surface aggregates (bilayer formation on solid surface known as admicelle) on the hydrophilic solid surface. The break between regions 1 and 2 represents the surfactant concentration where

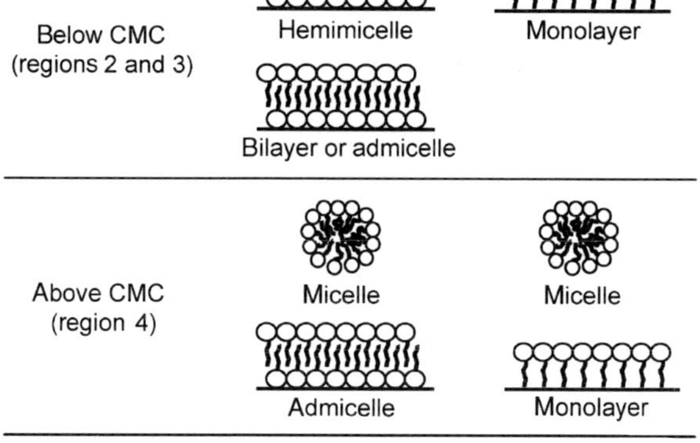

Surfactant concentration	Water–hydrophilic surface	Water–hydrophobic surface
Well below CMC (region 1)		
Below CMC (regions 2 and 3)	Hemimicelle Bilayer or admicelle	Monolayer
Above CMC (region 4)	Micelle Admicelle	Micelle Monolayer

Figure 2.5 Surfactant adsorption on solid surfaces.

the first hemimicelle or admicelle is formed. This concentration is well below the CMC of the surfactant. In this region, the charge of the solid surface is neutralised by the adsorption of oppositely charged ionic surfactants and reverses to the sign of the surfactant. In region 3, the slope of the isotherm decreases since the rate of bilayer formation decreases and adsorption occurs on the least energetic part of the solid surface [60–62]. In region 4, the plateau adsorption occurs due to micelle formation in the bulk liquid and slope of the isotherm levels off.

If the hydrophobic tails are short, the hydrophilic heads are composed of more than one similar charge ionic group and the ionic strength of the solution is low; then the hydrophobic bonding between tails does not occur and region 2 does not exist. Adsorption in region 1 proceeds until the charge of solid surface is neutralised and the slope of the isotherm reduces to that of region 3. If the ionic strength of the solution is high, the slopes of regions 1 and 3 become equal.

The adsorption of surfactants at the liquid–solid surface is affected by the nature of the solid surface (surface charge, polarity and non-polarity), the molecular structure of surfactant molecules (head group charge and characteristics of hydrophobic tail) and the characteristics of the liquid phase (pH, electrolyte concentration, presence of additives and temperature).

If the solid surface is composed of strongly charged sites, the adsorption of oppositely charged surfactant head groups on the solid surface is strong due to electrostatic interaction (solid and surfactant are in water). Eventually, the hydrophobic tails will orient towards the aqueous solution (region 2) and this will make the solid surface hydrophobic (zeta potential reduces to zero and the contact angle increases at the solid–water–air interface). Increase of surfactant concentration in water increases the surfactant adsorption on the solid surface due to hydrophobic bonding between surfactant tails. Hence, the zeta potential values reverse to the surfactant head group sign and contact angle decreases (regions 2 and 3). If the solid surface is non-polar (hydrophobic), surfactant molecules adsorb on the solid surface tail down. This makes the solid surface hydrophilic. If the surfactant is an ionic surfactant, the surface charge of the solid increases and it can be wetted and dispersed easily in the aqueous solution. If the solid surface contains polar groups, surfactant adsorbs on the solid surface head down in non-polar solvents which makes the solid dispersible in non-polar solvents.

The increase in the hydrophilic head group size reduces the amount of adsorbed surfactant at surface saturation. On the other hand, increasing the hydrophobic tail length may increase, decrease or maintain the surfactant adsorption. If the surfactant molecules are not closely packed, the increase in the chain length of the tail increases surfactant adsorption on solid surfaces. If the adsorption of surfactant on the solid surface is due to polarisation of π electrons, the amount of surfactant adsorbed on the surface reduces at surface saturation. If the adsorbed surfactants are closely packed on the solid surface, increasing the chain length of the surfactant tail will have no effect on the surfactant adsorption.

The change in solution pH affects the adsorption of ionic surfactants on the charged solid surface because the solid surface charge changes with pH. The solid surface becomes more negative with increase in pH and this decreases the adsorption of anionic surfactants but increases the cationic surfactant adsorption on the solid surface. Changing the pH also changes the ionic groups in the amphoteric surfactant structure making it either positively or negatively charged or neutral.

The increase in temperature increases adsorption of non-ionic surfactants on solid surfaces since the solubility of non-ionic surfactants in water decreases with increased temperature. On the other hand, increasing temperature decreases the adsorption of ionic surfactants on solid surfaces because the solubility of ionic surfactant increases with increased temperature. Furthermore, the presence of electrolytes increases the adsorption of ionic surfactants if the solid surface has the same charge as the surfactant head groups.

Acknowledgement

The author would like to thank Erdogan Ergican for the creation of the figures.

References

1. Porter, M.R. (1994) *Handbook of Surfactants*, 2nd edn. Chapman & Hall, London.
2. Rosen, M.J. (2004) *Surfactant and Interfacial Phenomena*, 3rd edn. Wiley, New York.
3. Domingo, X. (1995) *A Guide to the Surfactant World*. Edicions Proa, Barcelona.
4. Tsujii, K. (1998) *Surface Activity: Principles, Phenomena, and Applications*. Academic, Tokyo.
5. Ash, M. and Ash, I. (1993) *Handbook of Industrial Surfactants*. Gower Publishing, Hants.
6. Myers, D. (1988) *Surfactant Science and Technology*. VCH, New York.
7. Karsa, K.L. and Bothorel, P. (1987) *Surfactants in Solutions*, vols. 4–6. Plenum, New York.
8. Lucassen Reynders, E.H. (1981) *Anionic Surfactants Physical Chemistry of Surfactant Action*. Dekker, New York.
9. Rieger, M.M. (1997) Surfactant chemistry and classification. In M.M. Rieger and L.D. Rhein (eds), *Surfactants in Cosmetics*, 2nd edn, vol. 68, Surfactant Science Series. Dekker, New York, pp. 1–28.
10. Mahler, H.R. and Cordes, E.H. (1971) *Biological Chemistry*, 2nd edn. Harper & Row, New York.
11. Adamson, A.W. (1982) *Physical Chemistry of Surfaces*, 4th edn. Wiley, New York.
12. Hua, X.Y. and Rosen, M.J. (1982) Calculation of the coefficient in the Gibbs equation for the adsorption of ionic surfactants from aqueous binary mixtures with nonionic surfactants. *J. Colloid Interface Sci.*, **87**, 469.
13. Matijevic, E. and Pethica, B.A. (1958) The properties of ionized monolayers. Part 1. Sodium dodecyl sulphate at the air/water interface. Part 2. The thermodynamics of the ionic double layer of sodium dodecyl sulphate. *Trans. Faraday Soc.*, **54**, 1382–99.
14. Tajima, K. (1971) Radiotracer studies on adsorption of surface active substance at, aqueous surface. III. The effects of salt on the adsorption of sodium dodecylsulfate. *Bull. Chem. Soc. Japan*, **44**(7), 1767–71.
15. Ikeda, S. (1977) On the Gibbs adsorption equation for electrolyte solutions. *Bull. Chem. Soc. Japan*, **50**(6), 1403–08.
16. Van Voorst Vader, F. (1960) Adsorption of detergents at the liquid–liquid interface. Part 1 and Part 2. *Trans. Faraday Soc.*, **56**, 1067–84.
17. ISO Standard 304. Determination of surface tension by drawing up liquid films.
18. ASTM D1331-89(2001). Standard test methods for surface and interfacial tension of solutions of surface active agents.
19. ASTM D3825-90(2000). Standard test method for dynamic surface tension by the fast-bubble technique.
20. ISO 6889. Determination of interfacial tension by drawing up liquid films.
21. ISO 9101. Determination of interfacial tension – drop volume method.

22. Clint, J.H. (1992) *Surfactant Aggregation*, Chapman & Hall, New York.
23. Mukerjee, P. and Mysels, K.J. (1971) *Critical Micelle Concentrations of Aqueous Surfactant Systems.* NSRDS-NBS 36, Washington, DC.
24. Evans, H.C. (1956) Alkyl sulphates. Part 1. Critical micelle concentrations of the sodium salts. *J. Chem. Soc.*, **78**, 579–86.
25. Schick, M. J. (1962) surface film of nonionic detergents. I. Surface tension study. *J. Colloid Sci.*, **17**, 801–13.
26. Ray, A. and Nemethy, G. (1971) Effects of ionic protein denaturants on micelle formation by nonionic detergents. *J. Am. Chem. Soc.*, **93**(25), 6787–93.
27. Schick, M.J. and Fowkes, F.M. (1957) Foam stabilizing additives for synthetic detergents. Interaction of additives and detergents in mixed micelles. *J. Phys. Chem.*, **61**, 1062–68.
28. Schick, M.J. and Gilbert, A.H. (1965) Effect of urea, guadinium chloride, and dioxane on the CMC of branched-chain nonionic detergents. *J. Colloid Sci.*, **20**, 464–72.
29. Herzfeld, S.H., Corrin, M.L. and Harkins, W.D. (1950) The effect of alcohols and of alcohols and salts on the critical micelle concentration of dodecylammonium chloride. *J. Phys. Chem.*, **54**, 271–83.
30. Hunter, A.J. (1987) *Foundations of Colloid Science*, vols I and II. Clarendon, Oxford.
31. Flochart, B.D. (1961) The effect of temperature on the critical micelle concentration of some paraffin-chain salts. *J. Colloid Sci.*, **16**, 484–92.
32. Crook, E.H. Fordyce, D.B. and Trebbi, G.F. (1967) Molecular weight distribution of nonionic surfactants. I. Surface and interfacial tension of normal distribution and homogeneous *p*, *t*-octylphenoxyethoxyethanols (OPE'S). *J. Phys. Chem.*, **67**, 1987–94.
33. Klevens, H.B. (1953) Structure and aggregation in dilute solutions of surface active agents. *J. Am. Oil Chem. Soc.*, **30**, 74–80.
34. Kreshech, G.C. (1975) *Surfactants in Water – A Comprehensive Treatise.* Plenum, New York.
35. Corrin, M.L. and Harkins, W.D. (1947) The effect of salt on the critical concentration for the formation by nonionic detergents. *J. Am. Chem. Soc.*, **69**(3), 683–88.
36. Shinoda, K. Yamaguchi, T. and Hori, R. (1961) The surface tension and the critical micelle concentration in aqueous solution of β-D-alkyl glucosides and their mixtures. *Bull. Chem. Soc. Japan*, **34**(2), 237–41.
37. Hobbs, M.E. (1951) The effect of salts on the critical concentration, size, and stability of soap micelles. *J. Phys. Colloid Chem.*, **55**(5), 675–83.
38. Shinoda, K. (1953) The effect of chain length, salts and alcohols on the critical micelle concentration. *Bull. Chem. Soc. Japan*, **26**(2), 101–05.
39. Molyneux, P., Rhodes, C.T. and Swarbrick, J. (1965) Thermodynamics of micellization of N-alkyl betaines. *Trans. Faraday Soc.*, **61**, 1043–52.
40. Kaler, E.W. (1994) Basic surfactant concepts. In K.R. Lange (ed.), *Detergents and Cleaners – A Handbook for Formulators.* Hanser, New York, pp. 1–28.
41. ISO Standard 4311. Determination of the critical micelle concentration method by measuring surface tension with a plate, stirrup or ring.
42. Scamehorn, J.F. (1986) *Phenomena in Mixed Surfactant Systems*, ACS Symp. Series 311. ACS, Washington, DC.
43. Tanford, C. (1980) *The Hydrophobic Effect.* Wiley, New York.
44. Israelachvili, J. (1992) *Intermolecular and Surface Forces*, 2nd edn. Academic, Orlandao, FL.
45. Holmberg, K., Jonsson, B., Kronberg, B. and Lindman, B. (2003) *Surfactants and Polymers in Aqueous Solution*, 2nd edn. Wiley, London.
46. McBain, J.W. and Swain, R.C. (1936) Measurements of adsorption at the air–water interface by the microtome method. *Proc. R. Soc.*, A **154**, 608–23.
47. Wilson, A., Epstein, M.B. and Ross, J. (1957) The adsorption of sodium lauryl sulfate and lauryl alcohol at the air–liquid interface. *J. Colloid Sci.*, **12**, 345–55.

48. Nilsson, G. (1957) The adsorption of tritiated sodium dodecyl sulfate at the solution surface measured with a windowless, high humidity gas flow proportional counter. *J. Phys. Chem.*, **61**, 1135–42.
49. Langmuir, I. (1917) The constitution and fundamental properties of solids and liquids. *J. Am. Chem. Soc.*, **39**, 1848–906.
50. Langmuir, I. (1918) The adsorption of gases on plane surfaces of glass, mica and platinum. *J. Am. Chem. Soc.*, **40**, 1361–1403.
51. Von Szyszkowski, B. (1908) Experimentelle Studien Über kapillare Eigenscchaften der Wässeriyen Lõsungen van Fettsäuren. *Z. Phys. Chem.*, **64**, 385–414.
52. Frumkin, A. (1925) Die Kapillarkurve der hoheren Fettsauren und die Zustandsgleichung der oberblachenschicht. *Z. Phys. Chem.*, **116**, 466–80.
53. Borwankar, R.P. and Wasan, D.T. (1988) Equilibrium and dynamics of adsorption of surfactants at fluid–fluid interfaces. *Chem. Eng. Sci.*, **43**, 1323–37.
54. Dick, S.G., Fuerstenau, D.W. and Healy, T.W. (1971) Adsorption of alkylbenzene sulfonate (A.B.S.) surfactants at the alumina–water interface. *J. Colloid Interface Sci.*, **37**, 595–602.
55. Giles, C.H., D'Silva, A.P. and Easton, I.A. (1974) A general treatment and classification of the solute adsorption isotherm. Part II. Experimental interpretation. *J. Colloid Interface Sci.*, **47**, 766–78.
56. Rupprecht, H. and Liebl, H. (1972) Einflub von Tensiden auf das Kolloidchemische Verhalten hochdisperser Kieselsäuren in polaren und unpolaren Lösungsmitteln. *Kolloid, Z.Z. Polym.*, **250**, 719–23.
57. Fowkes, F.M. (1987) Role of acid-base interfacial bonding in adhesion. *J. Adhes. Sci. Technol.*, **1**, 7–27.
58. Snyder, L.R. (1968) Interactions responsible for the selective adsorption of nonionic organic compounds on alumina. Comparisons with adsorption on silica. *J. Phys. Chem.*, **72**, 489–94.
59. Law, J.P. and Kunze, G.W. (1966) Reactions of surfactants with montmorillonite: adsorption mechanisms. *Soil Sci. Soc. Am. Proc.*, **30**, 321–27.
60. Scamehorn, J.F., Schecter, R.S. and Wade, W.H. (1982) Adsorption of surfactants on mineral oxide surfaces from aqueous solutions. I. Isomerically pure anionic surfactants. *J. Colloid Interface Sci.*, **85**, 463–78.
61. Scamehorn, J.F., Schecter, R.S. and Wade, W.H. (1982) Adsorption of surfactants on mineral oxide surfaces from aqueous solutions. II. Binary mixtures of anionic surfactants. *J. Colloid Interface Sci.*, **85**, 479–93.
62. Scamehorn, J.F., Schecter, R.S. and Wade, W.H. (1982) Adsorption of surfactants on mineral oxide surfaces from aqueous solutions. III. Binary mixtures of anionic and nonionic surfactants. *J. Colloid Interface Sci.*, **85**, 494–501.

Chapter 3
Applied Theory of Surfactants

Peter Schmiedel and Wolfgang von Rybinski

3.1 Introduction

Surfactants are substances of outstanding importance both in nature and in technology. In a vast number of technical processes surfactants play a decisive role. The application that comes first to everybody's mind is detergency. Certainly surfactants are not the only active ingredient in a detergent; actually they are the most important one and other active ingredients, e.g. bleach or enzymes, can only have an optimum effect in the presence of surfactants. This will be discussed in detail in Section 3.2. In this field of application several interfacial effects come into play on the different interfaces involved. Adsorption of the surfactant molecules occurs on the interfaces and leads to various effects. At the liquid–solid interface it causes the wetting of the solid substrate e.g. a textile or a hard surface. At the liquid–liquid interface the interfacial tension is reduced which enables emulsification of oils. At the liquid–gas interface the surface tension is reduced and a disjoining pressure in thin films can arise. Thus, the generation of foam can occur. The latter two effects are relevant also in fields of application other than detergency and sections in this chapter are dedicated to them. Even if cleaning applications – laundry detergents as well as cleansers, hair shampoos or only simple bar soaps – are the first to come to most people's mind when thinking about surfactants, the range of possible applications is much wider. In many natural and synthetic foods they make oily and aqueous phases compatible. In agricultural or pharmaceutical preparations they enable the stable formulation of hydrophobic insoluble actives. In water-based paints they stabilise the latex particles and pigments by electrostatic or steric repulsion and in rolling oils they allow the combination of cooling and lubrication properties in one fluid.

For both the processing and the application of surfactant-containing products, further properties of these substances play an important role. Due to the association of surfactant molecules, micelles are formed and, at higher concentrations, lyotropic liquid crystals (mesophases) arise. Particularly for surfactant mixtures and surfactants of technical grade, the phase diagrams may become very complex. Sometimes only small variations in the composition may change the properties dramatically. Electrical and optical properties as well as flow behaviour can change over orders of magnitude and show discontinuities at the phase boundaries. Section 3.6.5 of this chapter is dedicated to this subject.

3.2 Detergency

3.2.1 Fundamental processes

Washing and cleansing are processes in which many interfacial effects are involved. Therefore, a fundamental description of detergency is very complex. The processes range from the adsorption of surfactants on the substrates, the wetting of fabrics or hard surfaces and the dissolution of stains from fabrics to the removal of ions from the washing liquor or the interaction of softeners with the fabric in the rinse cycle [1]. In many of these important processes, ingredients of the detergent other than surfactants, e.g. builder, enzymes or bleach, are involved. A comprehensive description of all the processes is beyond the scope of this chapter. Therefore only the processes directly linked to surfactants are discussed in more detail. Table 3.1 shows the different types of interfacial processes involved in the washing process.

In addition, the components involved in the washing process can be very different including a variety of fabrics to be cleaned, liquid or solid stains with different structure and the ingredients of the detergent [2]. Clustering of the different processes with focus on surfactants leads on the following main steps in washing or cleaning:

- Dissolution of the detergent formulation
- Wetting of the substrate to be cleaned or washed by the washing liquor
- Interaction of the detergent or cleanser with the stains
- Removal of the stains from fabric
- Stabilisation of the soil in the washing liquor
- Modification of the substrate (e.g. by softener in the rinse cycle)

All these processes occur consecutively or simultaneously and are influenced by the different interfacial parameters.

Table 3.1 Interfacial processes in detergency

Air–water interface	Solid–solid interface
Surface tension	Adhesion
Film elasticity	Flocculation
Film viscosity	Heterocoagulation
Foam generation	Sedimentation
Liquid–liquid interface	Interfaces in multicomponent systems
Interfacial tension	Wetting
Interfacial viscosity	Rolling-up processes
Emulsification	
Electric charge	
Active ingredient penetration	
Solid–liquid interface	
Adsorption	
Dispersion	
Electric charge	

Table 3.2 Substrates and soils in the washing process

Water-soluble materials	Pigments
Inorganic salts	Metal oxides
Sugar	Carbonates
Urea	Silicates
Perspiration	Humus
Fats	Carbon black (soot)
Animal fat	Proteins from the following
Vegetable fat	Blood
Sebum	Egg
Mineral oil	Milk
Wax	Skin residues
Bleachable dyes from the following	Carbohydrates
Fruit	Starch
Vegetables	
Wine	
Coffee	
Tea	

Table 3.2 gives an overview on the different substrates and soils [3] which can be either solid pigments or a liquid phase such as oils and fats but, usually, they occur in mixtures. The removal of soils can be carried out either by mechanical force or by chemical degradation, e.g. by enzymes or bleaching agents.

3.2.2 Basic formulae of detergents and cleansers

The composition of a modern heavy duty detergent (HDD) may be very complex, containing different types of substances. Table 3.3 shows the typical major components of detergents and cleansers [4].

In addition to this complex formulation, the components themselves are mixtures as they are usually of technical grade and this makes the description and interpretation of the interfacial processes even more complex.

In the following sections, the major characteristics of the single interfacial processes of the washing process are summarised, concentrating on the more general features applicable to different detergent types.

3.2.3 Adsorption at the solid–liquid interface

The physical separation of the soil from the fabrics is based on the adsorption of surfactants and ions on the fabric and soil surfaces. For a pigment soil the separation is caused by an increased electrostatic charge due to the adsorption (Figure 3.1) [5].

In the aqueous washing liquor the fabric surface and the pigment soil are charged negatively due to the adsorption of OH^- ions and anionic surfactants and this leads to an electrostatic repulsion. In addition to this effect, a disjoining pressure occurs in the adsorbed

Table 3.3 Major components of powder detergents

Ingredients	Composition (%)			
	United States, Canada, Australia	South America, Middle East, Africa	Europe	Japan
Surfactants	8–20	17–32	8–20	19–25
Foam boosters	0–2		0–3	
Foam depressants			0.3–5	1–4
Builders				
Sodium triphosphate	25–35	20–30	20–35	0–15
Mixed or nonphosphate	15–30	25–30	20–45	0–20
Sodium carbonate	0–50	0–60		5–20
Antiredeposition agents	0.1–0.9	0.2–1	0.4–1.5	1–2
Anticorrosion agents	5–10	5–12	5–9	5–15
Optical brighteners	0.1–0.75	0.08–0.5	0.1–0.75	0.1–0.8
Bleach			15–30	0–5
Enzymes			0–0.75	0–0.5
Water	6–20	6–13	4–20	5–10
Fillers	20–45	10–35	5–45	30–45

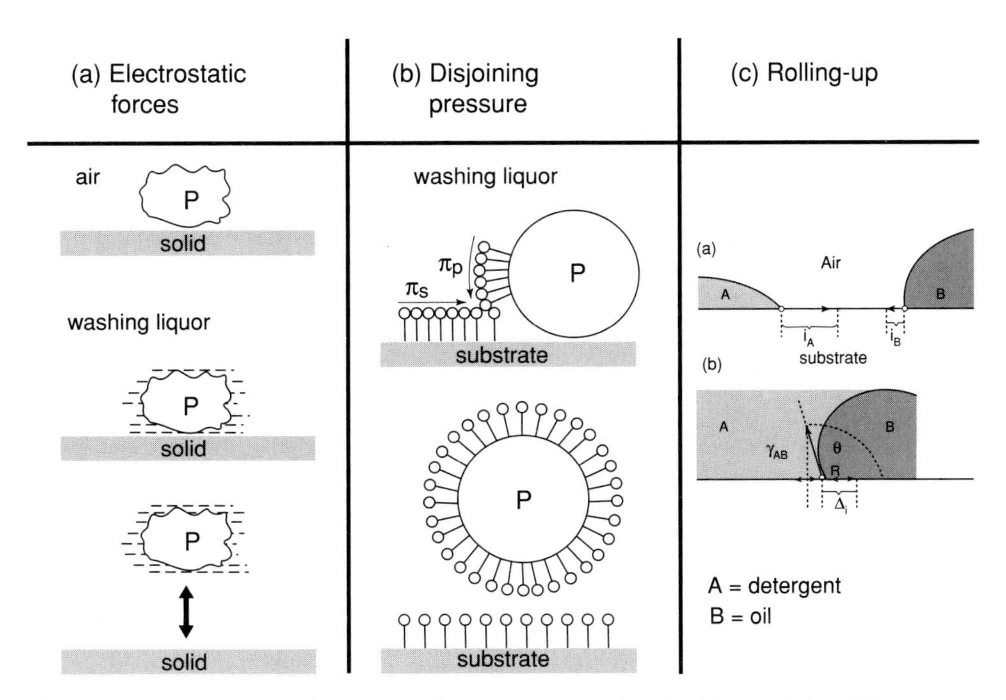

Figure 3.1 Separation mechanisms of detergency (reproduced with permission [5]).

layer which supports the lift-off process of the soil from the surface. For a spherical particle with a radius r the separation force is described by [5]:

$$f_d = 2\pi r(\pi_s + \pi_p) \quad (3.1)$$

with

π_s = disjoining pressure in the adsorption layer of the substrate
π_p = disjoining pressure in the adsorption layer of the particle

The non-specific adsorption of surfactants is based on the interaction of the hydrophilic headgroup and the hydrophobic alkyl chain with the pigment and substrate surfaces as well as the solvent. For the adsorption of surfactants, different models have been developed which take into account different types of interactions. A simple model which excludes lateral interactions of the adsorbed molecules is the Langmuir equation:

$$\frac{1}{Q_\infty} = \frac{1}{bQ_m}\frac{1}{c} + \frac{1}{Q_m} \quad (3.2)$$

with

Q_∞ = equilibrium adsorbed amounts
Q_m = adsorbed amounts in a fully covered monolayer
c = equilibrium concentration in solution
b = constant

This model is restricted to only very few systems. A more widely applicable model is presented in Figure 3.2 with a visualisation of the structure of the adsorbed molecules dependent on surface coverage [6].

Three different ranges are to be distinguished. In the low concentration range, single molecules are adsorbed on the surface with no interaction between the molecules which preferably are arranged on the surface in a flat structure or with a certain tilt angle. For ionic surfactants the adsorption sites on the surface are determined by the location of surface charge. When the surfactant concentration increases, a strong rise in the adsorbed amounts is observed by the lateral interaction of the hydrophobic parts of the surfactant molecules. The surfactant molecules have a perpendicular arrangement to the surface. There are different models for the structure of the adsorbed layer in this concentration range either assuming a flat monolayer or a hemimicellar structure, depending on the structure of the surfactants and the charge distribution on the solid surface. The hydrophilic groups of the surfactants can be directed towards either the surface of the solid or the solution, depending on the polarity of the solid surface. In the third part of the adsorption isotherm a plateau value is observed and, during a further increase of the surfactant concentration, a rise in the adsorbed amounts occurs. In this range of the adsorption isotherm a fully covered monolayer or double layer is adsorbed onto the surface, making the surface either hydrophilic or hydrophobic. Depending on the type of the surface, in some cases micellar structures of the adsorbed surfactants have been postulated instead of flat double layers. Typical examples of adsorption isotherms of sodium dodecyl sulphate onto different surfaces are shown in Figure 3.3 [5].

The adsorption isotherms for carbon black and graphitised carbon black (graphon) are completely different. For graphitised carbon black a step-like adsorption isotherm is

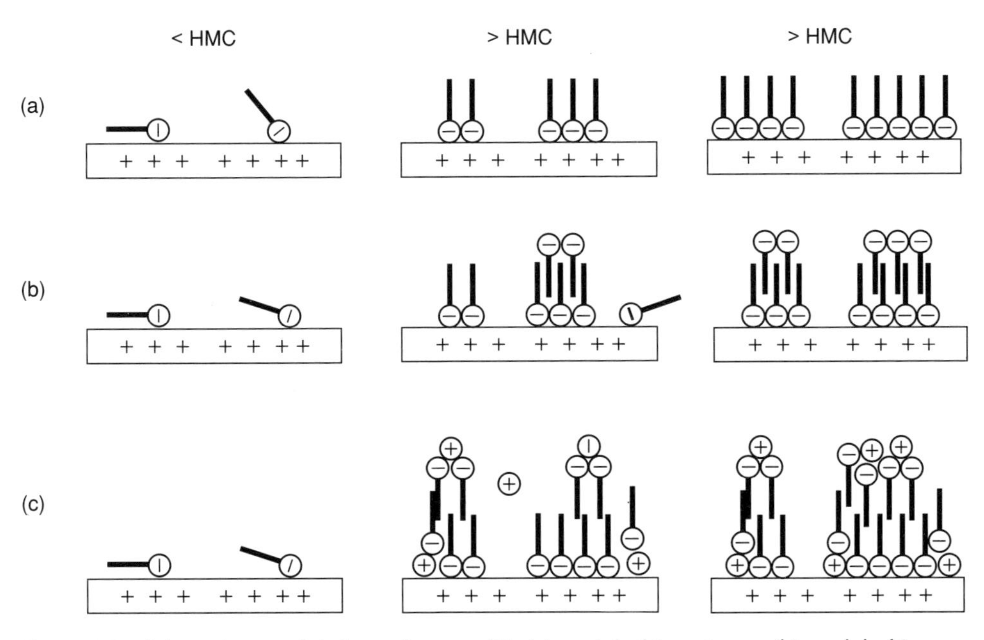

Figure 3.2 Adsorption models for surfactants [6]: (a) model of Fuerstenau, (b) model of Scamehorn, Chandar, Dobias and (c) model of Harwell *et al.*

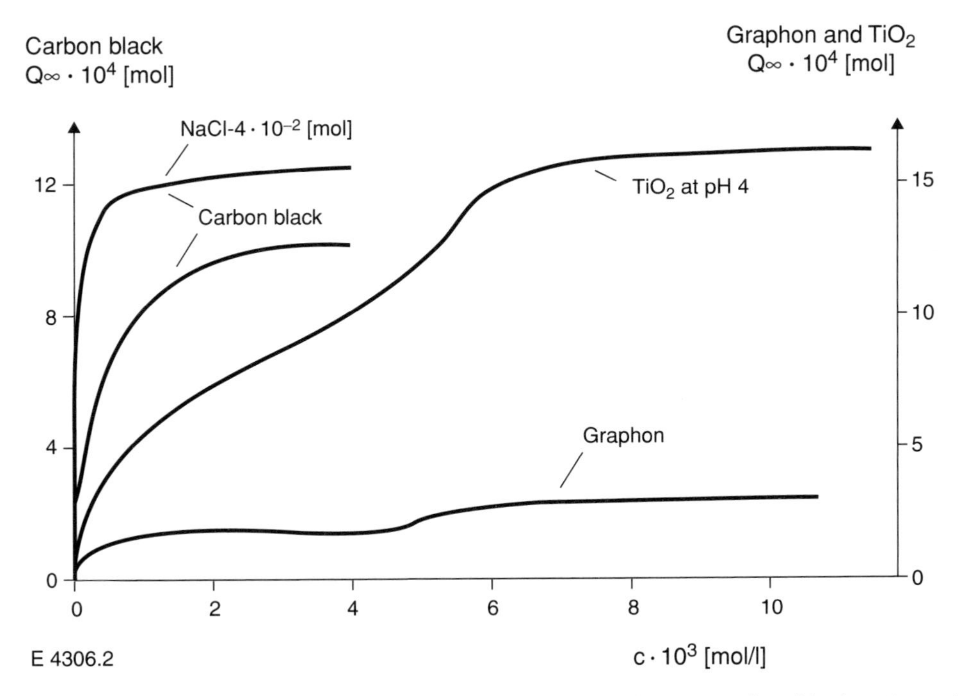

Figure 3.3 Equilibrium adsorption of sodium n-dodecyl sulphate on carbon black, TiO_2 and graphon at room temperature (reproduced with permission [5]).

observed which indicates flat arrangement of the surfactant molecules on the surface at low concentrations with a perpendicular structure at higher concentrations (see Figure 3.2). The adsorption process is exothermic with an adsorption enthalpy of about −128 to −36 kJ mol^{-1}. The adsorption of sodium dodecylsulphate on titanium dioxide is an example for the specific adsorption via the hydrophilic group onto the polar pigment surface. A second adsorption layer is formed via hydrophobic interaction with the first adsorption layer which makes the pigment surface hydrophilic again in the range of the plateau of the adsorption isotherm. Figure 3.3 also demonstrates the effect of the addition of electrolytes which take part in a washing process. In the presence of ions the amounts of the anionic surfactant adsorbed are increased. This is due to a decreased electrostatic repulsion of the negatively charged hydrophilic groups of the anionic surfactant in the presence of electrolytes. Therefore, the adsorption density in equilibrium can be enhanced significantly. A similar effect can be observed in a comparison of an anionic and non-ionic surfactant with the same alkyl chain length adsorbed onto a hydrophobic solid (Figure 3.4) [5].

The non-ionic surfactant gives higher adsorbed amounts at the same concentration than the anionic surfactants. This is especially valid at low concentrations, whereas at very high concentrations both surfactants reach the same plateau value. For a hydrophilic solid surface this effect can be just the opposite due to a higher affinity of anionic surfactant to the surface via specific interactions.

The electrolyte effect for the adsorption of anionic surfactants which leads to an enhancement of soil removal is valid only for low water hardness, i.e. low concentration of calcium ions. High concentration of calcium ions can lead to a precipitation of calcium surfactant

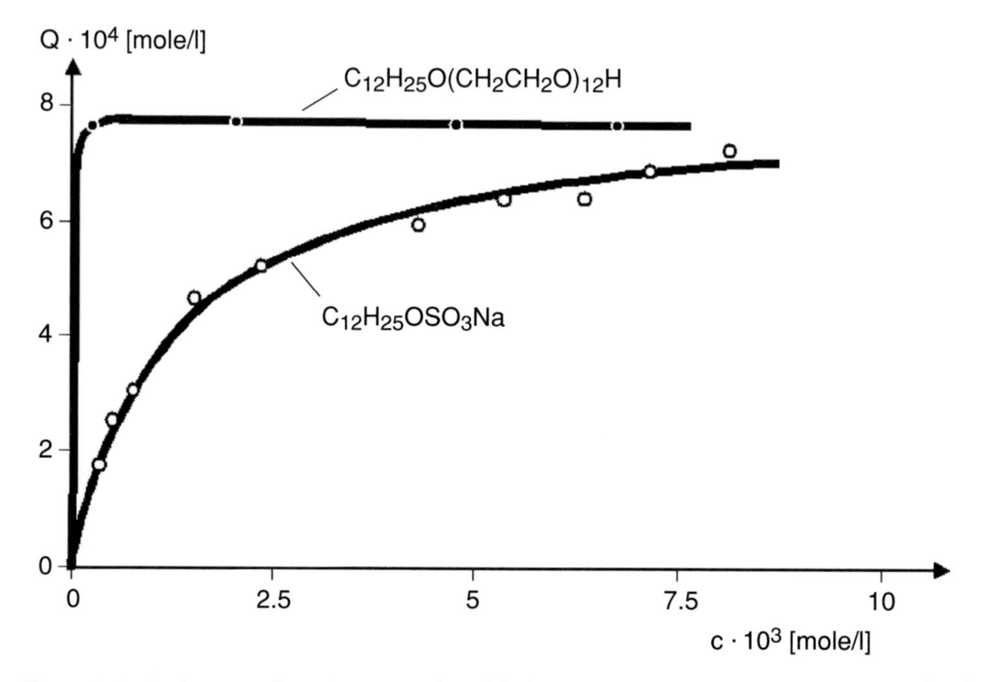

Figure 3.4 Surfactant adsorption on carbon black, $T = 298$ K, surface area = 1150 m^2 g^{-1} (BET) (reproduced with permission [5]).

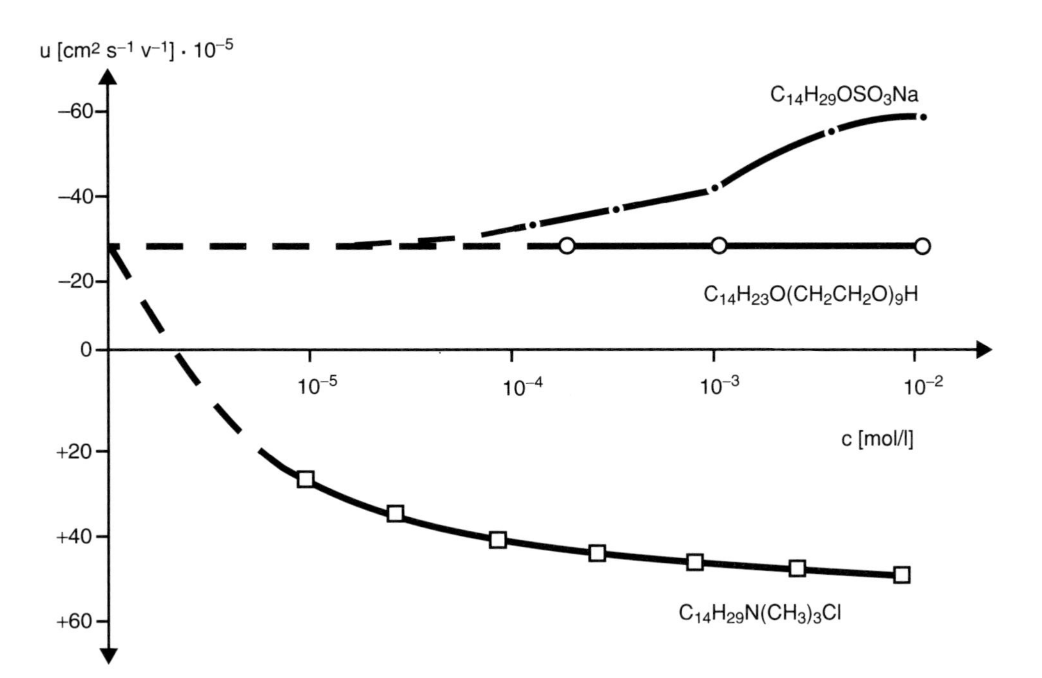

Figure 3.5 Electrophoretic mobility *u* of carbon black in solutions of different surfactants at 308 K (reproduced with permission [5]).

salts and therefore to a reduction of concentration of active molecules. In addition to this, the electrical double layer is compressed so much that the electrostatic repulsion between pigment soil and surface is reduced. Therefore, for many anionic surfactants the washing performance deteriorates with lower temperatures in the presence of calcium ions. This effect can be compensated by the addition of complexing agents or ion exchangers.

The characteristic change of the surface charge of the solid which depends on the nature of the hydrophilic groups of the surfactant is a consequence of the non-specific adsorption of the surfactants on pigments and fabrics or hard surfaces. This can be shown in aqueous solutions of different surfactants with the same alkyl chain length by the change of electrophoretic mobility of pigments which is a measure for the surface charge (Figure 3.5) [5].

The carbon black shown as an example has a negative surface charge in water at an alkaline pH value and, as for most pigments present in the washing process, the isoelectric point is below pH 10. The non-ionic surfactant shows no influence on the electrophoretic mobility, whereas the anionic surfactant increases the negative surface charge of the pigment due to the adsorption. By the adsorption of cationic surfactant the surface charge can be changed from a negative to a positive value during the adsorption process. This picture explains quite well the mode of action of different surfactant types for pigment removal in the washing process. As non-ionic surfactants do not influence the electrostatic repulsion of pigment and fabric, their washing efficiency mainly is caused by the disjoining pressure of the adsorption layer. In addition to this effect anionic surfactants increase the electrostatic repulsion, but usually have lower amounts adsorbed than the non-ionic surfactants. Cationic surfactants show effects similar to anionic surfactants in the washing process, but in spite of this they are not suited for most washing processes due to their adverse effects in the rinse cycles. In the rinse cycles the positively charged surfaces (due to the adsorption of cationic surfactants) are recharged to negative values due to the dilution of the washing liquor and the consecutive desorption of cationic surfactants. As the different fabrics and pigment soils have different isoelectric points, positively and negatively charged surfaces are present in the washing liquor which leads to heterocoagulation processes and a redeposition of the already removed soil onto the fabric. Therefore cationic surfactants are not used in washing processes, only as softeners in the rinse cycle when soil is no longer present and a strong adsorption of cationic softener on the negatively charged fabric is desired.

3.2.4 Surface tension and wetting

The characteristic effect of surfactants is their ability to adsorb onto surfaces and to modify the surface properties. Both at gas/liquid and at liquid/liquid interfaces, this leads to a reduction of the surface tension and the interfacial tension, respectively. Generally, non-ionic surfactants have a lower surface tension than ionic surfactants for the same alkyl chain length and concentration. The reason for this is the repulsive interaction of ionic surfactants within the charged adsorption layer which leads to a lower surface coverage than for the non-ionic surfactants. In detergent formulations, this repulsive interaction can be reduced by the presence of electrolytes which compress the electrical double layer and therefore increase the adsorption density of the anionic surfactants. Beyond a certain concentration, termed the critical micelle concentration (cmc), the formation of thermodynamically stable micellar aggregates can be observed in the bulk phase. These micelles are thermodynamically stable and in equilibrium with the monomers in the solution. They are characteristic of the ability of surfactants to solubilise hydrophobic substances.

In Figure 3.6, examples are given for the dependence of the surface tension of several surfactants on the concentration. Above the cmc the surface tension is minimum and remains constant so that, in a washing liquor, the concentration of surfactant has at least to be right above the cmc. Typical application concentrations of surfactants in washing liquors lie in the order of magnitude of $1\ \mathrm{g\ l^{-1}}$.

Most detergents contain electrolytes, e.g. sulphate, bicarbonate, carbonate or citrate and the presence of these electrolytes increases the adsorption of anionic surfactants at the gas/liquid interface as already mentioned. This leads to a reduction of the surface tension at an equal solution concentration [7] and to a strong decrease of the cmc. The effect can be of several orders of magnitude. Similar to this are the effects of mixtures of surfactants with the same hydrophilic group and different alkyl chain length or mixtures of anionic and non-ionic surfactants as they are mostly used in detergency [8]. Mixtures of anionic and non-ionic surfactants follow the mixing rule (eqn. 3) in the ideal case:

$$\frac{1}{\mathrm{cmc_{mix}}} = \frac{\alpha}{\mathrm{cmc_1}} + \frac{1-\alpha}{\mathrm{cmc_2}} \tag{3.3}$$

with

$\mathrm{cmc_{mix}}$ = cmc of surfactant mixtures
$\mathrm{cmc_1}$ = cmc of surfactant 1
$\mathrm{cmc_2}$ = cmc of surfactant 2
α = mole fraction of surfactant in bulk solution

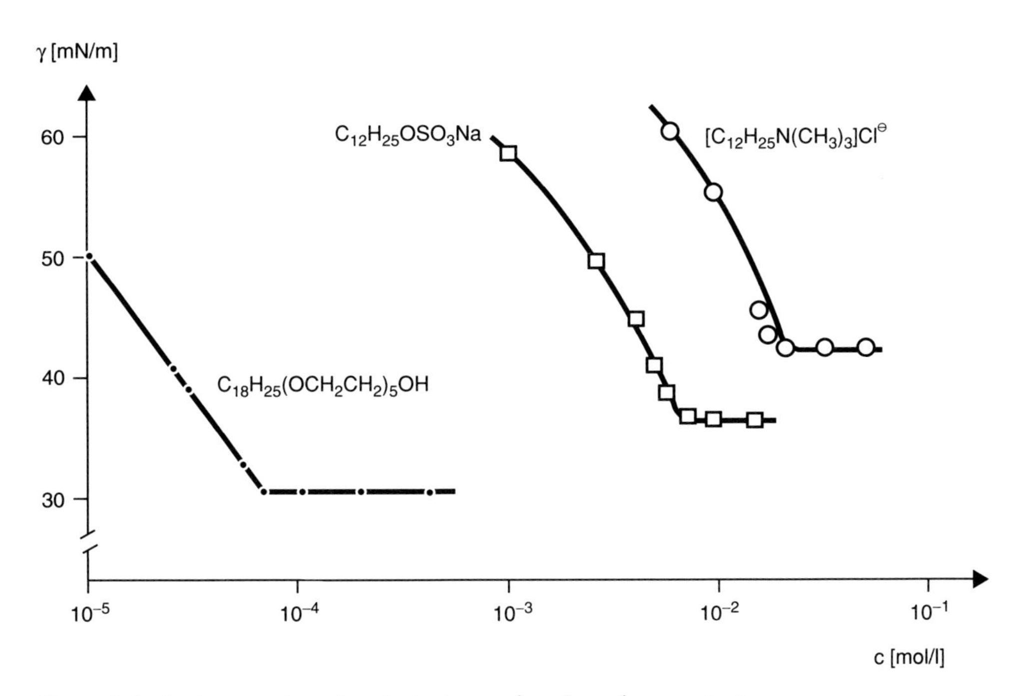

Figure 3.6 Surface tension of surfactants as a function of concentration.

According to a theory, based on the regular solution theory, a deviation from ideal behaviour can be described by the introduction of the activity coefficients f_1 and f_2:

$$\frac{1}{\mathrm{cmc_{mix}}} = \frac{\alpha}{f_1 \mathrm{cmc}_1} + \frac{1-\alpha}{f_2 \mathrm{cmc}_2} \tag{3.4}$$

$$f_1 = \exp\beta[1-x]^2 \tag{3.5}$$

$$f_2 = \exp\beta x^2 \tag{3.6}$$

$$\Delta H_{\mathrm{m}} = \beta RTx\,[1-x] \tag{3.7}$$

with

f_1 = activity coefficient of component 1
f_2 = activity coefficient of component 2
β = interaction parameter
x = mole fraction of component 1 in the micelle
ΔH_{m} = micellisation enthalpy

The interaction parameter β characterises the deviation from ideal behaviour. If β has negative values, there is an attractive interaction between the surfactants and the cmc of the mixture is lower than expected for ideal behaviour whereas for $\beta > 0$, there is a repulsive interaction and the cmc is higher than that for ideal behaviour. For highly negative values of β and when the cmcs of the surfactants are quite similar, the cmc of the mixture is even lower than that of the single surfactants. The strongest interaction is observed for mixtures

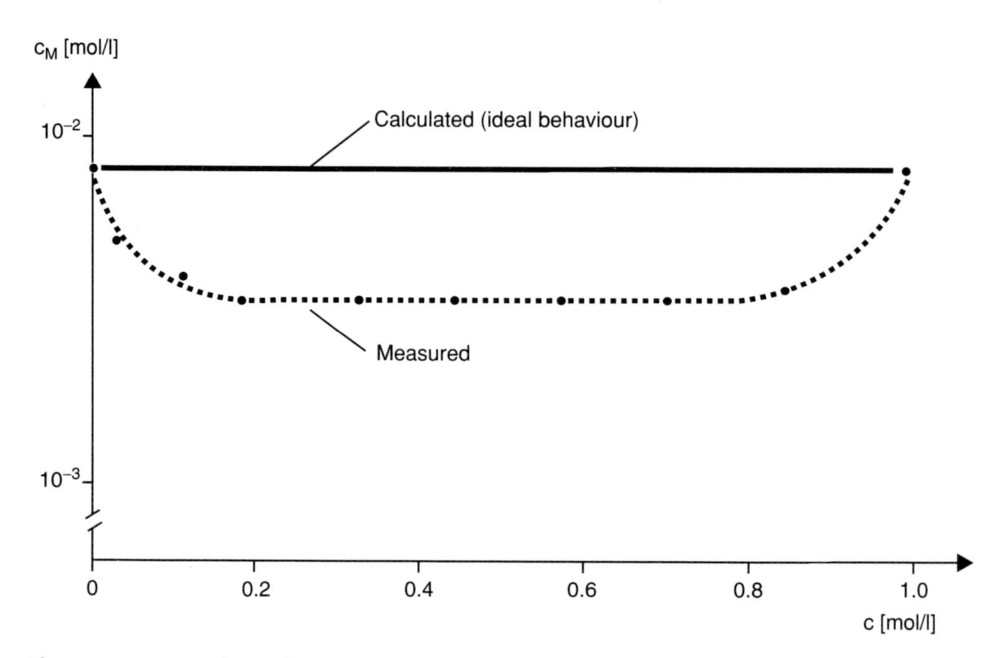

Figure 3.7 Critical micelle concentration of mixtures of sodium n-dodecyl sulphonate and n-octylnonaglycolether.

of anionic and cationic surfactants due to the electrostatic forces between the headgroups. An example of the influence of the interaction of the surfactant molecules on the cmc is shown in Figure 3.7.

The interaction between the surfactants has an influence not only on the cmc but also on different properties which are relevant for washing and cleaning. So a synergistic effect has been observed for foaming, emulsification and dispersing properties and even washing and cleaning efficiency for negative β parameters [8].

The kinetics of surface effects is an aspect which has been underestimated for a long time regarding the mechanisms of washing and cleaning. Especially at lower concentrations there might be a strong influence of time on the surface and interfacial tension.

Figure 3.8 shows the dynamic surface tension of a pure anionic and a non-ionic surfactant dependent on the absorption time after the creation of new surface for different concentrations [9]. For both surfactants, the time dependence of the surface tension is greatly reduced when the concentration increases and this effect is especially pronounced when the critical micelle concentration is reached. The reason for this dependence is the diffusion of surfactant molecules and micellar aggregates to the surface which influences the surface tension on newly generated surfaces. This dynamic effect of surface tension can probably be attributed to the observation that an optimum of the washing efficiency usually occurs well above the critical micelle concentration. The effect is an important factor for cleaning and institutional washing where short process times are common.

Connected with the parameter surface tension is the wetting process of the surface, e.g. fabrics or hard surfaces.

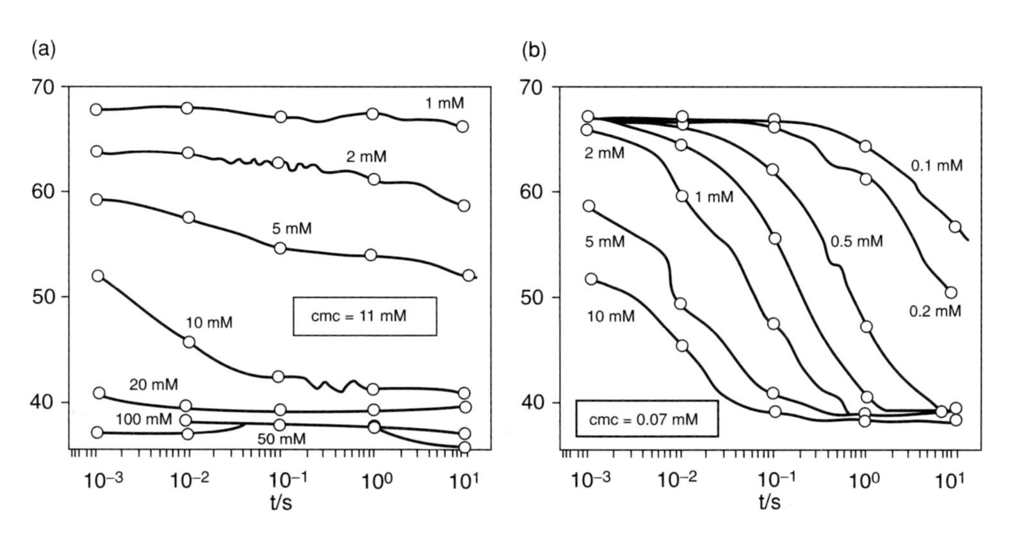

Figure 3.8 Dynamic surface tension of (a) $C_{12}SO_3Na$ and (b) $C_{12}E_6$ as a function of concentration at 40°C (reproduced with permission [9]).

The wetting can be described by the Young equation (see Figure 3.9):

$$\gamma_s = \gamma_{sl} + \gamma_l \cos\theta \tag{3.8}$$

γ_s = interfacial tension of the solid/gas interface
γ_{sl} = interfacial tension of the solid/liquid interface
γ_l = surface tension liquid/gas
θ = contact angle

The so-called wetting tension j can be defined from the following equation:

$$j = \gamma_s - \gamma_{sl} = \gamma_l \cos\theta \tag{3.9}$$

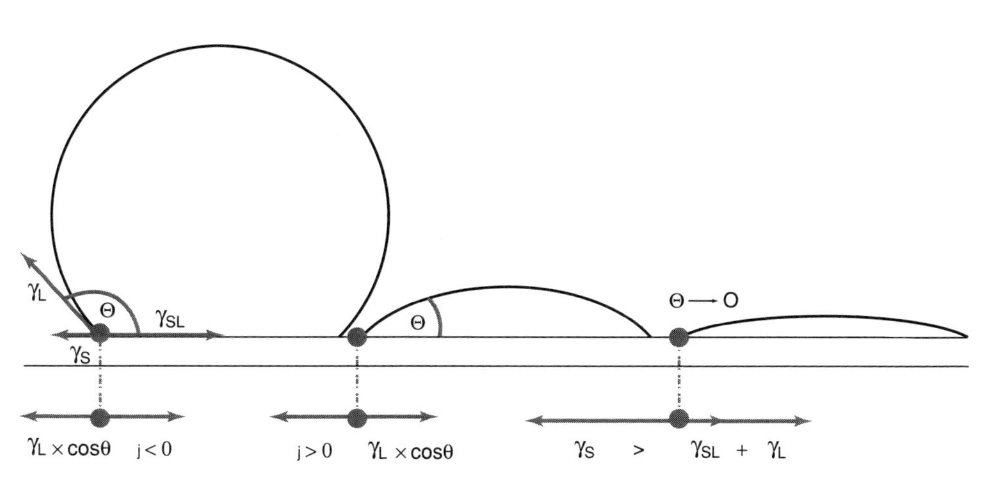

Figure 3.9 Schematic of the wetting of solid surfaces.

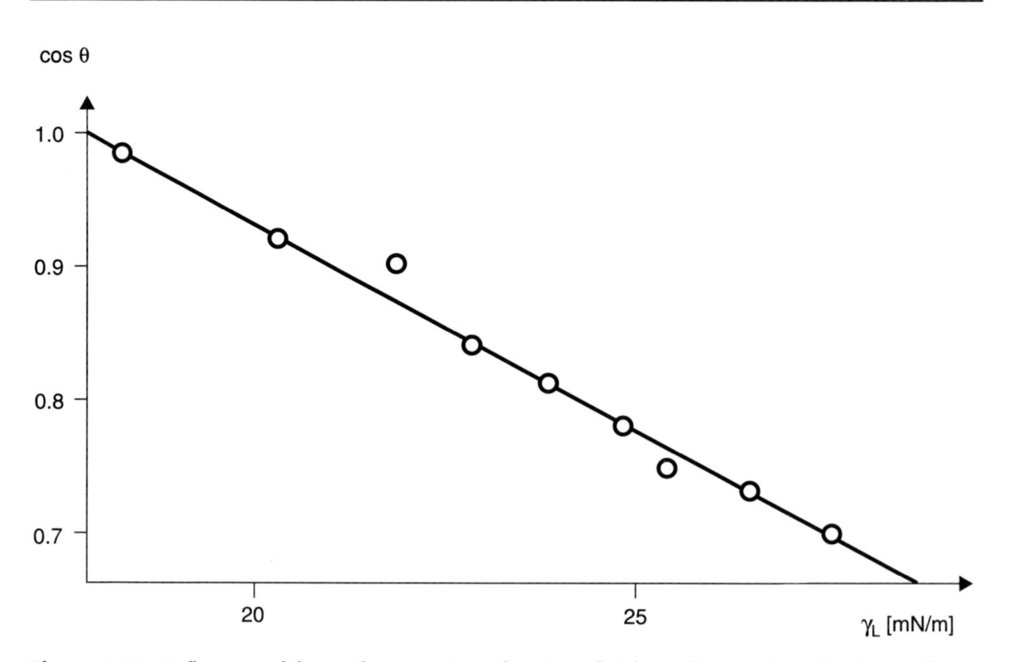

Figure 3.10 Influence of the surface tension of various fluids on the wetting of polytetrafluoro ethylene.

A complete wetting of a solid is only possible for spontaneous spreading of a drop of liquid at the surface, i.e. for $\theta = 0$ or $\cos\theta = 1$. For a specific solid surface of low surface energy, a linear correlation is observed between $\cos\theta$ and the surface tension. This is demonstrated for polytetrafluoro ethylene in Figure 3.10.

The limiting value for $\cos\theta = 1$ is a constant for a solid and is named critical surface tension of a solid γ_c. Therefore, only liquids with $\gamma_l \leq \gamma_c$ are able to spread spontaneously on surfaces and to wet them completely.

Table 3.4 gives an overview of critical surface tension values of different polymer surfaces [10]. From these data it is obvious that polytetrafluoro ethylene surfaces can only be wetted by specific surfactants with a very low surface tension, e.g. fluoro surfactants.

Figure 3.11 shows the wetting tension of two all-purpose cleaners for different surfaces [11]. As the wetting tension is in very good agreement with the surface tension of the cleaners, a spreading of the cleaner solution on the surfaces and therefore a good wetting can be assumed. Only on polytetrafluoro ethylene surfaces is an incomplete wetting observed.

In cleaning and washing, the situation becomes more complicated due to the presence of oily or fatty soil on the surface. In this case there is a competition of the wetting by the surfactant solution and that of the oily soil (see Figure 3.12).

When two droplets – one of surfactant solution and the other of oily soil – are set on a solid surface, on the basal plane two wetting tensions j_A and j_B will act [3]. When the two droplets approach each other, so that a common interface is formed, at the contact line the difference of the wetting tension will act. This parameter is called oil displacement tension:

$$\Delta j = j_A + j_B \quad (3.10)$$

Table 3.4 Critical surface tension of polymer solids [10]

Polymer	γ_c at 20°C (mN m^{-1})
Polytetrafluoro ethylene	18
Polytrifluoro ethylene	22
Poly(vinyl fluoride)	28
Polyethylene	31
Polystyrene	33
Poly(vinyl alcohol)	37
Poly(vinyl chloride)	39
Poly(ethylene terephthalate)	43
Poly(hexamethylene adipamide)	46

By adsorption of the surfactant from phase A, j_A is increased and thus Δj becomes larger. In addition to this a fraction of the interfacial tension γ_{AB} acts on the horizontal plane having a value of $\gamma_{AB} \cos\theta$ with θ being the contact angle in B, i.e. the oily phase. The resulting force R is called contact tension and is defined as:

$$R = \Delta j + \gamma_{AB} \cos\theta \qquad (3.11)$$

When R becomes zero, equilibrium is reached. For the washing and cleaning process the complete removal of the oil B by the surfactant solution A is the important step. This process is schematically shown in Figure 3.13 [12].

The interfacial tension γ_{AB} supports for $90° > \theta > 0°$ the contraction of the oil drop in the first step. For a contact angle $\theta > 90°$ this changes and the interfacial tension acts in

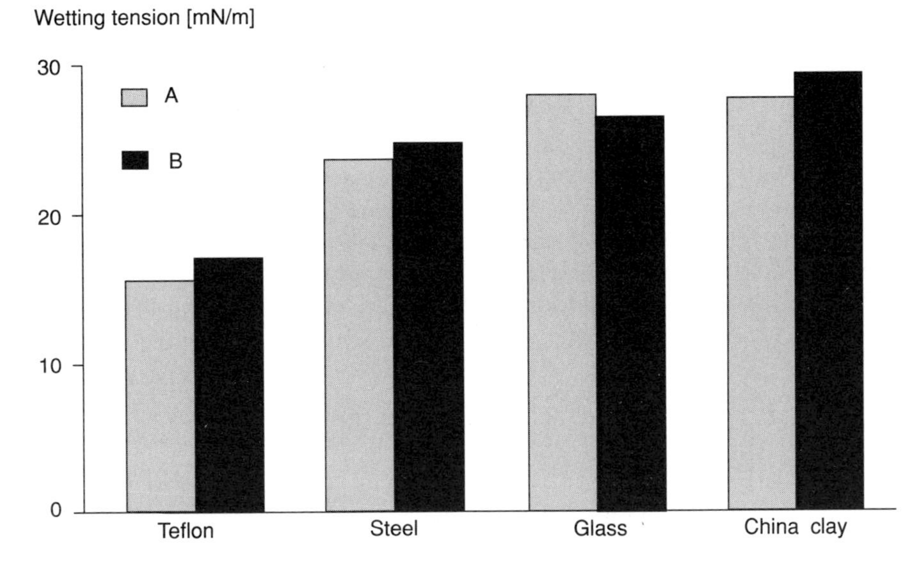

Figure 3.11 Wetting tension of two all-purpose cleaners versus different surfaces [11].

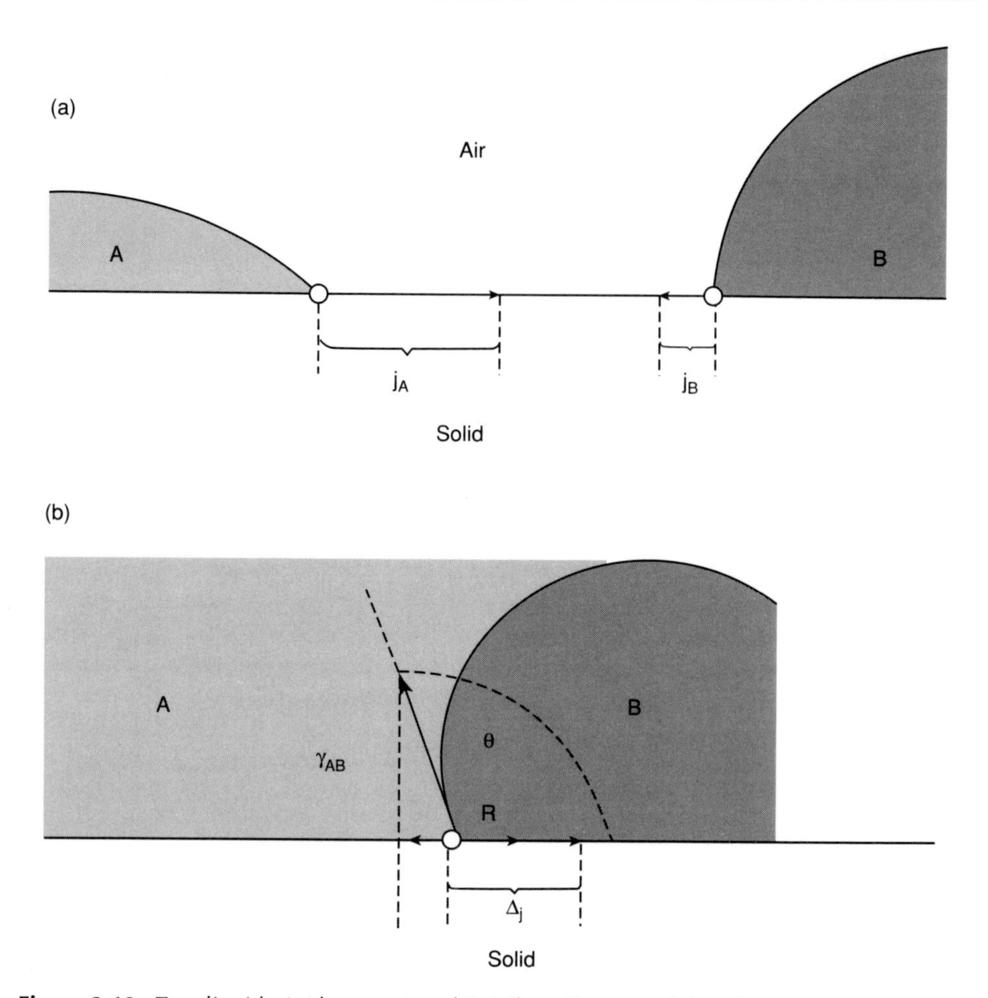

Figure 3.12 Two liquids A (detergent) and B (oily soil) on a solid surface: (a) separated and (b) in contact, j_A and j_B = wetting tensions, γ_{AB} = interfacial tension, R = interfacial wetting tension [3].

an opposite way. Depending on Δj and γ_{AB}, a complete removal of the oil can occur. In practice, the rolling-up is never complete, so that a support of the removal of the oil drop from a solid surface by mechanical forces is necessary for the washing and cleaning step.

3.2.5 Interplay of surfactants with other detergent ingredients

The presence of surfactants also influences the overall efficiency of other active ingredients of detergents, e.g. bleach and enzymes. A first requirement for the efficacy of these ingredients is a wetting of the textile substrate and hydrophobic or oily soils by the washing liquor. Besides this wetting function, all the other effects of soil removal of surfactants which have been

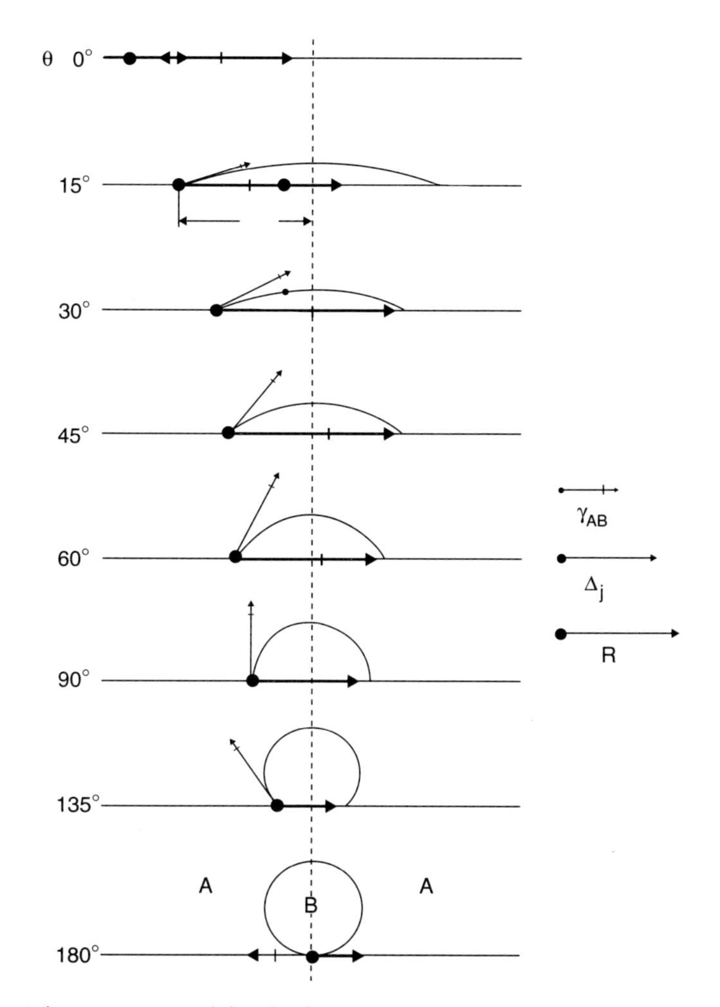

Figure 3.13 Schematic view of the displacement phases of an oily drop B by a cleanser A (reproduced with permission [12]).

discussed above also come into play. Bleaching agents in detergency usually are oxidative systems on the basis of chlorine or oxygen. The major effects of a bleaching agent are:

1 Oxidation and, hence, discolouration of chromophores in coloured soils such as tea, red wine or fruit juice
2 Hydrophilisation of hydrophobic soils by partial oxidation. In this way the wettability and the detachment of these soils is promoted
3 Cleavage of polymeric structures of soils so that the fragments of these soils can be more easily removed by surfactants
4 Disinfection by killing germs

As can be easily seen, effects 2 and 3 can work better in the presence of surfactants because the chemical fragments of the soil have still to be removed.

The general mode of action of detergent enzymes is quite similar. Detergent enzymes usually belong to the class of so-called hydrolases. These enzymes are able to split polymeric structures of stubborn soils such as proteins (e.g. blood, egg or starch) by hydrolysis and the fragments of the polymeric structures have to be subsequently detached by the surfactant system.

3.3 Phase behaviour of surfactants

3.3.1 Introduction

The phase behaviour of the surfactant systems is decisive for the formulation of liquid and solid products and the mode of action of the surfactants in soil removal during the washing and cleaning process. Due to the different phases of surfactant systems e.g. the flow properties can vary very strongly depending on the concentration and type of surfactants. This is of crucial importance for the production and the handling of liquid products. The dependence of the rheological behaviour of surfactant solutions on their microstructure is discussed in detail in Section 3.6. In addition to this the phase behaviour influences the dissolution properties of solid-surfactant-containing products when water is added, forming or preventing high-viscous phases. One can distinguish between the phase behaviour of surfactant–water systems and multi-component systems including an additional oil phase which occurs when, for example, soil is released from surfaces or in emulsification processes. This is discussed in detail in Section 3.4.

3.3.2 Surfactant phases

Surfactants form micelles beyond a certain cmc. The dimension of these aggregates can vary over several orders of magnitude from some nanometres to several hundred nanometres or even micrometres. As a rule of thumb, small globular micelles simply have the diameter of two times the chain length of the surfactant. With increasing surfactant concentration, the size or the concentration of the micelles increases while the concentration of the monomeric surfactant molecules in the solution remains constant, approximately equal to the cmc. It is important to notice that these micelles are no static structures. Rather they show a continuous exchange of surfactant molecules with the monomers in the solution and, in the same way, complete micelles can disintegrate and re-form. On dilution below the cmc, the micelles disappear completely. The time constants of this exchange process can range over several orders of magnitude from milliseconds to many seconds. Usual techniques for their detection are temperature jump or pressure jump experiments [13].

The shape of the micelles formed in a solution can be illustrated by a geometric consideration. A surfactant molecule consists of the hydrophilic headgroup including the solvation shell and the hydrophobic chain. In an aggregate, the hydrophilic headgroups are oriented to the outside so the shape of the aggregate is determined by the area required by the solvated headgroup, A_h, on the micellar surface and the cross section of the hydrophobic chain. It can be described by the packing parameter P [14]:

$$P = A_h l_c / V_c \tag{3.12}$$

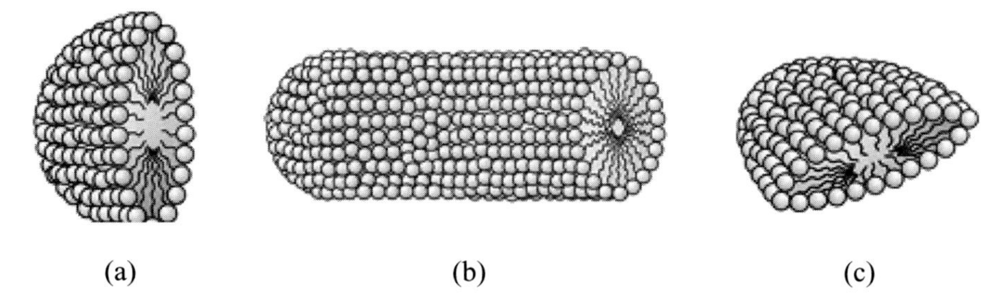

Figure 3.14 Schematic of surfactant aggregates (cut open): (a) spherical micelle, (b) rod-like micelle and (c) disc-like micelle.

l_c is the length of the hydrocarbon chain and V_c is the volume of the chain. The headgroup area, however, is not only a geometric issue since, besides the hydrate layers, it is also determined by electrostatic repulsion in the case of ionic surfactants. Consequently, it is dependent on the concentration of the surfactant in the solution, the degree of dissociation, the temperature and on the ionic strength of excess electrolyte which is present in many applications. For $P > 3$ spherical micelles (sometimes also termed globular micelles) are formed because the large headgroups have most space on the surface of a sphere. For $3 > P > 2$ rod-like or cylindrical micelles are formed and for $2 > P > 1$ lamellar bilayers or disk-like micelles arise. For $P < 1$ the headgroup area is smaller than that of the hydrocarbon chain and reverse aggregates occur. According to this consideration the sequence of micellar shapes in many cases is spherical, cylindrical, lamellar and reverse, but this is not necessarily the case: it is also possible that a surfactant forms rod-like micelles directly above the cmc.

The diameter of a spherical micelle and a rod-like or cylindrical micelle and the thickness of a bilayer are given by the length of the hydrocarbon chain of the surfactant and do not vary with concentration. The length of rod-like micelles, however, can increase with increasing surfactant concentration. Here the rods can become flexible and form entanglement networks similar to polymers. The interesting rheological properties of these systems are discussed in Section 3.6. The three types of aggregates are schematically represented in Figure 3.14.

At higher surfactant concentrations, a packing of the micellar aggregates induces the formation of lyotropic liquid crystalline phases, sometimes also called mesophases. These phases are crucial in the manufacture and mode of action of detergents and have an important role in cosmetics. A dense packing of spherical micelles in a cubic lattice leads to cubic phases. Rod-like micelles form hexagonal phases in which the long axis of the rods is packed on a hexagonal lattice. Bilayers form lamellar phases. At high concentrations inverse phases can also follow. Thus, the sequence of phases with increasing surfactant concentration is usually isotropic micellar (L_1), cubic, hexagonal (H_1), lamellar (L_α) and reverse. The transition from one phase to another occurs owing to a change in concentration but, of course, the temperature can also cause phase transitions. The single-phase regions are often separated by two-phase regions. It should be emphasised that a certain surfactant does not necessarily exhibit all the mentioned liquid crystalline phases. Rather many systems show, for example, a hexagonal or a lamellar phase as the first liquid crystalline phase. Moreover systems are known with a packing parameter of the surfactant close to unity and these systems can form lamellar phases at very low concentration (ca 1%). There may be stacked lamellar phases

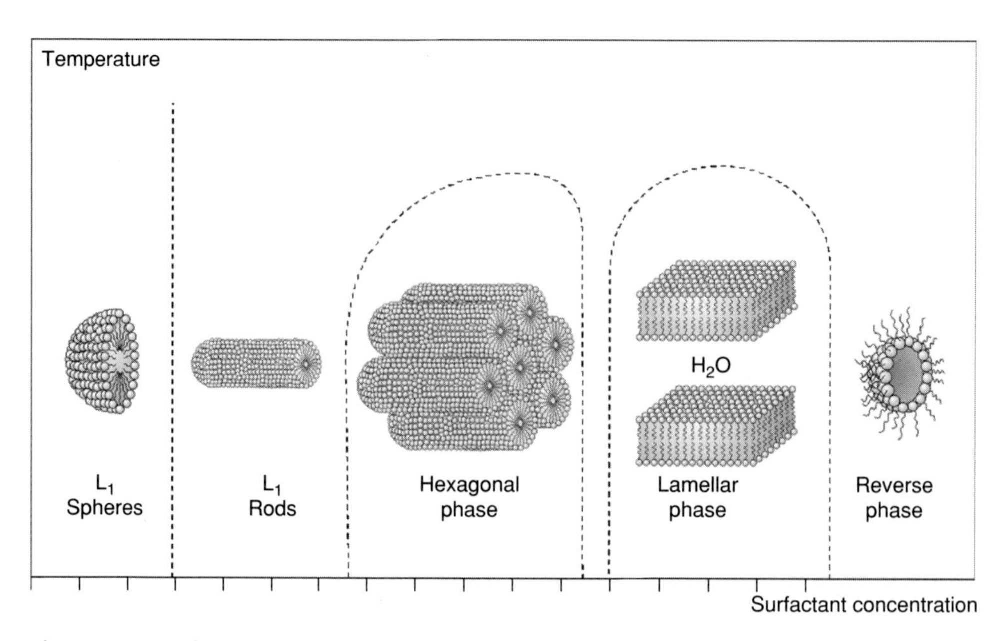

Figure 3.15 Schematic phase diagram of a surfactant: a typical sequence of phases with increasing concentration is micellar isotropic (L_1), hexagonal (H_1), lamellar (L_α) and reverse.

or phases of vesicles; a schematic phase diagram with a typical phase sequence is shown in Figure 3.15.

Phases with flat lamellae can show fascinating iridescent effects if the distance of the lamellae is in the range of the wavelength of the visible light [15–17]. Then Bragg reflection occurs and the solution plays in various colours when it is illuminated with white light against a black background. The colour depends on the angle of observation and there are samples which show the full rainbow spectrum. Vesicles may be unilamellar spheres or multi-lamellar, a kind of onion-like arrangement of bilayers. These systems show interesting rheological properties that are discussed more in detail below. Sometimes so-called sponge phases (L_3 phases) are observed which consist of bicontinuous bilayers, neither flat nor with a uniform curvature-like vesicles but with saddle point-like structures or branched tubes [18]. A schematic sketch of these structures is given in Figure 3.16.

The most common and easily applicable method of characterising liquid crystalline mesophases is polarisation microscopy. In this method, thin samples of the surfactant solution are viewed under a microscope between crossed polarisation filters. Due to optical anisotropy of liquid crystals they are birefringent. Hence, they give rise to a brightness in the microscope and show patterns that are very characteristic for the specific phases: examples are shown in Figure 3.17.

It should be emphasised that the micellar structures themselves are still too small to be seen under an optical microscope but they form domains of uniform orientation that can be observed. Other methods for the characterisation of mesophases are scattering methods, e.g. neutron scattering, x-ray diffraction or rheology as pointed out in Section 3.6. For a more detailed description see the literature relevant to this subject [19].

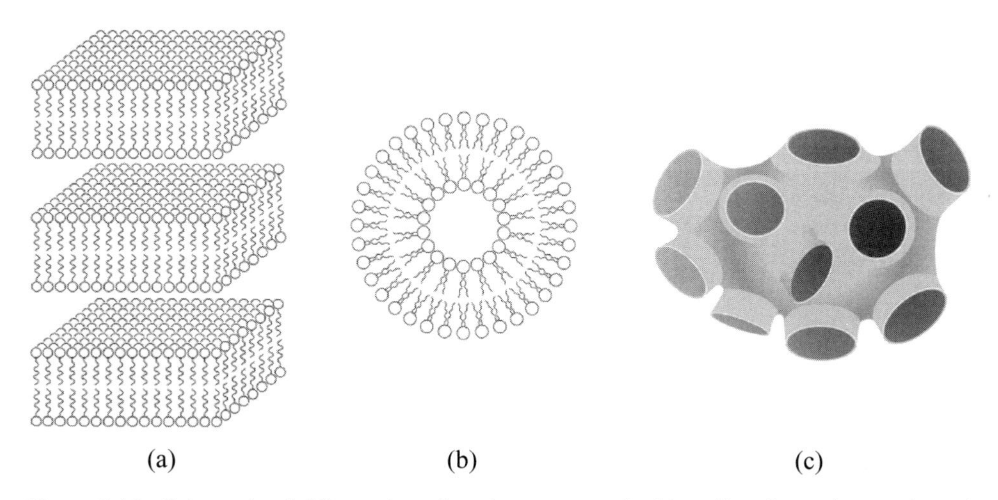

Figure 3.16 Schematic of different lamellar phases: (a) stacked lamellar phase, (b) vesicle and (c) L_3 phase; the grey area consists of a surfactant double layer similar to those in (a).

As a practical example for the phase behaviour of surfactants, Figure 3.18 shows the phase diagram of a pure non-ionic surfactant of the alkyl polyglycol ether type C_nE_m. n denotes the length of the hydrocarbon chain and m the degree of ethoxylation [20].

However, the phase behaviour of non-ionic surfactants with a low degree of ethoxylation m may be more complex than the schematic description above. As the lower consolute boundary is shifted to lower temperatures with decreasing EO (ethylene oxide) number of the molecule, an overlapping of this boundary with the mesophase region may result, as depicted in Figure 3.18. At low surfactant concentrations in such systems, several two-phase areas are observed in addition to the single-phase isotropic L_1 range: there are two coexisting liquid phases ($W + L_1$), a dispersion of liquid crystals ($W + L_\alpha$) and a two-phase region of water in equilibrium with a surfactant liquid ($W + L_2$). The temperature of the phase separation is called the, 'cloud point' because the solution gets turbid on exceeding this temperature.

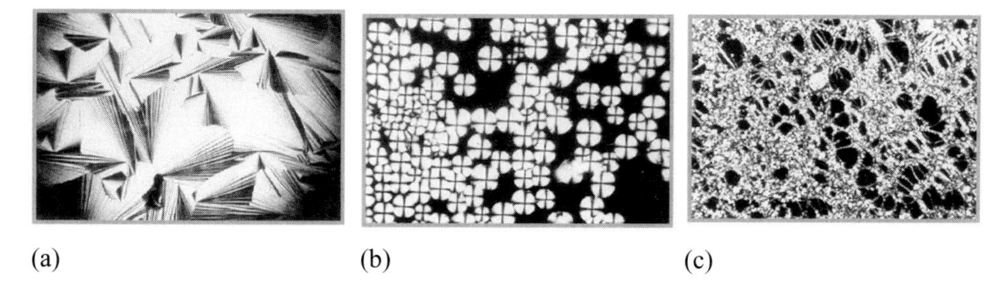

Figure 3.17 Patterns of liquid crystalline phases of surfactants under the polarisation microscope: (a) hexagonal phase; the typical fan-like structure can be seen, (b) lamellar droplets with typical Maltese crosses and (c) lamellar phase.

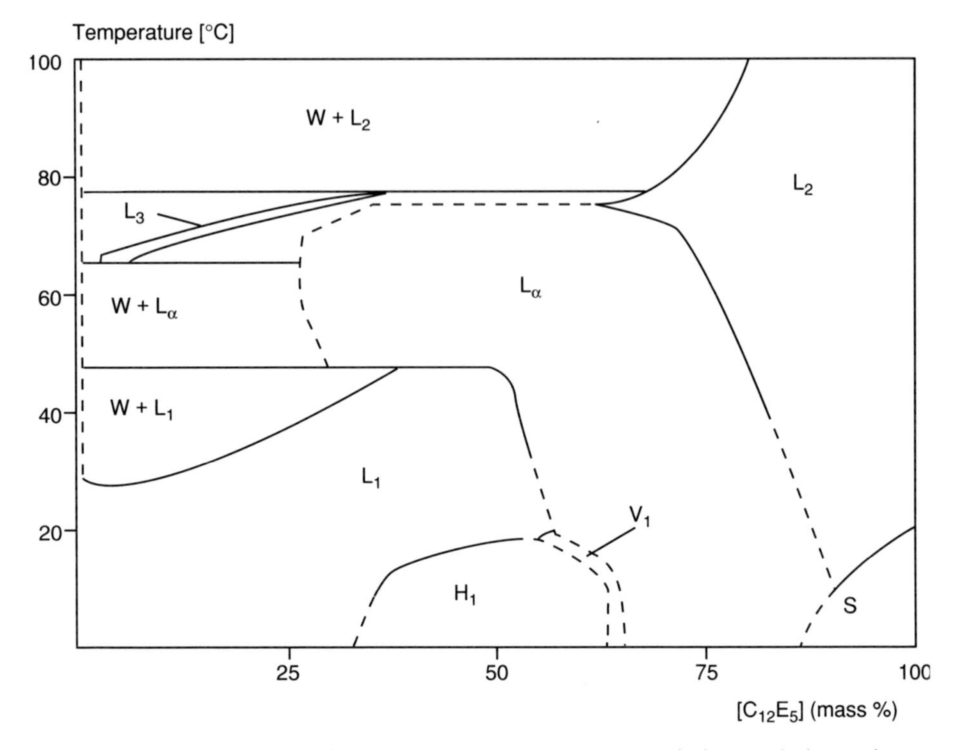

Figure 3.18 Phase diagram of the binary system water-pentaoxyethylene n-dodecanol ($C_{12}E_5$) (reproduced with permission [21]).

3.3.3 Impact of the phase behaviour on detergency

The phase behaviour can have a significant impact on detergency [21] but, if there is no phase change for the surfactant–water system, a linear dependence of detergency on temperature is observed (Figure 3.19).

The surfactant is in an isotropic micellar solution at all temperatures. The cloud point of the surfactant used here is 85°C at the given concentration (2 g l^{-1}), i.e., above the highest washing temperature.

Tests with other pure ethoxylated surfactants have revealed that a discontinuity is observed with respect to oil removal versus temperature in cases where there exist dispersions of liquid crystals in the binary system water/surfactant. Figure 3.20 shows that the detergency values for mineral oil and olive oil, i.e. two oils with significantly different polarities, are at different levels.

It also demonstrates that in both cases a similar reflectance vs temperature curve exists. In the region of the liquid crystal dispersion, i.e. between 20°C and 40°C, the oil removal increases significantly. Above the phase transition $W + L_\alpha \rightarrow W + L_3$, between 40°C and 70°C, no further increase in oil removal takes place. For olive oil, a small decrease in detergent performance is observed. The interfacial tensions between aqueous solutions of $C_{12}E_3$ and mineral oil lie at about 5 mN m^{-1} at 30°C and 50°C and these relatively high values indicate that, in this system, the interfacial activity is not the decisive factor in oil removal from fabrics.

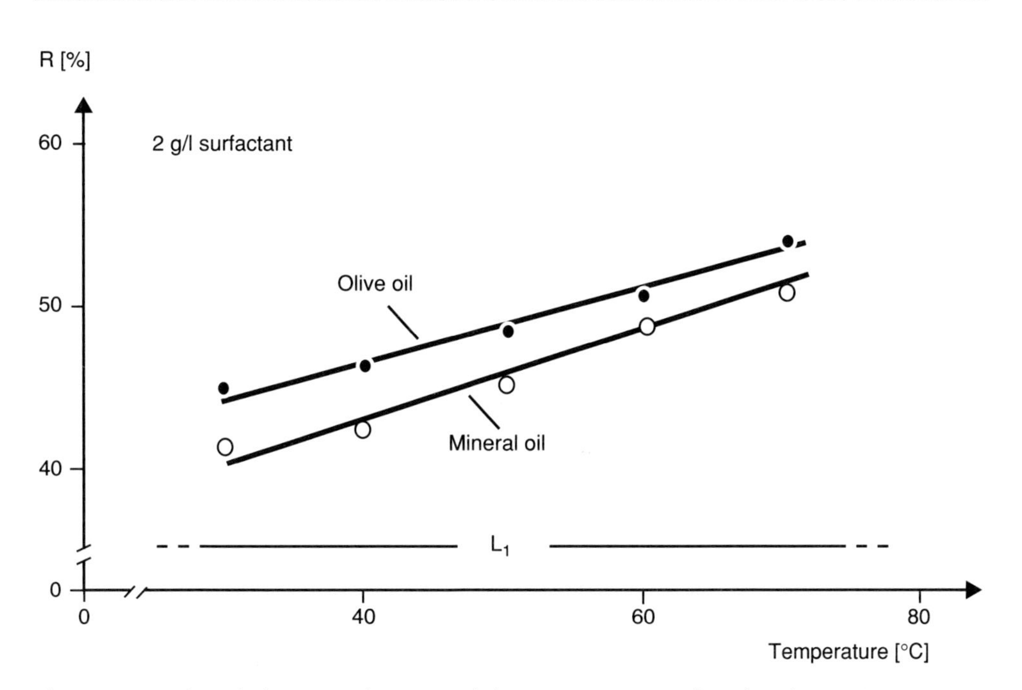

Figure 3.19 Phase behaviour of $C_{12}E_9$ and detergency R (reproduced with permission [22]).

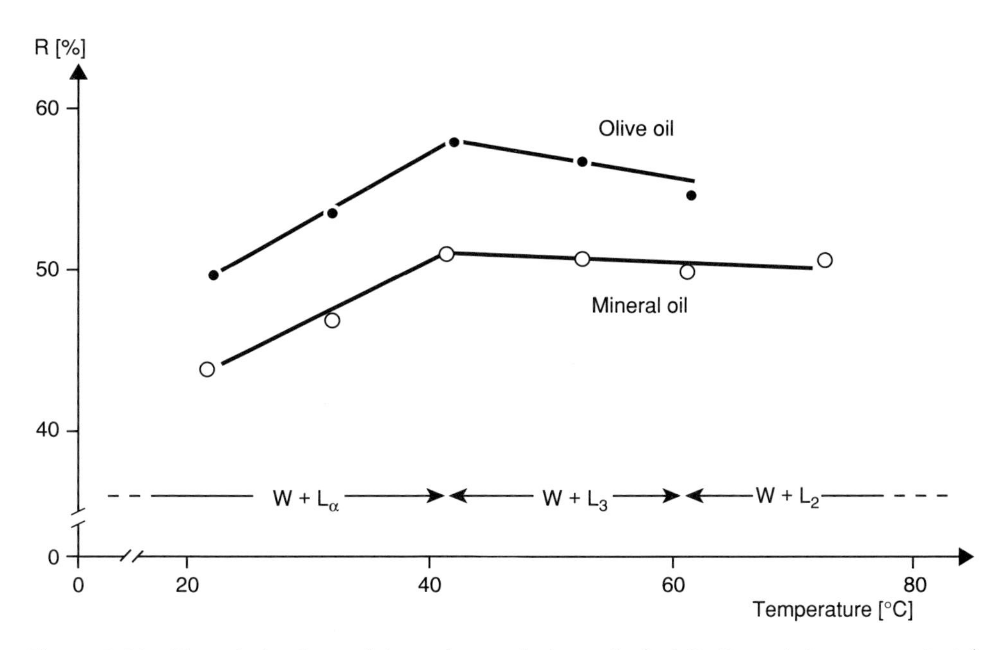

Figure 3.20 Phase behaviour of the polyoxyethylene alcohol $C_{12}E_3$ and detergency, $2\,gl^{-1}$ surfactant (reproduced with permission [22]).

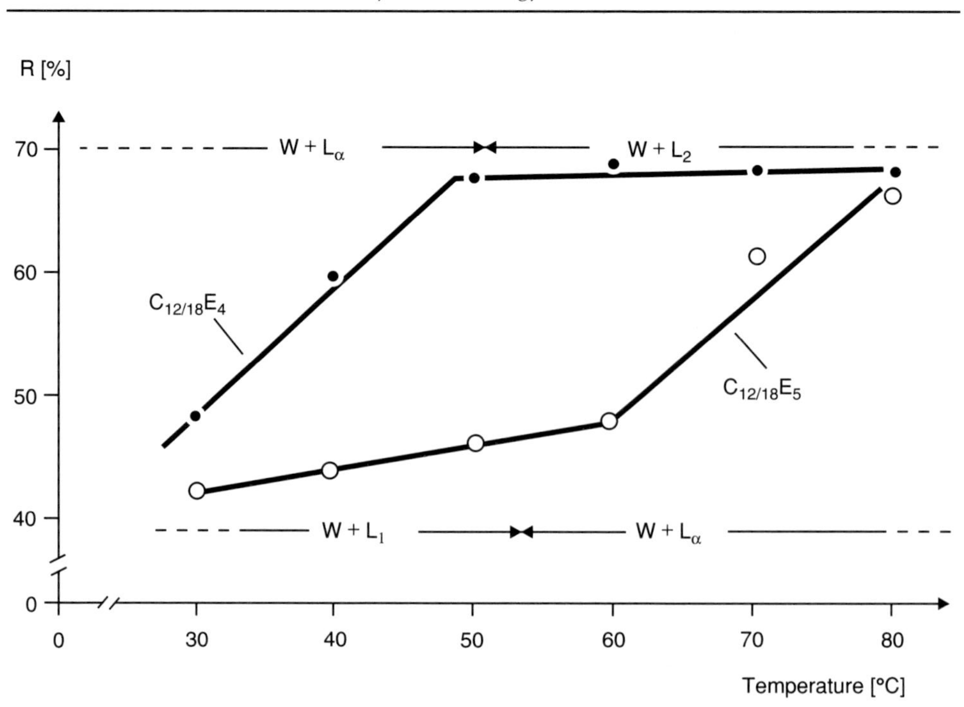

Figure 3.21 Phase behaviour of the polyoxyethylene alcohols $C_{12/18}E_4$ and $C_{12/18}E_5$ and detergency (reproduced with permission [22]).

The macroscopic properties of the liquid crystal dispersion seem to be responsible for the strong temperature dependence and it can be assumed that fragments of liquid crystals are adsorbed onto fabric and oily soil in the $W + L_\alpha$ range during washing. The local surfactant concentration is, therefore, substantially higher in comparison to the molecular surfactant layer that forms when surfactant monomers adsorb. As the viscosity of liquid crystals in the single-phase range is strongly temperature dependent, it can be assumed that the viscosity of a fragment of a liquid crystal deposited on a fabric also significantly decreases with increasing temperature. Thus the penetration of surfactant into the oil phase and removal of oily soil are promoted. Technical grade surfactants are of specific interest for applications. As in the case of pure non-ionic surfactants, definite ranges exist in which there is only a slight dependence of oil removal on the temperature (see Figure 3.21).

For $C_{12/18}E_5$, this is in the range of the two coexisting liquid phases ($W + L_1$) and for $C_{12/18}E_4$ it is in the range of the surfactant liquid phase ($W + L_2$), and an unusually strong increase of oil removal with increasing temperature occurs in the region of the liquid crystal dispersion ($W + L_\alpha$). At 30°C and 50°C the interfacial tensions between aqueous surfactant solutions and mineral oil and the contact angles on glass and polyester were determined for $C_{12/18}E_4$. Whereas the values of interfacial tensions are practically identical (approximately 10^{-1} mN m^{-1}), the contact angles on both substrates are slightly less advantageous at higher temperatures. Hence, the increased oil removal between 30°C and 50°C cannot be attributed to an increase in the adsorbed amounts of surfactants. Rather, in both cases, the decisive part is probably played by the macroscopic properties of the liquid crystal dispersion and their temperature dependence.

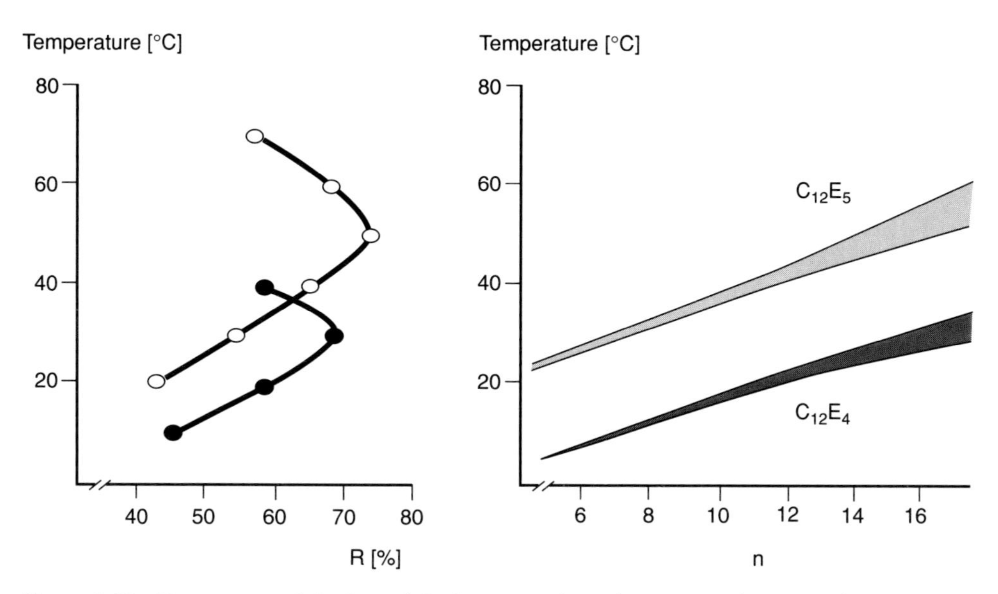

Figure 3.22 Detergency of $C_{12}E_4$ and $C_{12}E_5$ against hexadecane as a function of temperature (left) and the corresponding three-phase ranges for these surfactants as a function of the number *n* of carbon atoms of alkanes (reproduced with permission [22]).

Figure 3.22 (right) represents the three-phase temperature intervals for $C_{12}E_4$ and $C_{12}E_5$ vs the number *n* of carbon atoms of n-alkanes (for the phase behaviour of ternary systems see Section 3.4.2, Figure 3.26). The left part of Figure 3.22 shows the detergency of these surfactants for hexadecane. Both parts of Figure 3.22 indicate that the maximum oil removal is in the three-phase interval of the oil used (n-hexadecane) [22]. This means that not only the solubilisation capacity of the concentrated surfactant phase, but probably also the minimum interfacial tension existing in the range of the three-phase body is responsible for the maximum oil removal. Further details about the influence of the polarity of the oil, the type of surfactant and the addition of salt are summarised in the review of Miller and Raney [23].

3.4 Emulsions

3.4.1 *Introduction*

Emulsions are formed when two immiscible liquids are mixed with each other. The most familiar types are oil-in-water emulsions (O/W emulsions), which consist of colloidal or microscopic oil droplets in water, and water-in-oil (W/O emulsion), where an aqueous solution is emulsified in an outer oil phase [24].

Emulsions are not a human invention. In living nature they play an especially important role in the absorption of fats with nutrients. The earliest known use of an emulsion by humans is certainly the exploitation of milk and milk products such as cream, butter and cheese for nutritional purposes. With increasing prosperity the advanced civilisations of antiquity began to use emulsions for cosmetic purposes. Nowadays emulsions are applied in a wide variety of technical processes, so that they play a role in many everyday products and processes.

For practical applications emulsions are so attractive because they consist of at least two phases, an oil and an aqueous phase, so that they are suitable solvents for both hydrophobic and hydrophilic active substances. The pharmaceutical and agricultural sectors as well as soil remediation take advantage of the good solubilisation capacity of emulsions for substances with different polarity. In other fields the interactions of emulsions with solid surfaces play a central role. Cooling lubricants, rolling oil emulsions, fibre and textile auxiliaries and other lubricants have the primary task of lowering undesirable frictional effects during machining processes. Here the oily phase, emulsifiers and other auxiliary substances are adsorbed on the treated materials and cause lubricating effect whereas water with its high thermal capacity is responsible for the cooling effect.

3.4.2 Emulsion types

Emulsions made by agitation of pure immiscible liquids are usually very unstable and break within a short time. Therefore, a surfactant, mostly termed emulsifier, is necessary for stabilisation. Emulsifiers reduce the interfacial tension and, hence, the total free energy of the interface between two immiscible phases. Furthermore, they initiate a steric or an electrostatic repulsion between the droplets and, thus, prevent coalescence. So-called macroemulsions are in general opaque and have a drop size > 400 nm. In specific cases, two immiscible liquids form transparent systems with submicroscopic droplets, and these are termed microemulsions. Generally speaking a microemulsion is formed when a micellar solution is in contact with hydrocarbon or another oil which is spontaneously solubilised. Then the micelles transform into microemulsion droplets which are thermodynamically stable and their typical size lies in the range of 5–50 nm. Furthermore bicontinuous microemulsions are also known and, sometimes, blue-white emulsions with an intermediate drop size are named miniemulsions. In certain cases they can have a quite uniform drop size distribution and only a small content of surfactant. An interesting application of this emulsion type is the encapsulation of active substances after a polymerisation step [25, 26].

There is a common rule, called Bancroft's rule, that is well known to people doing practical work with emulsions: if they want to prepare an O/W emulsion they have to choose a hydrophilic emulsifier which is preferably soluble in water. If a W/O emulsion is to be produced, a more hydrophobic emulsifier predominantly soluble in oil has to be selected. This means that the emulsifier has to be soluble to a higher extent in the continuous phase. This rule often holds but there are restrictions and limitations since the solubilities in the ternary system may differ from the binary system surfactant/oil or surfactant/water. Further determining variables on the emulsion type are the ratios of the two phases, the electrolyte concentration or the temperature.

The solubility of the surfactant of polyethyleneglycol type in different phases can be described by the HLB (hydrophilic-lipophilic-balance) concept [27]. This concept attributes to the molecule a HLB number that represents the geometric ratios of the hydrophilic and the hydrophobic moieties. It should, however, be emphasized that the HLB does not represent a fundamental property of the system but is based on experience. For fatty alcohol ethoxylates the HLB can be calculated as follows:

$$HLB = E/5 \tag{3.13}$$

E denotes the wt. percentage of polyoxyethylene in the molecule.

As a rule of thumb, surfactants with $3 < HLB < 6$ are suitable for the preparation of W/O emulsions while surfactants with $8 < HLB < 18$ are O/W emulsifiers.

The HLB concept assumes that the emulsion type is mainly governed by the curvature of the interface. Large headgroups may need considerable space on the outside of oil droplets in a continuous water phase and cause a positive curvature of the interface. On the other hand, small hydrophilic headgroups can be forced together inside a water droplet whereas large hydrophobic moieties extend into the continuous oil phase. The interface now has a negative curvature.

This concept is, however, quite simplified and takes no account of the real conformation of the surfactant molecules adsorbed at the interface, which depends on variables such as electrolyte concentration, particularly the temperature or effects of further ingredients. The significance of the temperature in influencing the emulsion type can be illustrated by a system of equal amounts of water and hydrocarbon containing a certain concentration of the surfactant $C_{12}E_5$ (Figure 3.23).

At low temperatures an O/W microemulsion (O/W_m) is formed which is in equilibrium with an excess oil phase. This condition is termed a Winsor I system. At high temperatures the headgroup requires less space on the interface and, thus, a negative curvature can result. A phase inversion occurs and a W/O microemulsion (W/O_m) is formed which is in equilibrium with an excess water phase. This situation is termed a Winsor II system. At intermediate temperatures three phases – a water phase, a microemulsion D and an oil phase – are in equilibrium. This is called a Winsor III system. Here the curvature of the interfaces is more or less zero. Hence, the interfacial tension is minimum as depicted in Figure 3.24 (right) for the system $C_{12}E_5$, tetradecane and water.

For this system the temperature of phase inversion (PIT) is between 45°C and 55°C. Variation of both the temperature and the surfactant concentration in a system with a fixed ratio of water and oil leads to a phase diagram that is called informally the 'Kahlweit fish' due to the shape of the phase boundaries that resemble a fish. In Figure 3.24 (left), this diagram is given for the system water/tetradecane/$C_{12}E_5$. For small surfactant concentrations (<15%), the phases already discussed occur but, at higher emulsifier concentrations, the surfactant is able to solubilise all the water and the hydrocarbon which results in a one-phase microemulsion D or a lamellar phase L_α.

The described phase inversion phenomenon can be used in practice to prepare very fine and stable emulsions, so called PIT-emulsions. An example for the procedure is given in Figure 3.25.

First a coarse O/W emulsion is prepared and, on heating, phase inversion occurs. After cooling down through the microemulsion zone, the finely dispersed nature of the microemulsion is partially retained and emulsions with drop sizes of about 100 nm result [28–30]. They show considerable long-term stability as a consequence of the Brownian motion of the oil droplets [31] and pump sprayable deodorants are one of the cosmetic products based on this technology.

The Kahlweit fish, however, is only a special case for a fixed water/oil ratio of an even more complex phase behaviour of the ternary system water/oil/surfactant. The more general

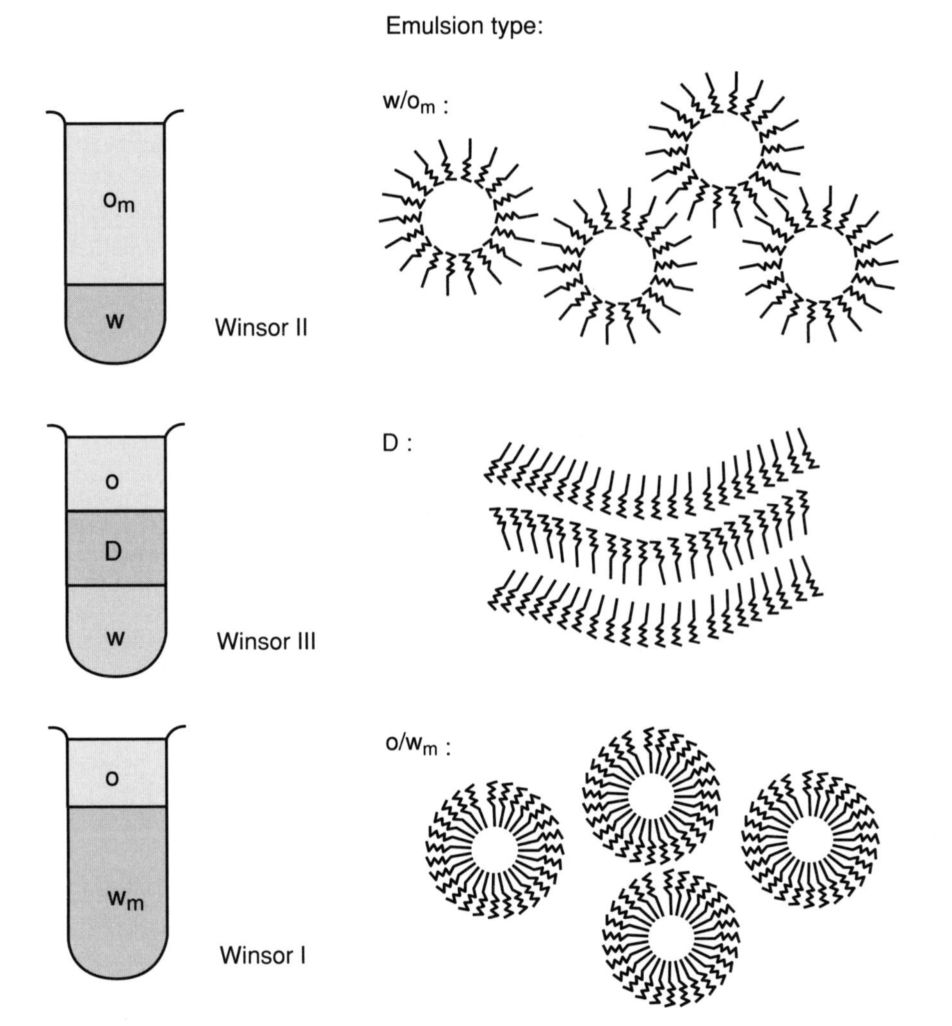

Figure 3.23 Schematic representation of the emulsion type depending on the temperature.

phase diagram for arbitrary water/oil ratios is depicted qualitatively in Figure 3.26 for alkyl polyglycol ethers as surfactants [31].

In the figure, the Kahlweit fish is represented by the bold line and the three-phase region is between T_l and T_u at low surfactant concentrations (in the foreground). The microemulsion zone is at high surfactant concentrations and extends over the whole temperature range (in the background).

This behaviour has a particular importance for the soil removal process in detergency. During the oil removal from stained fabrics or hard surfaces, ternary systems occur where three phases coexist in equilibrium. As already pointed out above, in this region the interfacial tension is particularly low. Because the interfacial tension is generally the restraining force,

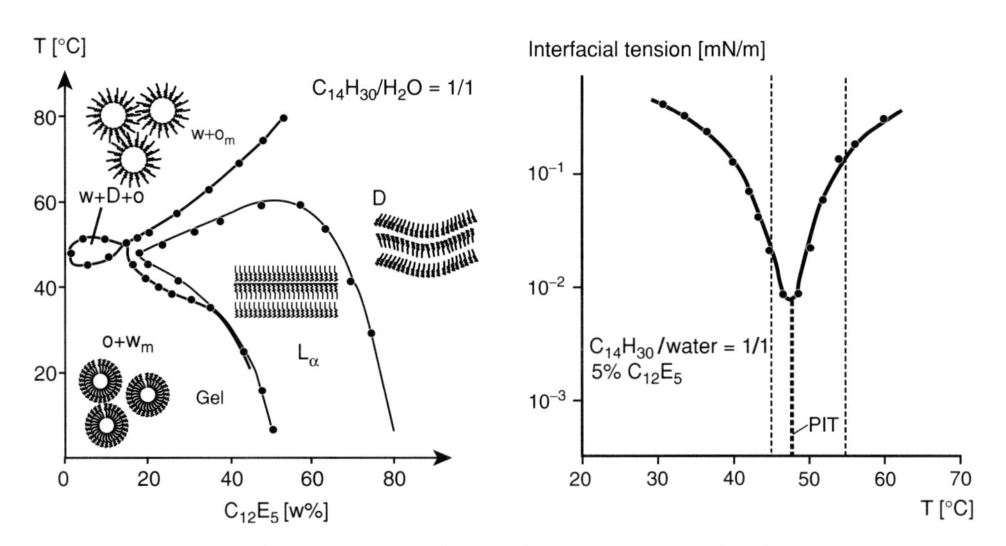

Figure 3.24 Left: emulsion type depending on the temperature and surfactant concentration ($C_{12}E_5$) for a constant tetradecane/water ratio of 1 : 1. Right: interfacial tension as a function of the temperature of the system tetradecane/water/$C_{12}E_5$.

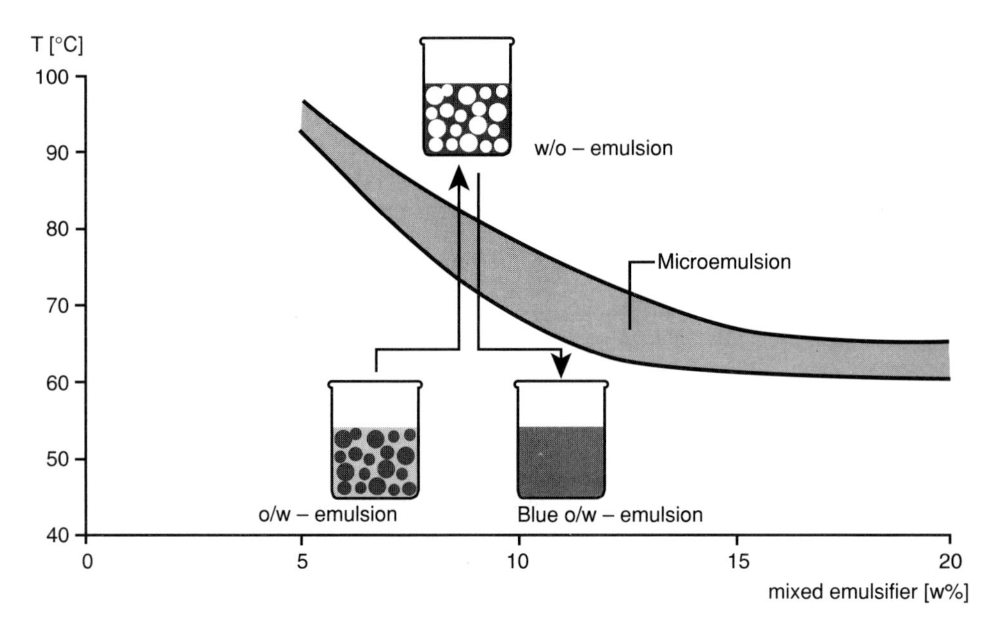

Figure 3.25 Procedure of the preparation of a PIT emulsion.

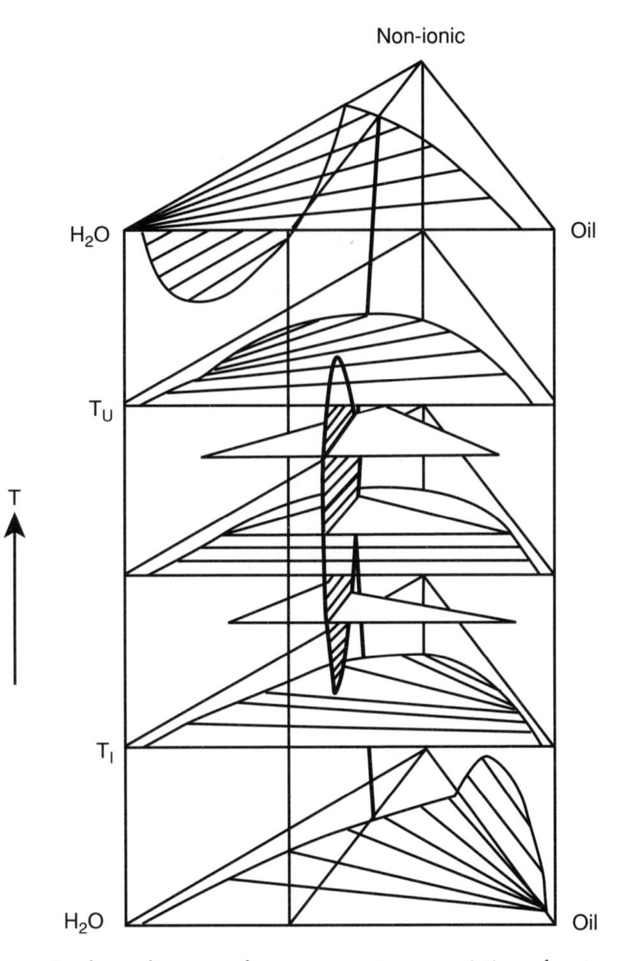

Figure 3.26 Schematic phase diagram of a ternary system consisting of water, oil and ethoxylated non-ionic surfactant.

with respect to the removal of liquid soil in the washing and cleaning process, it should be as low as possible for optimal soil removal.

Besides W/O and O/W emulsions there are so-called multiple emulsions of the W/O/W type. These emulsions can be produced in a one-stage modified PIT process [32] or by emulsification of a primary W/O emulsion in an outer water phase. These systems are an approach to protect sensitive active substances such as vitamins or enzymes in a formulation.

3.4.3 Breakdown of emulsions

The breakdown of emulsions can either be desirable or unwanted. Of course, cosmetic emulsions such as creams or cleansers have to be stable and become useless if separated. On the other hand, in processes such as enhanced oil recovery emulsions may be formed that are considerably stable and a notable effort is necessary for their separation.

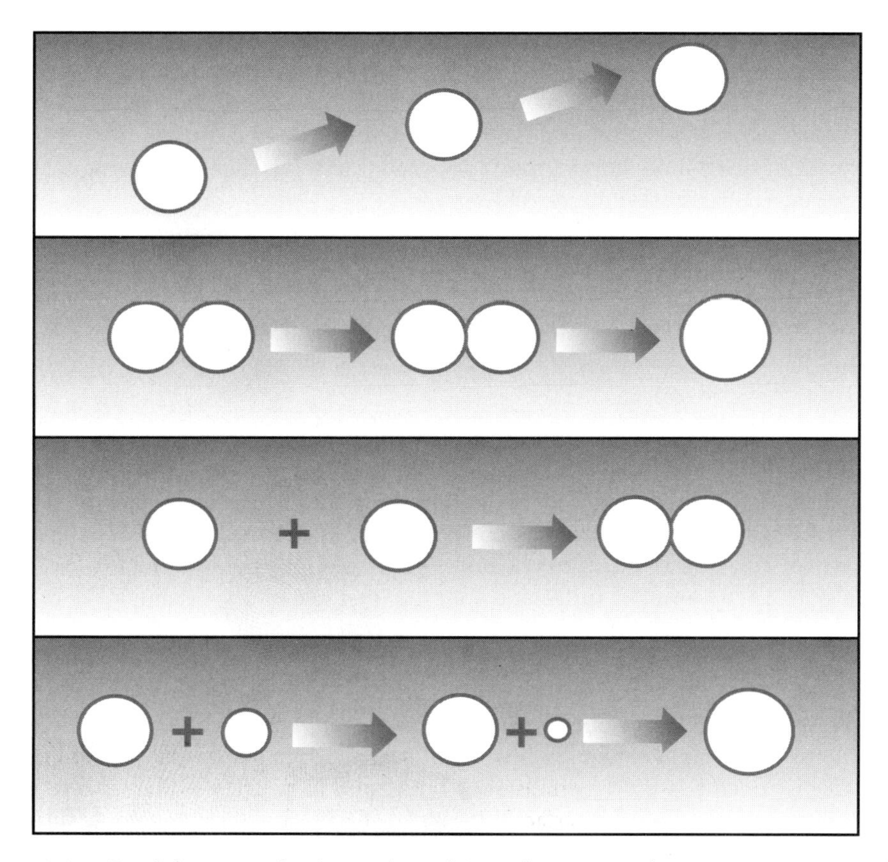

Figure 3.27 Breakdown mechanisms of emulsions (from top to bottom: creaming, coalescence, flocculation and Ostwald ripening).

Apart from microemulsions, all types of emulsions are thermodynamically unstable and their stability is solely a kinetic issue. The relevant timescale can vary between seconds and years. The following mechanisms are responsible for the breakdown of an emulsion. They are depicted schematically in Figure 3.27.

- *Creaming and sedimentation.* Creaming means the floating of the oil droplets of an O/W emulsion due to buoyancy – a consequence of the different densities of oil and water. On the topside of the sample a more concentrated oil phase is formed and on the bottom more or less pure water is left and so a concentration gradient develops. The distribution of the oil drop sizes may not yet be affected in an early stage of creaming. The equivalent process in a W/O emulsion is called sedimentation. In this case water droplets settle down to the bottom of the sample. In practice, creaming can be slowed down by reducing the density difference of the two phases, thickening of the continuous phase, e.g. by polymeric thickeners or by reduction of the drop size.
- *Coalescence.* In the case of coalescence, the separating film of the continuous phase between the droplets breaks and an irreversible fusion of emulsion droplets occurs.

- *Flocculation.* Flocculation means an aggregation of emulsion droplets but, in contrast to coalescence, the films of the continuous phase between the droplets survive. Hence, the process may be partially reversible. Both processes, flocculation and coalescence, speed up the creaming of an emulsion due to the increase of the drop size. The process of flocculation is even more important for dispersions of solids than for emulsions because in this case a coalescence is not possible.
- *Ostwald ripening.* The small residual solubility of oil in water gives rise to a process called Ostwald ripening. This solubility of the oil is further increased with decreasing drop size. Single oil molecules leave the smaller droplets into the water phase and, here, they can diffuse around and recondense in a larger droplet so the larger droplets grow at the cost of the smaller ones. The driving force of the process is the decrease of interfacial area and, thus, interfacial energy.

Generally speaking, for a stable emulsion a densely packed surfactant film is necessary at the interfaces of the water and the oil phase in order to reduce the interfacial tension to a minimum. To this end, the solubility of the surfactant must not be too high in both phases since, if it is increased, the interfacial activity is reduced and the stability of an emulsion breaks down. This process either can be undesirable or can be used specifically to separate an emulsion. The removal of surfactant from the interface can, for example, be achieved by raising the temperature. By this measure, the water solubility of ionic surfactants is increased, the water solubility of non-ionic emulsifiers is decreased whereas its solubility in oil increases. Thus, the packing density of the interfacial film is changed and this can result in a destabilisation of the emulsion. The same effect can happen in the presence of electrolyte which decreases the water solubility mainly of ionic surfactants due to the compression of the electric double layer: the emulsion is salted out. Also, other processes can remove surfactant from the water–oil interface – for instance a precipitation of anionic surfactant by cationic surfactant or condensing counterions.

3.5 Foaming and defoaming

3.5.1 Introduction

Foaming and the control of foam is an important factor in the application of surfactants, particularly for detergents and cleansers [33]. For some applications, a high foam is desired e.g. for manual dish-washing detergents, hair shampoos or detergents for manual textile washing. In these applications, foam is understood as an important measure of washing performance by the consumer who expects the product to generate voluminous and dense foam. In other cases, only a low foam is acceptable e.g. for use in textile washing machines – especially those with a horizontal axis of the drum – or dish washing machines. In these devices, the use of a high foaming detergent would lead to an overfoaming and furthermore to a decrease of the washing performance due to a damping of the mechanical action in the washing drum. Also, in institutional laundry or technical cleaning processes, high foam is not acceptable. Besides detergency there are many more applications in which surfactant foams play a role. They are used, for example, in enhanced oil recovery, for drilling operations, for flotation in mineral processes for fire fighting purposes and for personal care applications [34].

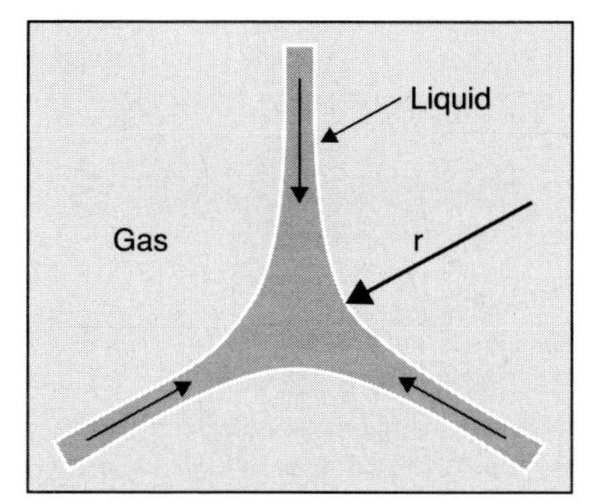

Figure 3.28 Sketch of the plateau border in a surfactant foam. Due to the curvature (radius *r*) a suction effect in the direction of the arrows results.

3.5.2 Stabilising effects in foams

A foam consists of a high volume fraction of gas dispersed in a liquid where the liquid forms a continuous phase. 'Wet' foams with a high water content, e.g. immediately after the formation, can have more or less spherical bubbles. As a consequence of a drainage process of the foam lamellae, the wet foam loses water with time. Due to the resulting high volume fraction of gas, the bubbles are no longer spherical but they are deformed into a polyhedral shape. The polyhedra are separated from each other by thin liquid films. The intersection lines of the lamella are termed plateau borders (see Figure 3.28).

Due to the radius of the plateau borders the pressure inside the plateau border is lower than that inside the adjoining lamella and this gives rise to a 'suction effect' of the plateau borders on the flat films between the foam bubbles. The plateau borders form a continuous network inside the foam and as long as the pressure difference between plateau border and lamella conveys liquid into the plateau border, the above mentioned drainage process of the foam occurs driven by gravity. As the film thickness falls under several tens of nanometres, a disjoining pressure of the surfaces comes into play. This disjoining pressure is caused by electrostatic and steric repulsion forces between the absorption layers on the surfaces and slows down the rate of film drainage. At a certain film thickness an equilibrium of the suction of the plateau border and the disjoining pressure is reached and the drainage process ceases.

In order to generate foam, surfaces of thin liquid films always have to be stabilised by layers of surfactants, polymers or particles. This is why pure liquids never foam. Foaming is always accompanied by an increase in the interfacial area and, hence, its free energy. Thus, in a thermodynamic sense foams are basically unstable and are, therefore, sooner or later destroyed. The lifetime of a foam can span a remarkable range from milliseconds to very long duration.

The situation in a foam lamella and another important stabilising mechanism of foam is depicted in Figure 3.29.

A surfactant film is adsorbed on the surfaces of the foam lamella and, depending on concentration, micelles may be present. If the foam lamella is stretched in the direction of the arrows, for instance by an external force, the surface film is depleted and the coverage is incomplete in this zone. This causes an increase of the surface energy in this zone and

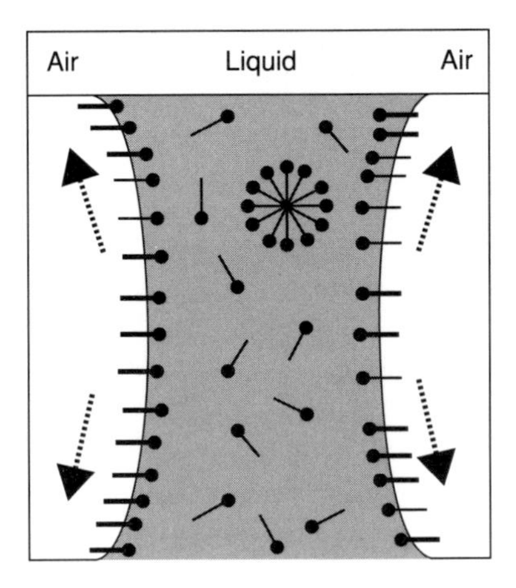

Figure 3.29 Situation in a foam lamella: the surface film, monomeric surfactant and a micelle are represented. A stretching of the lamella in the direction of the arrow causes a surface tension gradient.

a surface tension gradient arises. The surfactant tries to compensate this surface tension gradient by diffusion within the surface from the densely packed into the depleted zone dragging along a layer of hydrate water in the bulk phase. This process causes a self-healing of stretched or thin zones in a film and it is referred to as the Marangoni effect. This is an essential stabilising effect in foams. For neutralisation of the surface tension gradient by the Marangoni effect, it is a necessary prerequisite that the depletion of surfactant in the surface is not compensated by a diffusion of surfactant molecules from the bulk phase. Therefore, the effect is particularly pronounced for non-ionic surfactants with a slow adsorption kinetic or anionic surfactants slightly below the cmc (cf. Figure 3.8).

3.5.3 Correlation of foamability with interfacial parameters

The foam properties of products are mainly governed by the surfactant system and the use of anti-foams discussed below. Besides this the chemical composition of the product or the washing liquor, for example electrolyte content and soil, strongly influences the foam properties. Physical parameters such as temperature and pH value or mechanical input in the system additionally have to be taken into account.

The basis for the foam properties is given by interfacial parameters. An overview of some interfacial parameters and the correlation to foam properties is shown in Figure 3.30 [9].

It can be considered from the scheme that one has to distinguish between the foam kinetics, i.e. the rate of generation of foam under well defined conditions (air input and mechanical treatment) and the stability and lifetime of a foam once generated. The foam kinetics is also sometimes termed foamability in the literature. These quantities can be related to interfacial parameters such as dynamic surface tension, i.e. the non-equilibrium surface tension of a newly generated surface, interfacial rheology, dynamic surface elasticity and interfacial potential. In the case of the presence of oily droplets (e.g. an antifoam, a

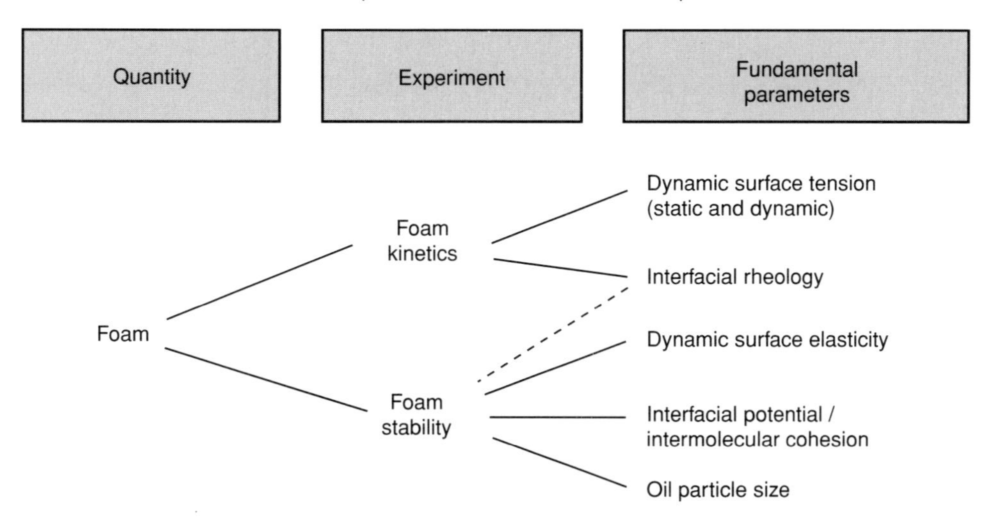

Figure 3.30 Interfacial parameters influencing the foam stability and foam kinetics (reproduced with permission [9]).

coacervate phase (above the cloud point of the surfactant system) or an oily soil) their drop sizes play a role. Small particles in foaming solution can have both effects, i.e. they can destroy the foam by entering and destabilising the surface film on a foam lamella or they can stabilise a foam by blocking the plateau borders and preventing the lamellae from drainage. All these parameters influence the foam properties in a complex way. Partially they have been studied in detail and, although correlations have been shown between a single parameter and foam properties, there is still a lack of a general correlation between interfacial properties and the foam behaviour of complex systems in applications. Here, a few examples should be given to illustrate the correlation of interfacial parameters and foam properties. The simplest approach to correlate interfacial parameters to foam properties is the comparison of the surface activity pronounced by the equilibrium surface tension of a surfactant system and the foam stability. This has been done for a series of a pure surfactant. Within a specific class of surfactants the surface tension directly correlates to the foam stability of the surfactant–water system but a more general approach of this concept is not possible due to the influence of other parameters summarised in Figure 3.30.

As foam generation and also foam stability are dynamic processes generating and reducing surface area, in a surfactant–water system the diffusion of the surfactant to the surface and the change in surface coverage, at least locally during bubble generation and drainage of the film, is a more useful way of explaining foam properties. If one distinguishes between foam formation and foam stability, a good correlation has been found between the relative (normalised) dynamic surface pressure $\Pi(t)/\Pi_{eq}$

$$\frac{\Pi(t)}{\Pi_{eq}} = \frac{\sigma_w - \sigma(t)}{\sigma_w - \sigma_{eq}} \tag{3.14}$$

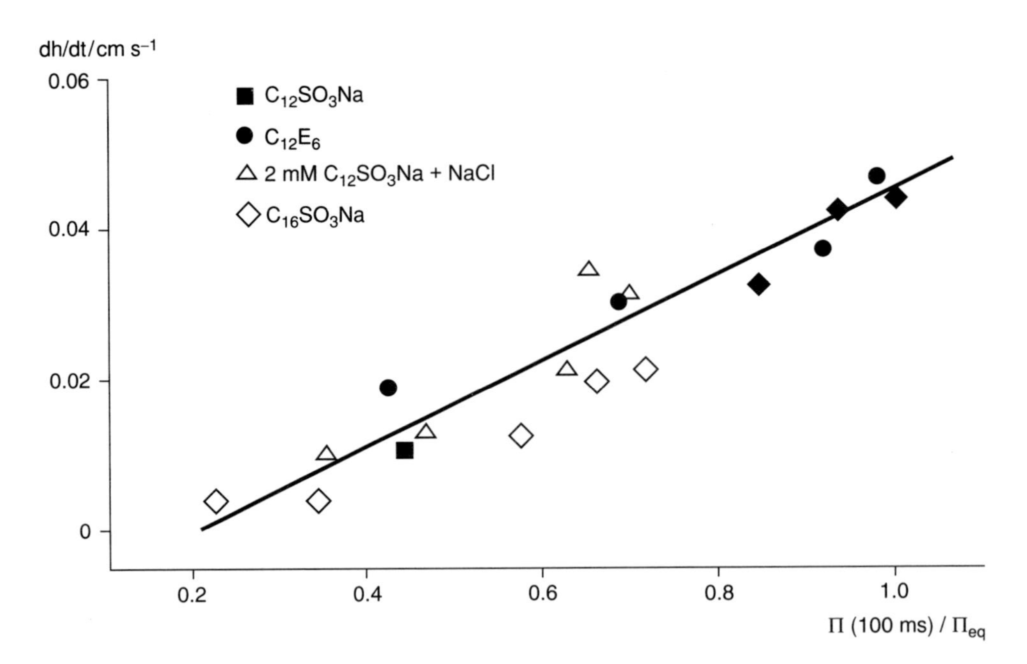

Figure 3.31 Correlation of the normalised dynamic surface pressure $\Pi(t)/\Pi_{eq}$ at a surface age of 100 ms with the rate of foam formation dh/dt. The correlation holds for various systems (reproduced with permission [9]).

derived from the non-equilibrium surface tension at a certain surface age (see Figure 3.8 in section 3.2.4) and the rate of foam formation (see Figure 3.31).

σ_w denotes the surface tension of water, $\sigma(t)$ is the surface tension of the surfactant solution after the generation of new surface and σ_{eq} is the equilibrium surface tension beyond the cmc.

The rate of foam formation is, for example, the growth of a foam column in an experimental device with a stirrer. In practice it may be the increase of the amount of foam in the drum of a washing machine or the amount of foam on the head of a person shampooing hair. The specific time for the relative dynamic surface pressure was chosen empirically to be 100 ms in these experiments and it may be attributed to typical time scales of the experiment. For instance, it can be the time between the generation of a new bubble and the moment this bubble reaches the surface of the foam for the first time. If enough surfactant diffuses from the bulk phase to the newly generated surface of the bubble in this time interval, this surface is stabilised and the bubble survives. Otherwise it bursts, releases its air content and, thus, does not contribute to the foam volume. The correlation of the two parameters is valid for different surfactants and the addition of electrolyte. Anionic surfactants of the alkyl sulphonate type with various chain lengths and non-ionic surfactants have been studied. The effect can be explained by the micellar kinetics of the surfactant solution and the diffusion of the molecules and micelles to the surface. The faster a surfactant is able to adsorb at a newly generated surface, the higher the rate of foam formation. Therefore, ionic

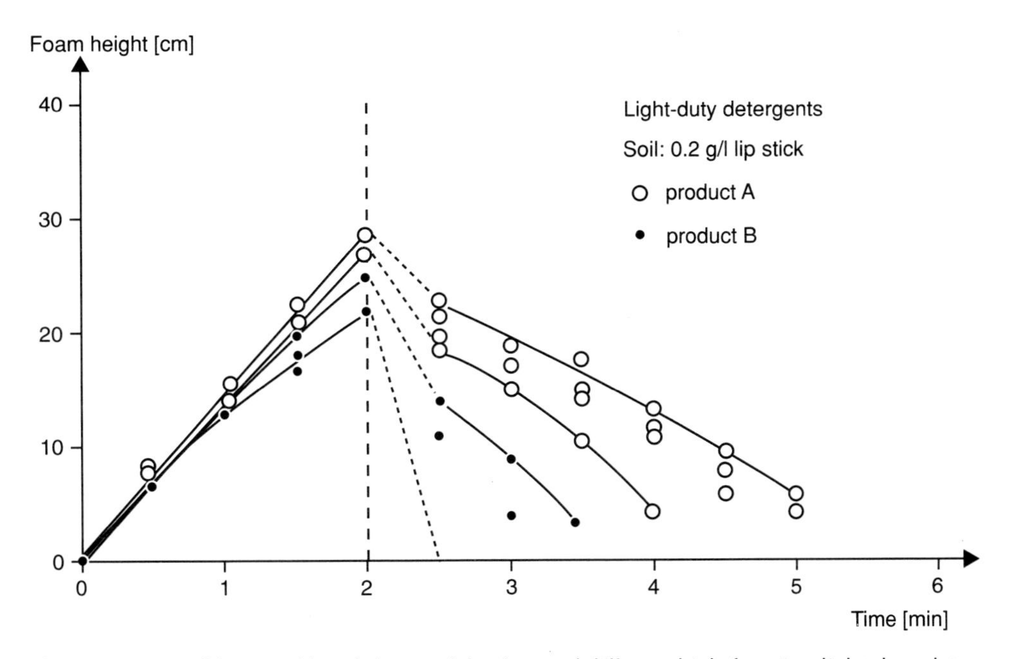

Figure 3.32 Build-up and breakdown of the foam of different high-foaming light-duty detergents in presence of an oily soil (reproduced with permission [9]).

surfactants in general show a higher rate of foam formation than non-ionic surfactants. In the case of the non-ionic surfactants the more hydrophilic ones, e.g. alkyl polyglycosides, exhibit a higher foam than the more hydrophobic ones. The anionic alkylethersulphates are surfactants with very high foamability and, therefore, they are frequently used in applications in which high foam is required. Other surfactants which show a strong foam tendency or a high foam stability are fatty alcohol sulphates, alkyl glucon and alkyl glucamides.

The importance of the described effects for finding the optimum surfactant system in detergents is shown in Figure 3.32. For high-foaming light duty detergents, the foam stability of the products is shown in the presence of oily soils which usually suppress the foam formation. It can be demonstrated that foam stability strongly depends on the formulation, i.e. the surfactant system and, in this case, can be adjusted at a high level for care aspects of the detergents towards sensitive textiles. This care effect is due to the reduction of the mechanical action of the foam in a washing drum.

3.5.4 Foam control

Sometimes, the developer of surfactant products is compelled to use high foaming surfactants for applications in which only little foam is acceptable, e.g. in detergency and, in these cases, antifoams are necessary. Antifoams are substances or mixtures of substances which are able to reduce the foaming tendency of a system or destroy an existing foam when added in small quantities [33]. These substances typically exhibit a high surface activity, i.e. a surface spreading pressure and they are scarcely soluble in the foaming liquid. As already mentioned, many oils and fatty substances are used as defoamers. In detergents, the defoaming effect of calcium soaps is often used as an efficient defoaming system. These calcium soaps are formed in situ by the precipitation of fatty acids with calcium originating either from water hardness or calcium-containing soil. A disadvantage of these systems may be that they have only a small effect in water with low hardness. In these cases overfoaming may result.

Silicone oils, primarily those of the polydimethylsiloxane-type, show a particularly prominent defoaming effect and, since they are chemically inert, they do not interfere with other components of the formulation.

The mode of action of antifoams is usually explained by assuming that they either force surfactant molecules away from the interfaces or they penetrate interfaces that are already occupied by surfactants, thereby creating defects. These defects weaken the mechanical strength of the foam lamellae and cause their rupture.

Another way of foam regulation is the use of special low foaming surfactants. This is, however, only possible for certain applications, e.g. automatic dishwashing or rinsing, because it is hardly possible to combine both properties in one surfactant molecule: low foam and optimum washing power. Low foaming surfactants are usually fatty alcohol alkoxylates or end-capped fatty alcohol alkoxylates with several moles of propylene oxide or butylene oxide (e.g. EO-PO adducts). Their defoaming effect arises because the temperature of application is above their cloud point. Contrary to common ethoxylates, the composition of the precipitating surfactant-rich coacervate phase of EO-PO surfactants is very different from that of the continuous phase. Hence, the droplets of the coacervate have a particularly strong defoaming effect. The defoaming effect of the coacervate of an ethoxylate is lower because the compositions of the two phases are quite similar [35].

3.6 Rheology of surfactant solutions

3.6.1 Introduction

The rheology of surfactant solutions and surfactant-containing products is of great importance for practical applications both from a technical point of view and in the perception of consumers. People often attribute special properties to a product, e.g. ‘richness’, on the basis of its flow behaviour, for instance for liquid detergents. Regarding cosmetic applications, the rheology is even more important for the acceptance of a product by the consumer. Here, for example, the feeling on the skin is determined by the rheological behaviour of the product. In practice polymeric thickeners are frequently used to tailor-make the flow behaviour of surfactant containing products. Sometimes, however, the intrinsic rheological behaviour of the surfactants is appropriate to design the flow properties. From a scientific point of view the rheological study of surfactant solutions is interesting because the rheological behaviour is directly linked to the microstructure, i.e. the micellar or liquid crystalline structure. The viscosities of surfactant systems range over several orders of magnitude from pure water for spherical micelles of low concentration to stiff pastes e.g. hexagonal phases.

3.6.2 Rheological terms

As the detailed description of rheological terms is not the subject of this volume, only a very brief introduction is given.

The shear rate $\dot{\gamma}$ is the rate of deformation, γ, of a body. A deformation or a shear rate occurs due to the application of a shear stress σ. The apparent viscosity is given by:

$$\eta = \sigma/\dot{\gamma} \tag{3.15}$$

If η is independent of the shear rate $\dot{\gamma}$ a liquid is called Newtonian. Water and other low molecular weight liquids typically are Newtonian. If η decreases with increasing $\dot{\gamma}$, a liquid is termed shear thinning. Examples for shear thinning liquids are entangled polymer solutions or surfactant solutions with long rod-like micelles. The zero shear viscosity is the value of the viscosity for small shear rates: $\eta_0 = \lim_{\dot{\gamma}\to 0} \eta(\dot{\gamma})$. The inverse case is also sometimes observed: η increases with increasing shear rate. This can be found for suspensions and sometimes for surfactant solutions. In surfactant solutions the viscosity can be a function of time. In this case one speaks of shear induced structures.

If a sample shows elastic, solid-like deformation below a certain shear stress σ_y and starts flowing above this value, σ_y is called a yield stress value. This phenomenon can occur even in solutions with quite low viscosity. A practical indication for the existence of a yield stress value is the trapping of bubbles in the liquid: Small air bubbles that are shaken into the sample do not rise for a long time whereas they climb up to the surface sooner or later in a liquid without yield stress even if their viscosity is much higher. A simple model for the description of a liquid with a yield stress is called Bingham's solid:

$$\sigma = \eta\dot{\gamma} + \sigma_y \tag{3.16}$$

These terms describe the flow behaviour of matter under steady shear flow. Elastic properties which provide more profound information about the structure of a sample can be taken from dynamic measurements with sinoidal deformation of the frequency ω. Here the storage modulus $G'(\omega)$ gives information about the elastic response of a sample and the loss modulus $G''(\omega)$ describes the dissipative loss of energy under oscillating shear. The two quantities are the real part and the imaginary part of the complex shear modulus G^*. If both properties can be found, a sample is called viscoelastic. The magnitude of the complex viscosity which is depicted in the dynamic rheograms below is given by:

$$|\eta^*| = \frac{1}{\omega}\sqrt{G'^2 + G''^2} \tag{3.17}$$

More detailed information can be found in the relevant literature e.g. [36].

3.6.3 Rheological behaviour of monomeric solutions and non-interacting micelles

Solutions below the cmc contain only monomeric surfactant molecules. To a first approximation these dissolved molecules do not influence the viscosity which is the same as for pure water (1 mPa s at room temperature) [37].

The influence of non-interacting micelles (i.e. spherical micelles or small rod-like micelles not far beyond the cmc) on the viscosity η of a solution can be described by the Einstein equation:

$$\eta = \eta_s(1 + 2.5\Phi) \tag{3.18}$$

where η_s is the viscosity of the solvent water and Φ is the volume fraction of the micelles. It can be seen from the equation that the viscosity is linearly increased with the concentration above the cmc but the flow behaviour is still Newtonian. For the calculation of Φ it has to be taken into account that the micelles are solvated and the effective volume fraction is larger than that calculated only from the concentration. For ionic micelles electroviscous effects can lead to a further increase of the viscosity.

In the case of non-interacting anisometric aggregates, the situation becomes more complex. Such aggregates rotate in a shear flow with the consequence that their effective volume is much larger than the volume fraction of the micelles. Hence, the viscosity as a function of concentration increases with a larger slope once rod-like micelles are formed.

3.6.4 Entanglement networks of rod-like micelles

Surfactant solutions with rod-like micelles can have notable viscosities up to six times higher than the water viscosity [37]. This can be explained by the presence of entangled rod-like micelles (often also called worm-like micelles or thread-like micelles) which arrange themselves in a supramolecular transient network [38–41]. Such solutions often have elastic properties but they do not show a yield stress. This means that even high viscous solutions flow under the influence of very small shear stress. In this situation they show a zero shear viscosity which is given by:

$$\eta_0 = G_0\tau \tag{3.19}$$

with a characteristic time constant τ and the shear modulus:

$$G_0 = \nu kT \tag{3.20}$$

ν is the density of the entanglements, k is Boltzmann's constant and T is the temperature. From these equations, it is clear that the viscosity of the undisturbed network is determined by its density and a characteristic time for dissolving the entanglements. The time constant is the result of two possible stress relaxation mechanisms. On the one hand the network can disentangle under flow by reptation. This process is similar to polymer solutions. On the other hand it has been pointed out above that micelles exchange monomers with the solution, disintegrate and re-form in a dynamic way. Therefore, a stress applied to the network can relax by the breakage and rearrangement of micelles. The relaxation time τ can be determined by dynamic (oscillating) rheological measurements. In some cases one discrete relaxation time can be found and this can be attributed to the breakage of the micelles. In Figure 3.33 an example is given for such a rheological behaviour. The figure shows a dynamic rheogram of the surfactant system 100 mM cetylpyridiniumchloride and 60 mM sodium salicylate. At low frequencies, the loss modulus G'' is higher than the storage modulus G' and the

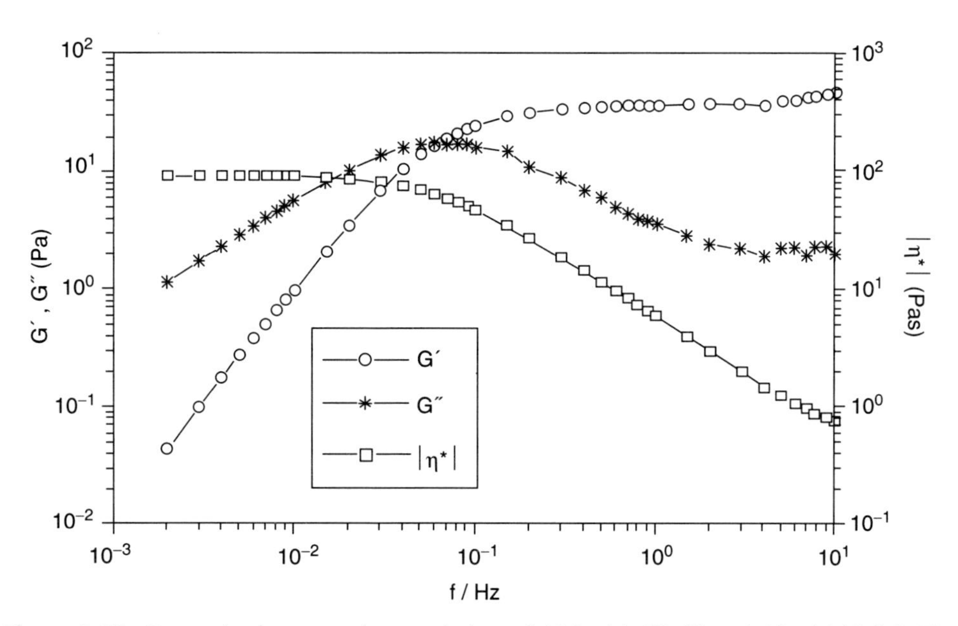

Figure 3.33 Dynamic rheogram for a solution of 100 mM CPyCl and 60 mM NaSal. The solution behaves like a Maxwell fluid with a single relaxation time.

sample shows viscous flow. At high frequencies the stress in the system cannot relax and the sample behaves as a rubber-like elastic. This type of rheological behaviour is also termed 'Maxwell fluid'. The typical time constant can be identified with the reciprocal frequency of the intersection point of the moduli G' and G''.

In other cases, several discrete relaxation times or distributions of relaxation times can be found [39]. This is typically the case if the stress relaxation is dominated by reptation processes [42]. The stress relaxation model can explain why surfactant solutions with worm-like micelles never show a yield stress: Even the smallest applied stress can relax either by reptation or by breakage of micelles. For higher shear rates those solutions typically show shear thinning behaviour and this can be understood by the disentanglement and the orientation of the rod-like micelles in the shear field.

While the network density ν is more or less given by the concentration of the surfactant, the relaxation time can depend on many parameters such as surfactant concentration, temperature, type of counterions or ionic strength.

The ionic strength is often used to adjust the viscosity of surfactant-containing products such as hair shampoos or dishwashing detergents. Here a certain amount of salt is added to a surfactant system that may contain spherical micelles of an anionic surfactant. The salt causes a screening of the charges of the surfactant headgroups. Consequently the electrostatic repulsion of the headgroups in the micellar surface is reduced and the effective headgroup area decreases. This induces a sphere-rod transition and a growth of the rods. Hence, the viscosity increases. On further addition of salt the viscosity may decrease again. An example of this behaviour is given in Figure 3.34 [43].

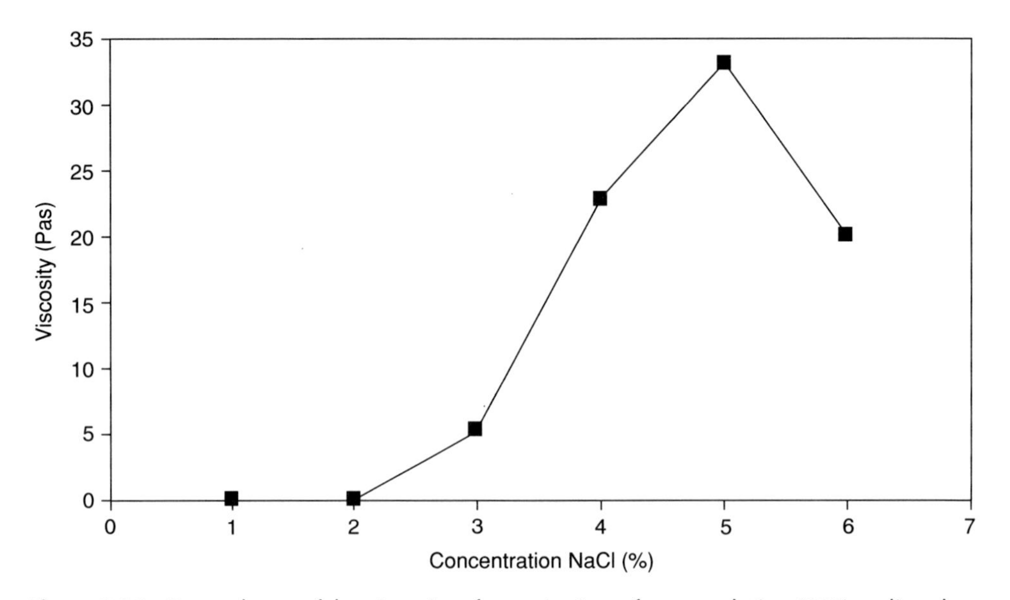

Figure 3.34 Dependence of the viscosity of an anionic surfactant solution (10% sodium laureth sulphate) on the concentration of excess salt.

3.6.5 The rheological behaviour of bilayer phases

As pointed out above, many surfactant systems can form bilayer phases. Most systems form bilayers at higher concentrations in the mesophase region. If the packing parameter of the system is approximately 1, however, systems can form lamellar phases even at very low concentration. Here, phases with flat bilayers, with vesicles or with sponge-like bicontinuous bilayers (L_3) can be found. The structure of these phases is determined by the spontaneous curvature of the lamellae and their flexibility and the latter can be influenced, e.g. by their charge. The appropriate packing parameter can be realised either by double chain surfactants or by a mixture of surfactant and a cosurfactant with a small headgroup, for example, a fatty alcohol. Of course, these structures determine the rheological behaviour of the solutions. L_3 phases typically show low viscous Newtonian flow behaviour [37]. This is due to the high flexibility and the rapid dynamics of the lamellae which permits a fast relaxation of shear stress. Phases with flat lamellae show a similar behaviour at low concentrations. The behaviour of higher concentrated lamellar phases in the mesophase region, however, can be much more complex due to the close packing of the lamellae and the resulting stronger interactions between them. Vesicle phases can be of low viscosity and Newtonian as long as they are dilute and the vesicles do not interact. On the other hand, vesicle phases exist which contain densely packed multi-lamellar vesicles with onion-like structures. These vesicles are highly polydisperse and the wedges between larger vesicles are filled with smaller ones. The structure of these solutions has been visualised by electron microscopy [44–48]. The systems often consist of non-ionic (e.g. $C_{12}E_6$) or zwitterionic surfactants

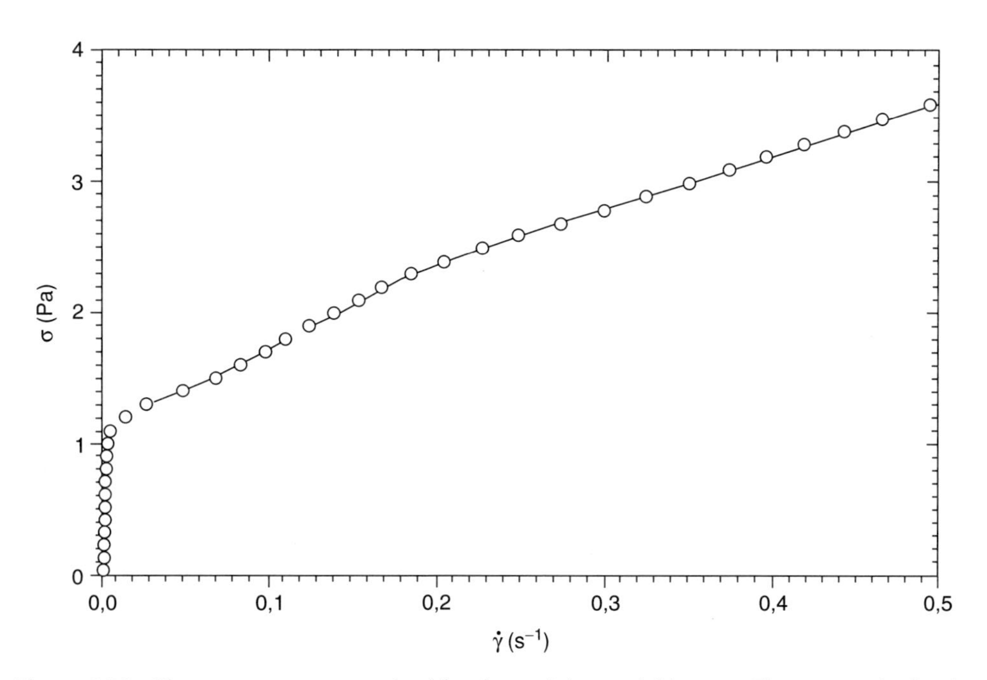

Figure 3.35 The most common method for determining a yield stress. The stress σ is slowly increased and plotted against the shear rate $\dot{\gamma}$. It is clearly visible when the sample (here 90 mM $C_{14}DMAO$, 10 mM $C_{14}TMABr$ (tetradedecyltrimethylammoniumbromide) and 220 mM C_6OH) begins to flow (reproduced with permission [44]).

(e.g. tetradecyldimethylaminoxide $C_{14}DMAO$) and a cosurfactant (e.g. fatty alcohol). On charging up the bilayers by an ionic surfactant or by protonation of the aminoxide, the vesicles form under the additional influence of shear.

It is easy to understand that these solutions must exhibit viscoelastic properties. Under shear flow the vesicles have to pass each other and, hence, they have to be deformed. On deformation, the distance of the lamellae is changed against the electrostatic forces between them and the lamellae leave their natural curvature. The macroscopic consequence is an elastic restoring force. If a small shear stress below the yield stress σ_y is applied, the vesicles cannot pass each other at all. The solution is only deformed elastically and behaves like Bingham's solid. This rheological behaviour is shown in Figure 3.35. which clearly reveals the yield stress value, beyond which the sample shows a quite low viscosity.

As pointed out above, small air bubbles in the liquid do not rise. This can be considered as an indication for the existence of a yield stress for a person doing practical work without using highly sophisticated instruments. In Figure 3.36 a dynamic rheogram of this system is depicted.

It demonstrates the highly elastic behaviour. At small deformation amplitudes the storage modulus G' is one order of magnitude larger than the loss modulus G'' and independent of the frequency. This is the behaviour of a solid body.

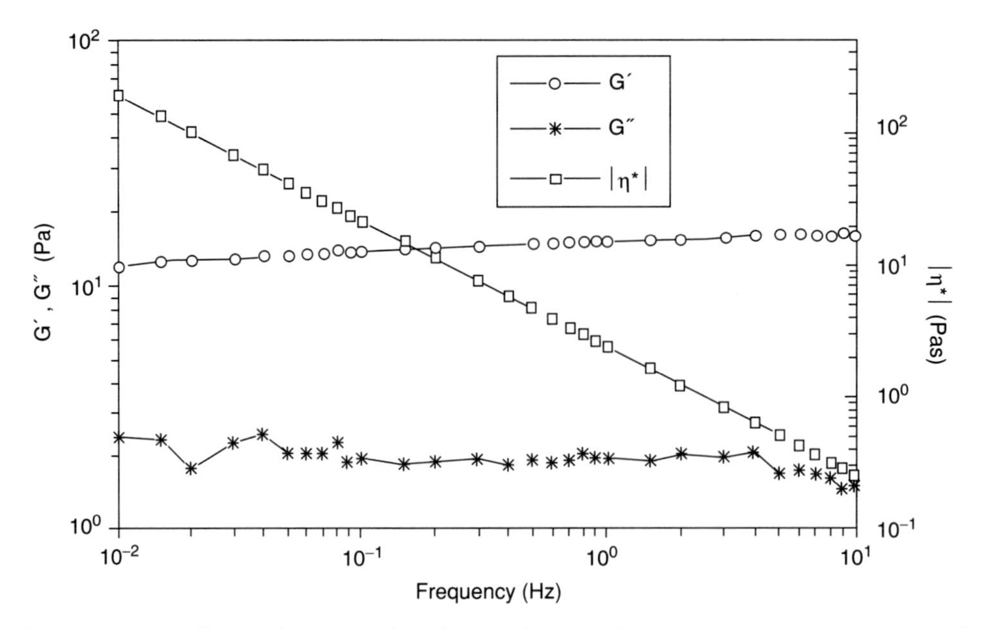

Figure 3.36 Oscillatory rheogram of a solution of 90 mM C_{14}DMAO, 10 mM C_{14}TMABr and 220 mM C_6OH. The moduli are almost independent of the frequency: *G*′ is one order of magnitude larger than *G*″ and does not vanish for low frequencies. This indicates a yield stress (reproduced with permission [44]).

References

1. Cutler, W.G. and Kissa, E. (1987) *Detergency – Theory and Applications.* Dekker, New York.
2. Lange, K.R. (1994) *Detergents and Cleaners.* Hanser, Munich.
3. Jakobi, G. and Löhr, A. (1986) *Detergents and Textile Washing.* VCH, Weinheim.
4. Showell, M.S. (1997) *Powdered Detergents.* Dekker, New York.
5. Schwuger, M.J. (1979) Washing and cleaning as a separation process. *Ber. Bunsenges. Phys. Chem.*, **83**(11), 1193–205.
6. Dobias, B., Qiu, X. and von Rybinski, W. (1999) *Solid–Liquid Dispersions.* Dekker, New York.
7. Berth, P. and Schwuger, M.J. (1979) Chemical aspects of washing and cleaning. *Tenside Det.* **16**(4), 175–84.
8. Jost, F., Leiter, H. and Schwuger, M.J. (1988) Synergisms in binary surfactant mixtures. *Colloids Polym. Sci.*, **266**(6), 554–61.
9. Engels, Th., von Rybinski, W. and Schmiedel, P. (1998) Structure and dynamics of surfactant-based foams. *Progr. Colloid Polym. Sci.*, **111**, 117–26.
10. Shafrin, E.G. and Zisman, W.A. (1960) Constitutive relations in the wetting of low-energy surfaces and the theory of the retraction method of preparing monolayers. *J. Phys. Chem.*, **64**, 519–24.
11. Nickel, D., Speckmann, H.D. and von Rybinski, W. (1995) Interfacial tension and wetting as parameters for product characterization. *Tenside Surfactants Det.* **32**(6), 470–4.
12. Kling, W. (1949) Der Waschvorgang als Umnetzung. *Kolloid-Z.*, **115**, 37–44.
13. Hoffmann, H. (1978) The dynamics of micelle formation. *Berichte der Bunsengesellschaft*, **82**(9), 988–1001.

14. Hoffmann, H. (2000) The micellar structures and macroscopic properties of surfactant solutions. In H. Hoffmann, M. Schwoerer and Th. Vogtmann (eds). *Macromolecular Systems: Microscopic Interactions and Macroscopic Properties.* Wiley-VCH, pp 199–250.
15. Hoffmann, H. (1994) Fascinating phenomena in surfactant chemistry. *Adv. Mater,* **6**(2), 116–29.
16. Thunig, C., Hoffmann, H. and Platz, G. (1998) Iridescent colors in surfactant solutions. *Prog. Colloid Polym. Sci.,* **79**, 297–307.
17. Platz, G., Thunig, C. and Hoffmann, H. (1990) Iridescent phases in aminoxide surfactant solutions. *Prog. Colloid Polym. Sci.,* **83**, 167–75.
18. Miller, C.A. and Ghosh, O. (1986) Possible mechanism for the origin of lamellar liquid crystalline phases of low surfactant content and their breakup to form isotropic phases. *Langmuir,* **2**(3), 321–9.
19. Stegemeyer, H. (ed.) (1999) *Lyotrope Flüssigkristalle.* Dr. Dietrich Steinkopff Verlag GmbH & Co. KG, Darmstadt.
20. Mitchell, D.J., Tiddy, G.J.T., Warring, L., Bostock, T. and Mc Donald, M.P. (1983) Phase behavior of polyoxyethylene surfactants with water. mesophase structures and partial miscibility (cloud points). *J. Chem. Soc. Faraday Trans.* I, **79**(4), 975–1000.
21. Schambil, F. and Schwuger, M.J. (1987) Correlation between the phase behavior of ternary systems and removal of oil in the washing process. *Colloid Polym. Sci.,* **265**(11), 1009–17.
22. Benson, H.L., Cox, K.R. and Zweig, J.E. (1985) Nonionic-based detergent systems for cold water cleaning. *Soap Cosmet. Chem. Specialties,* **61**(3), 35–47.
23. Miller, C.A. and Raney K.H. (1993) Solubilization - emulsification mechanisms of detergency. *Colloids Surf.* A, **74**, 169.
24. Binks, B.P. (ed.) (1998) *Modern Aspects of Emulsion Science.* The Royal Society of Chemistry, Cambridge.
25. Landfester, K., Bechthold, N., Tiarks, F. and Antonietti, M. (1999) Formulation and stability mechanisms of polymerizable miniemulsions. *Macromolecules,* **32**(16), 5222–8.
26. Antonietti, M. and Landfester, K. (2001) Single molecule chemistry with polymers and colloids: a way to handle complex reactions and physical processes? *Chem. Phys. Chem.* **2**(4), 207–10.
27. Griffin, W.C. (1955) Calculation of "HLB" values of nonionic surfactants. *Am. Perfumer Essential Oil Rev.,* **65**(5), 26–9.
28. Engels, T. and von Rybinski, W. (1998) Liquid crystalline surfactant phases in chemical applications. *J. Mater. Chem,* **8**(6), 1313–20.
29. Förster, T., Schambil, F. and von Rybinski, W. (1992) Production of fine disperse and long-term stable oil-in-water emulsions by the phase inversion temperature method. *J. Disp. Sci. Technol.,* **13**(2), 183–93.
30. Förster, Th., von Rybinski, W., Tesmann, H. and Wadle, A. (1994) Calculation of optimum emulsifier mixtures for phase inversion emulsification. *Int. J. Cosmet. Sci.,* **16**(2), 84–92.
31. Kahlweit, M. (1993) Microemulsions. *Tenside Surfactants Det.,* **30**(2), 83–9
32. Gohla, S.H. and Nielsen, J. (1995) Partial phase solu-inversion technology (PPSIT). A novel process to manufacture long term stable multiple emulsions by an in situ one step procedure. *Seife Öle Fette Wachse,* **121**(10), 707–10.
33. Smulders, E., Rähse, W., von Rybinski, W., Steber, J., Sung, E. and Wiebel, F. (2002) *Laundry Detergents.* Wiley-VCH, Verlag GmbH.
34. Prud'homme, R. and Khan, S.A. (eds) (1996) *Foams: Theory, Measurements and Applications.* Surfactant Science Series 57, Dekker, New York, Basel, Hong Kong.
35. Jacobs, B., Breitzke, B., Stolz, M. and Verzellino, R. (2004) 51. SEPAWA Congress, Conference Proceedings, pp 24–9.
36. Barnes, H.A., Hutton, J.F. and Walters, K. (1989) *An Introduction to Rheology.* Elsevier, Amsterdam.

37. Hoffmann, H. (2002) Rheological effects in surfactant solutions. In C. Holmberg, D.O. Shah and M.J. Schwuger (eds). *Handbook of Applied Surface and Colloid Chemistry*, vol. 2. Wiley, New York, pp 189–214.
38. Hoffmann, H. and Ulbricht, W. (1997) Viscoelastic surfactant solutions. In K. Esumi and R. Ueno (eds). *Structure-performance relationships in surfactants.* Decker, New York, pp 285–324.
39. Schmiedel, P. (1995) Weiterentwicklung eines dynamischen Rheometers für Frequenzen bis 1 kHz und rheologisches Verhalten viskoelastischer Tensidlösungen im Frequenzbereich bis 1 kHz. Ph.D. Dissertation, Universität Bayreuth.
40. Cates, M.E. and Candau, S.J. (1990) Statics and dynamics of wormlike micelles. *J. Phys., Condens. Matter*, **2**, 6869–80.
41. Granek, R. and Cates, M.E. (1992) Stress relaxation in living polymers: results from a poisson renewal model. *J. Chem. Phys.*, **96**, 4758–69.
42. Ferry, J.D. (1980) *Viscoelastic Properties of Polymers.* Wiley, New York.
43. Tesmann, H., Kahre, J., Hensen, H. and Salka, B.A. (1997) Alkyl polyglycosides in personal care products. In K.H. Hill, von W. Rybinski and G. Stoll, (eds). *Alkyl Polyglycosides.* VCH Verlagsgesellschaft mbH; Weinheim, pp 71–98.
44. Hoffmann, H., Thunig, C., Schmiedel, P. and Munkert, U. (1994) Surfactant systems with charged multilamellar vesicles and their rheological properties. *Langmuir*, **10**(11), 3972–81.
45. Hoffmann, H., Thunig, C., Schmiedel, P. and Munkert, U. (1994) Complex fluids with a yield value; their microstructures and rheological properties: multilamellar vesicle systems with a yield stress value. *Il Nuovo Cimento*, **16D**(9), 1373–90.
46. Hoffmann, H., Thunig, C., Schmiedel, P. and Munkert, U. (1995) Gels from surfactant solutions with densely packed multilamellar vesicles. *Faraday Discuss.*, **101**, 319–33.
47. Hoffmann, H., Thunig, C., Schmiedel, P. and Munkert, U. (1994) The rheological behavior of different viscoelastic surfactant solutions: systems with and without a yield stress value. *Tenside Surf. Det.*, **31**(6), 389–400.
48. Hoffmann, H. and Ulbricht, W. (1998) Vesicle phases and their macroscopic properties. *Recent Res. Devel. Phys. Chem.* **2**, 113–58.

Chapter 4
Anionic Surfactants

John Hibbs

Anionic surfactants are the most commonly used class of surfactants in cleansing applications. These surfactants, in addition to their ability to emulsify oily soils into wash solutions, can lift soils, including particulates, from surfaces. This is because the negatively charged head group is repelled from most surfaces, which also tend to be slightly negatively charged – the reverse action to a cationic surfactant, where the positively charged head group is adsorbed onto a surface, giving an antistatic and conditioning effect.

The great majority of anionic surfactants will generate significant foaming in solutions above their critical micelle concentration (CMC), which is a desirable attribute in most cleansing applications, but can restrict the use of anionic surfactants in areas where foam is a problem.

Anionics can be classified according to the polar group and the following will be considered:

- Sulphonates

  Aromatic – alkylbenzene, alkyltoluene, alkylxylene, alkylnaphthalene
  Aliphatic – α-olefin sulphonates, alkane sulphonates, sulphosuccinates

- Sulphates

  Alkyl sulphates e.g. sodium lauryl sulphate (SLS)
  Alkyl ethoxy sulphates e.g. sodium laureth sulphate

- Phosphate esters
- Carboxylates
- Soaps, isethionates, taurates

When specifying an anionic surfactant for an application, it is important to understand how the composition of the raw material (especially that of the hydrophobe) influences the performance of the surfactant and the properties of the formulated product. In looking at the properties of each surfactant type, the basic chemistry will be considered together with sources of hydrophobe and the manufacturing process used to functionalise them. How the composition of the surfactant affects its performance and physical properties will be examined together with how these properties lead to the applications of the surfactant.

4.1 Sulphonates

Sulph(on)ation processes. Since many of the anionic surfactants to be discussed are made by the addition of SO_3 to an organic substrate, it is appropriate to consider, in overview, the main processes used and the contribution of the sulphonation process to the quality and performance of the surfactant. (The term 'sulphonation' is used here generically to describe reaction of an organic with SO_3, regardless of the nature of the substrate.)

Sulphuric acid or oleum is probably the simplest and oldest sulphonating agent. Sulphuric acid is made by the reaction of gaseous sulphur trioxide with water, which is a very exothermic reaction. The ratio of SO_3 to water determines the acid strength, 96% or 98% being common commercial grades. Once the molar ratio of SO_3:water is >1, the product is oleum, or fuming sulphuric acid. The strength of oleum is described in terms of the percentage of SO_3 added to 100% sulphuric acid. A product nominally consisting of 80% H_2SO_4 and 20% SO_3 would be called Oleum 20 but, in fact, the excess SO_3 reacts with the H_2SO_4 to form pyrosulphuric acid $HO(SO_2)O(SO_2)OH$.

Sulphonation with sulphuric acid or oleum is initially rapid and exothermic. The rate of reaction is highly dependent on the concentration of the sulphuric acid, so the reaction slows significantly as the sulphonating agent is consumed and further diluted by the water which is a by-product of the reaction. Processes based on sulphuric acid use large excesses, resulting in high levels of waste acid or residual sulphate in the neutralised product. An alternative is to remove the water from the reaction mixture, usually by two-phase distillation, with excess substrate as a carrier.

Chlorosulphonic acid (CSA), HSO_3Cl, has also been used as an effective sulphonating agent. The effectiveness of chloride as a leaving group and the absence of water as a by-product mean that chlorosulphonation can be run at stoichiometry close to 1:1 and with efficient conversion of the organic substrate. The reaction temperature can be controlled by addition rate of the CSA and some very good product colours can be achieved. The by-product of chlorosulphation is HCl, or NaCl after neutralisation. The salt level in the surfactant is typically $< 0.5\%$ and this would need to be accounted for in formulation since it could affect the viscosity.

The most common and cost effective sulphonating agent in use today is sulphur trioxide itself which has the benefit of being a highly aggressive sulphonating agent, with no direct by-products of sulphonation. SO_3 can be bought as a liquid, which must be maintained at $\sim 35°C$ since if the temperature is too low, the product can freeze and at higher temperatures, polymers can form over time, which will foul storage tanks and reactors. The liquid SO_3 is normally diluted prior to use to help to moderate the reaction. In cold sulphonation, SO_3 is dissolved in liquid SO_2 at $-10°C$ and the resulting solution used as the sulphonating agent. This technique is used on a very large scale to manufacture synthetic petroleum sulphonates in which the reaction scheme is very elegant. The concentrations of the reactants are such that the heat of sulphonation is sufficient to vaporise the SO_2 solvent, leaving a neat suphonic acid. This process is also operated by Pilot Chemical Company of New Jersey, USA, where the low temperatures used during the sulphonation of linear alkyl benzene (LAB) give a very pure sulphonate with exceptionally low colour.

The dominant sulphonation technology is undoubtedly falling film reaction. Here, a solution of SO_3 in air (normally 3–10% SO_3) flows concurrently with a thin film of the organic substrate down the inner surface of a tube, with sulphonation taking place on the walls of the tube. Efficient removal of the heat of reaction is essential in order to produce a good quality sulphonate normally defined by the colour, degree of conversion of the organic and the levels of impurities. Sulphonators are essentially efficient heat exchangers, with headgear to give efficient distribution of the gas and liquid feeds. The two most popular designs are the multi-tube reactor, similar in design to a shell and tube heat exchanger (Ballestra SpA) and the annular reactor (Chemithon Corporation). Both reactors give controlled film formation on the reactor walls, consistent distribution of the gas feed and very efficient removal of heat. Most modern sulphonation plants include equipment to generate the large volumes of very dry air needed for dilution and often systems to produce SO_3 in situ, by catalytic oxidation of SO_2, produced from burning elemental sulphur.

While it is true that the nature of a sulphonated surfactant is largely determined by the choice of hydrophobe, sulphonation process can have a significant impact on the quality and performance of the surfactant. 'Good' sulphonation achieves a balance between conversion of the substrate and formation of undesirable by-products. Efficient conversion of the feedstock is important for both economics and quality reasons. The cost of the organic substrate is the dominant cost driver of the surfactant and if significant quantities are left unconverted, then the manufacturing costs are increased. Additionally, high levels of unsulphonated organic matter (UOM) will affect the solution properties and formulation properties. Common problems from undersulphonated surfactants include cloudy solutions and very high viscosities, and possibly even the formation of solid gels at low concentrations. Oversulphonation is as problematic as undersulphonation. If a significant excess of SO_3 is used, or reaction temperatures are too high, or if the mixing and heat removal are inadequate, then the sulphonate will be dark in colour and undesirable impurities such as polysulphonates, sulphones and degradation products (such as 1,4 dioxane) will be formed. In summary, good sulphonation will give the best possible surfactant properties from a raw material: low colour, low impurities, high conversion and consistent solution properties.

4.1.1 Alkylbenzene sulphonates

Alkylbenzene sulphonates are one of the most important classes of anionic surfactants. The surfactants based on LAB are used in detergent formulations in most regions of the world, usually as the primary surfactant. Surfactants based on branched alkylbenzene (BAB), while no longer used in detergent products in most developed regions, are still important in certain agrochemical and industrial applications, where rapid biodegradability is of reduced importance.

4.1.1.1 Linear alkylbenzene sulphonate

Global consumption of LAB is estimated at approximately 3 million tonnes per annum, making it the most commonly used anionic surfactant [1]. Its popularity can be attributed to its relative ease of manufacture, ability to be easily stored, transported and handled in a highly concentrated form (the 96% active sulphonic acid) and its efficiency as a detergent. Due to this near-universal application, it is also probably the most researched and documented, especially in terms of its fate in the environment.

Chemistry and general properties. Alkylbenzene sulphonate is made from the reaction of an alkylbenzene with a sulphonating agent, to add a SO_3^- group to the aromatic ring, which forms the polar head group of the surfactant molecule. By far the most common sulphonating agent is gaseous sulphur trioxide which gives a clean reaction with minimal by-products. There is still some use of sulphuric acid and chlorosulphonic acid processes but high levels of impurities (sulphates and chlorides respectively) have made these products much less favoured. Reaction of the LAB with a dilute solution of SO_3 in air (typically 10% SO_3) in a falling film sulphonator is the most common method of production.

The resulting sulphonic acid is typically a low viscosity liquid, even at manufacturing concentrations of 96%. It can be stored for extended periods, due to the stability of the carbon–sulphur bond which makes the use of the concentrated sulphonic acid a highly cost-effective way to deliver the surfactant into a formulation. It is common for commercial sulphonic acids to contain a small amount of water, typically 0.5–1%, and this is added to sulphonic acid after an 'ageing' period in a process called quenching. When the LAB is sulphonated, a number of impurities can form; these include sulphones and anhydrides. Once formed, the sulphones are very stable and can usually be detected with the unreacted starting material (UOM, also called free oil) in ether extractions or by HPLC analysis. The anhydrides will react further with LAB to make more sulphonic acid, or they can be hydrolysed with water. To obtain maximum conversion to sulphonic acid, it is good practice to allow the acid from the reactor to age, i.e. to stand for a period between 30 min and 2 h, to allow rearrangement. During this time, the active matter of the acid increases and often the colour improves. If the reactions are allowed to continue indefinitely, very high levels of conversion can be reached, but often the remaining active SO_3 species will cause some oxidation of the alkylate, leading to dark colours. The ageing process is terminated by the addition of quench water to the acid which hydrolyses the anhydrides and converts any active SO_3 to sulphuric acid. The composition of the product is stable after quenching and it is interesting to see that the addition of water leads to an increase in the active matter of sulphonic acid.

The acid must be neutralised prior to use and the sodium salt is prevalent in most applications but amine salts can be used to give additional solubility and detergency. Sodium salts give clear, stable, low viscosity solutions up to 30% active matter but above 30%, the surfactant solution becomes paste-like in nature, with rapidly increasing viscosity. A 50% active sodium salt is a flowable paste which can separate on standing into a low active supernatant liquor and a higher active very high viscosity paste which can be difficult to homogenise without efficient mixing. Products of concentrations between 60% and 85% are not usual, since they would be practically solid at ambient temperatures and remain very highly viscous even at elevated temperatures. Solid products with surfactant concentrations between 80% and 90% are available and these are made by drying a mixture of the sodium salt of the sulphonate with a 'builder'. This is necessary because dry alkylbenzene sulphonates are very hygroscopic and rapidly absorb water to become a sticky mass. The addition of a crystalline inorganic (typically a phosphate or carbonate) improves the powder properties and greatly reduces hygroscopicity.

Raw materials. LAB is made by Friedel–Kraft alkylation of benzene with an n-alkene. The choice of catalyst and the composition of the alkene have a very significant effect on the performance of the surfactant, especially on its physical characteristics and that of the overall formulation.

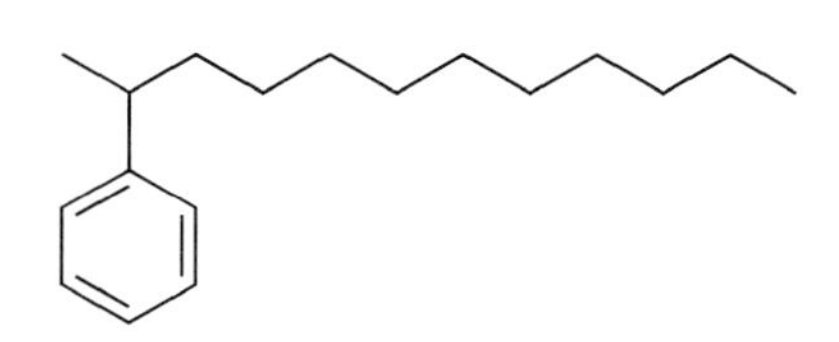

Figure 4.1 2-phenyl dodecane.

The bond between the alkyl chain and the benzene ring can form at any position, except the terminal carbon which means that the LAB is a mixture of isomers, with the phenyl group at positions 2-, 3-, 4-, etc. Figure 4.1 shows the structure of 2-phenyl dodecane.

The catalyst used is a major factor in determining the isomer distribution and such is its importance that the LAB and LAS surfactants are classified as either high 2-phenyl, (typically >30%) or low 2-phenyl (<20%). The physical properties and the performance of the two types of LAS are significantly different and the correct selection of either H2P or L2P surfactant can be key to a successful LAS-based formulation. The choice of catalyst can also influence the level of impurities commonly referred to as tetralins (more correctly, dialkyl tetralins; see Figure 4.2). These bicyclic compounds form when the alkyl group reacts with the benzene ring a second time to form a second ring, typically tetralin (six membered ring) or indane (five membered ring). These compounds can be sulphonated and influence the surfactant performance.

Polyalkylation is possible, and two impurities are the dialkylbenzene (Figure 4.3) and diphenylalkane (Figure 4.4).

Due to their high molecular weights, these compounds are easily removed by distillation. There is a market for the dialkylbenzenes which can be sulphonated to produce synthetic petroleum sulphonates.

Figure 4.5 shows a summary of the most common variants of LAB that are produced.

The chloroparaffin/$AlCl_3$ route is the longest established, but now represents only a minority (~10%) of LAB supply. The product is H2P, with 2-phenyl isomer levels of 28–30%. The reaction can be more difficult to control and yields not only a high level of dialkyl tetralin (DAT) impurities, but also high levels of dialkylbenzene and diphenylalkanes. The spent catalyst has to be removed by filtration.

Most of the world's LAB supply is made using the HF process. Here the DAT impurities are significantly reduced (from ~7% to <1%), but the product is low 2-phenyl, typically 17% 2-phenyl isomer. The Detal process developed by UOP is the newest technology, commercialised in the late 1990s. The Detal catalyst is a fixed bed, solid catalyst which gives high levels of 2-phenyl isomer (~25%), <1% DAT and very good colours of both the alkylate and the sulphonate. In 2004 there were a small number of plants operating, with five additional

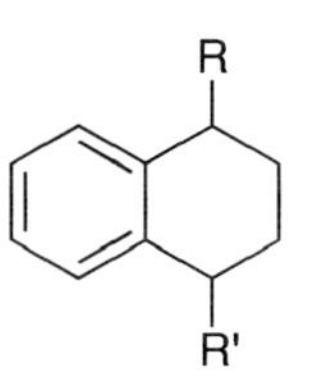

Figure 4.2 Dialkyl tetralin.

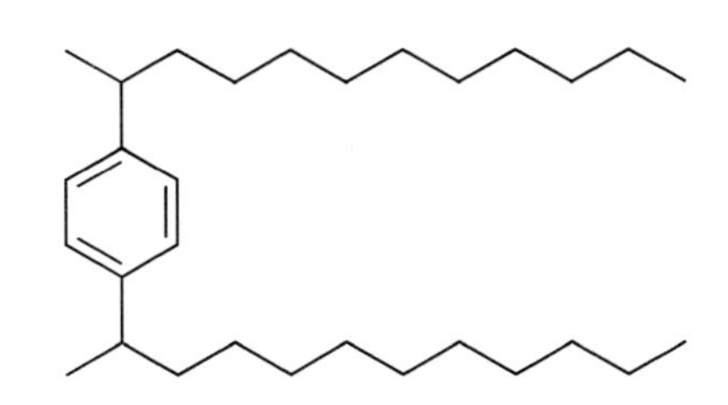

Figure 4.3 Dialkyl benzene.

plants under construction. The surfactants produced via the Detal process are in increasing demand, since the combination of high 2-phenyl, low DAT and very low colours of both alkylate and the sulphonated product are very attractive in certain applications.

Composition vs. performance and properties. The key parameters controlling the physical properties of surfactants based on LAB are molecular weight, 2-phenyl isomer composition and the level of DAT impurities. Molecular weight is often described in terms of the mean of the carbon chain distribution and most commercial alkylates contain carbon chains between C9 and C14, so the mean is usually ~12. North American applications tend to favour slightly lower molecular weights than Europeans, with a typical carbon number of 11.3, while 11.6 might be more typical in Europe. Increasing molecular weight leads to reduced solubility, lower foaming (and increased sensitivity to hard water) and higher formulation viscosities. With the relatively narrow range of molecular weight used in common detergent application, the importance of molecular weight is generally less than the other factors discussed below.

LAB is often described as high or low 2-phenyl. As already seen, the level of 2-phenyl isomer is determined by the catalyst choice, with low 2-phenyl product made by HF catalyst being dominant. Isomer distribution has a very significant effect on solubility. Figure 4.6 shows how the cloud point (the temperature below which the formulation becomes turbid) varies with LAS type and concentration.

The cloud point of the high 2-phenyl product remains below 0°C up to 25% LAS whereas the low 2-phenyl cloud point is ~15°C. In practice, this means that a formulation based on L2P surfactant would need higher additions of solvent or hydrotrope to keep the solution clear and stable at lower temperatures.

Dialkyl tetralins in LAB feedstocks are readily sulphonated and act as hydrotropes. High DAT levels give surfactants with high solubility and low viscosity. This effect is very significant in formulations. For example the salt curve of 15% active H2P LAS with 3% cocodiethanolamide (a common thickener and foam stabiliser) can give a maximum viscosity of ~600 cPs with a high DAT LAB, but over 1300 cPs with a low DAT LAB.

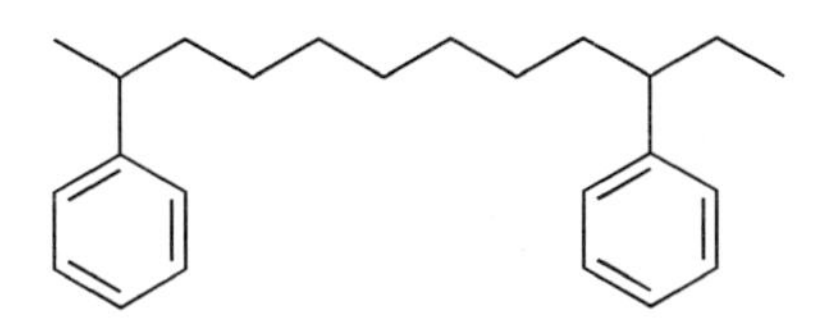

Figure 4.4 Diphenyl alkane.

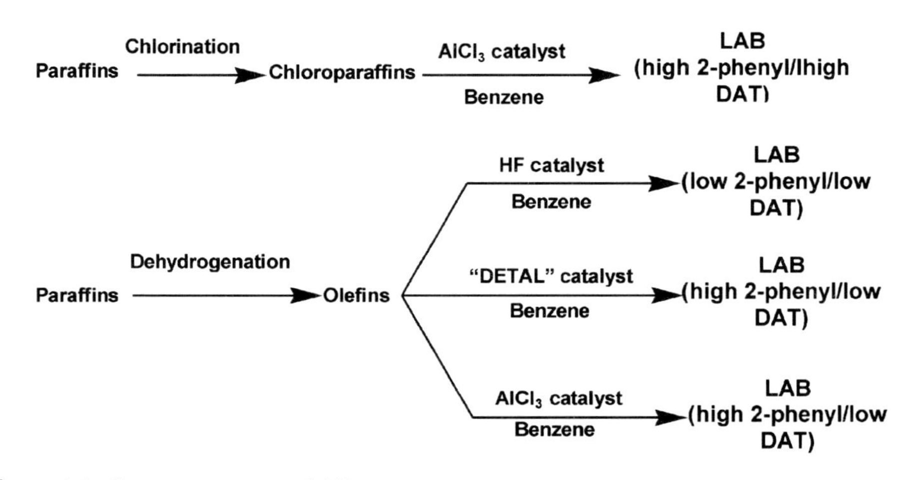

Figure 4.5 Common routes to LAB.

Applications. LAS has many uses, but is used predominantly in cleansing applications, mainly laundry products and hard surface cleaners. The use of LAS in personal care applications was almost totally phased out by 2004.

Laundry formulations are the greatest consumer of LAS, being used as the primary surfactant in powder, tablet and liquid formulations providing good degreasing and soil removal properties. Some soap is normally added to the formulation to control the foaming

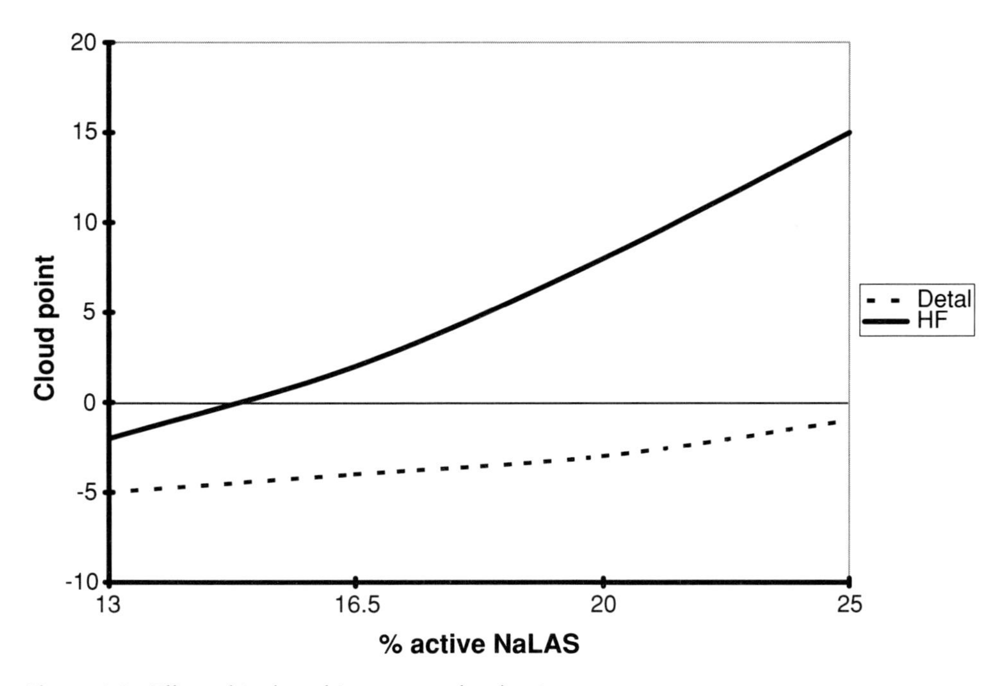

Figure 4.6 Effect of 2-phenyl isomer on cloud point.

for horizontal axis (European) machines. In low density powders, sulphonic acid is normally neutralised with caustic soda, and other ingredients such as inorganic builders and fillers are added. The base slurry is spray dried and agglomerated to give a 'fluffy' powder of ~10% active surfactant and a bulk density of 200–500 g l^{-1}. While these products remain popular in Southern Europe, higher density powders are becoming increasingly popular and here the amount of filler and builder is reduced and surfactant active levels are raised to ~15–20% or higher. These powders, produced by 'non-tower' processes, have bulk densities of 600–900 g l^{-1} and the techniques for the production of such powders normally involve some form of high shear mixing. In some cases, the sulphonic acid is neutralised and agglomerated with sodium carbonate in a single step. In detergent tablets, the surfactant loading is even higher and surfactant concentrations approach the acceptable limits to provide a desirable solid form (without stickiness) and ease of solution, without forming a gel phase. A wide range of LAB types is used in laundry powders, although H2P variants are thought to give better dissolution properties. In laundry liquids solubility is important, so salts other than sodium such as amine may be used to improve solubility. Incorporating high levels of builders such as STPP into a liquid formulation is difficult but Huntsman corporation has developed techniques to suspend builders into structured liquid formulations.

Manual dishwash liquids (also called light duty liquids, LDLs) are another important area of application for LAS. These formulations use a combination of LAS and an alkyl ether sulphate to give both good detergency (often measured by the ability to emulsify grease) and a high level of foam. Although not strictly necessary for good cleansing, foam is often used as a visual cue by the consumer, and formulators try to preserve foam in the presence of high levels of soil. Currently L2P type LAS is most common in these formulations as it thickens well with salt to give a viscosity of 500–600 cPS with actives as low as 10%. Since consumers also use viscosity as an indicator of quality, formulators of 'economy' LDLs value the ability to thicken low active formulations. In the premium sector of the market, the reverse is the case and formulations which may contain as much as 30% surfactant usually need a hydrotrope and/or a solvent to reduce their viscosity and keep the cloud point above 0°C. The current trend towards concentrated, pale or pastel coloured LDLs in clear bottles presents some real challenges to a formulator using LAS.

LAS is also used in degreasing preparations such as domestic and industrial hand cleansers. Here the triisopropylamine salt of LAS is used to emulsify paraffin and form a gel with the combination of the hydrocarbon and the LAS being highly effective at removing oil from skin.

LAS is also used in emulsion polymerisation.

4.1.1.2 Petroleum sulphonates

The so-called petroleum sulphonates are close cousins to LAB in that they are sulphonated alkyl aromatics but the molecular weight of the alkylate is higher than that of detergent LAB and the product is somewhat less water soluble. The aromatic portion of the surfactant may also be toluene (methylbenzene) or xylene (dimethylbenzene). Principal applications are as oil emulsifiers, lubricant additives and corrosion inhibitors.

Raw materials. Petroleum sulphonates were originally obtained as by-products of white oil or lubricant oil production. In this process, the oil was treated with sulphuric acid, sulphonating aromatic compounds which could then be separated from the non-polar oils. These were further separated into oil soluble ('mahogany') and water soluble ('green') sulphonates. These materials were a complex mixture of alkyl benzene and alkyl naphthalene sulphonates with a wide distribution of molecular weight. This complexity and variability of the natural petroleum sulphonates, coupled with increasing demand, drove the development of synthetic petroleum sulphonates. The synthetics are made from alkylates produced in a similar manner to that described for LAB although the carbon chains are typically longer and a higher degree of branching is acceptable, or even preferred in some applications. Toluene and xylene (methyl- and dimethyl benzene) are also used as aromatic feed as are alkyl naphthalenes. Here the carbon chain can be shorter (as low as C3), due to the increased hydrophobicity from the naphthalene but it is common for naphthalene-based products to be di-alkyl, as in diisopropyl or dinonyl naphthalene.

The high molecular weight fractions from LAB manufacture can also be used as raw materials for petroleum sulphonates. The preferred feed is the dialkylbenzene which can give products of similar performance to a custom-made C20+ alkylbenzene. The alkylate still bottoms (as these materials are often called) also contain diphenyl alkanes. When sulphonated, these become highly water soluble and should be removed, either by distillation of the organic feed, or separation from the oil soluble sulphonates.

Composition vs. performance and properties. The petroleum sulphonates vary in composition from being very similar to LAS (molecular weights of ~370 compared to a typical European LAS of 340) to far removed from any detergent material (calcium dinonylnaphthalene sulphonate, molecular weight 708). The main effect of molecular weight is to influence the solubility of the product, lower molecular weights tending to better water solubility, higher weights to oil solubility. This general trend will apply within a structural type but should be treated with caution across differing systems, for example, alkylbenzene vs alkyl naphthalene.

Alkylates made specifically as feedstocks for synthetic petroleum sulphonates are typically long chain (average C chain > 16) and may use propylene oligomers which result in branched chains. Naphthalene products use the same propylene technology but tend to shorter chains (di-isopropylnaphthalene, di-nonylnaphthalene).

The method of sulphonation will influence composition. Large scale commercial sulphonation of petroleum sulphonates is carried out using SO_3/liquid SO_2 technology to give a 'clean' product with good conversion and low levels of inorganic sulphates. Lower molecular weight (~350–400) acids have been manufactured using standard falling film reactors. As molecular weight increases, the products become more viscous and the reaction becomes difficult to control, leading to darker colours and increasing levels of sulphates. High molecular weight products are more difficult to manufacture and may use air/SO_3 reactors equipped to deal with higher temperatures and viscosity, or may use batch reaction with oleum. In the latter case, excess sulphuric acid leads to high levels of sodium sulphate on neutralisation and, since it is insoluble in oil, it must be removed either by separation of waste sulphuric acid from the sulphonic acid or by a difficult filtration after neutralisation.

Applications. Some of the very low molecular weight sulphonates (such as sodium diisopropylnaphthalene sulphonates) are used as dispersants/coupling agents in agrochemical formulations and they are also used to improve wetting and leaf penetration.

Shorter chain sodium alkylbenzene/alkylxylene sulphonates (sodium salts) are used as emulsifiers for oil in water systems. Cutting fluids are often made by diluting an oil-based concentrate containing a petroleum sulphonate with water, at the point of use. The petroleum sulphonate aids the formation of the emulsion and provides some 'detergency', helping to remove grease and swarf from the metal surface. As molecular weight increases and water solubility decreases, petroleum sulphonates, especially as dibasic metal salts, can be used as detergents and corrosion inhibitors in lubricant oils. The mechanism of corrosion inhibition is by displacement of water from the metal surface and the development of a barrier film, preventing further attack. Petroleum sulphonates such as calcium didodecylbenzene sulphonate can be used as detergents in oils where they prevent deposition of tars and carbon particles on pistons. Soot particles are formed from microscopic particles of (mostly) carbon from incomplete combustion of the hydrocarbon fuel and these have a very strong tendency to aggregate, due to their high surface energy. The result would normally be the deposit of soot particles, especially around the piston 'O' ring. The sulphonate helps to disperse the soot particles (the sulphonate adsorbing onto the carbon particle and the hydrophobe stabilising the dispersed particle). Partially combusted fuels can also oxidise in the presence of sulphuric acid (from combustion products) to form tars or resins which deposit onto the metal surface and the petroleum sulphonates also help to solubilise the oil insoluble resins. An interesting variation of the petroleum sulphonates is the so-called overbased products which are often used in engine lubricants since overbased sulphonates provide a 'reservoir' of alkalinity to neutralise combustion gases. During the neutralisation of the sulphonate, excess neutralising agent is suspended as a stable dispersion, approaching a colloidal suspension. This can be achieved by physical means (shearing of the solid base) or more usually by chemical means to form an insoluble base in situ – by forming a carbonate from CO_2 for example. The overbased sulphonates are incorporated into a lubricant formulation. If combustion products from an engine pass into the lubricant, they can form highly acidic sulphates and chlorides, which would rapidly damage engine components, and also catalyse the formation of tars. The free alkalinity in the overbased sulphonate neutralises the acidic contaminants and provides an additional degree of protection to the metal surface.

Higher molecular weight products, such as calcium dinonylnaphthalene sulphonate, can also be used as demulsifiers, to remove water from oil systems. Examples of the include sump oils in ships. If sea water penetrates into the lubricant system, the sulphonate will remove the water by first including it in a micelle which leads to subsequent aggregation into droplets which are deposited into the sump. The metal of the sump is further protected by the corrosion inhibiting properties of the sulphonate.

4.1.1.3 Alkyl diphenyloxide disulphonates

Alkyl diphenyloxide disulphonates (ADPODSs) are a variant of an alkylaromatic sulphonate. Their structure is more complex and, in practice, the surfactants may consist of a mixture of different species. They have a number of high performance attributes which enable them to be used in various demanding application areas.

Chemistry and general properties. The surfactants are made by reacting an olefin with diphenyloxide and SO_3. The reaction can produce the species shown in Figure 4.7.

The properties of the surfactant can be varied by changing the length of the alkyl chains and commercial products are available with alkyl groups of average length between C6 and C16. The description of their properties is complex, since the solution behaviour and detergency are determined not simply by the length of the alkyl chain, but also by the degree of alkylation and the degree of sulphonation. It is reasonable to assume that the isomer

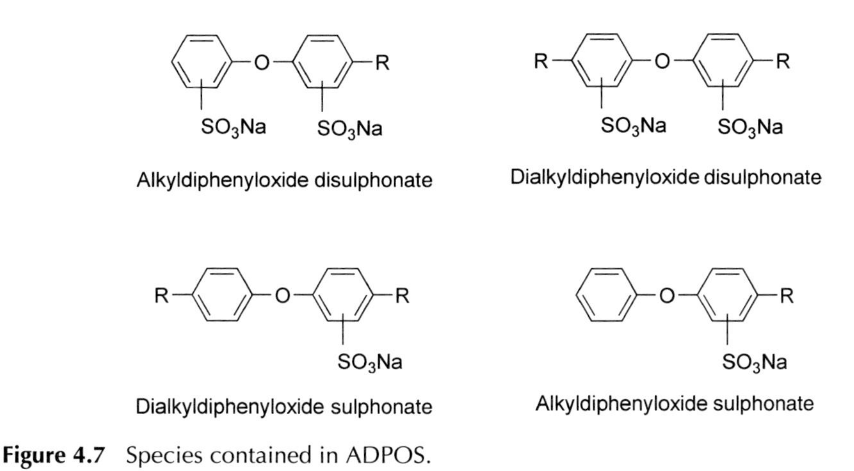

Figure 4.7 Species contained in ADPOS.

distribution will also play some role in determining properties, though little work on this appears to have been published.

As a group of surfactants, ADPODSs all show very good chemical stability and high solubility, generally better than LAS. They are sold as per cent active solutions, the Dowfax series being the best known.

Composition vs. performance. The properties and performance of the different species shown in Figure 4.7 have been studied in some detail [2] although it should be borne in mind that commercial products are a mixture of species. It is difficult to draw overall trends, with the combined effect of alkyl chain length and degree of sulphonation playing a role but if a fixed alkyl chain length is considered, then trends emerge. The lowest CMC (often linked to low irritancy potential) can be obtained from a dialkyl disulphonate. Changing from a dialkylate to a monoalkylate reduces the CMC by a factor of 3–10. Wetting and foaming follow the trend:

$$\text{mono (alkyl):mono (sulphonate)} > \text{di:di} > \text{mono:di}$$

All the surfactants in the class tend to show better solubility than a comparable LAS, especially in the presence of calcium ions or electrolytes. Detergency of ADPODS is reported to be improved in hard water, with maximum detergency obtained in water with 100 ppm hardness and it has been observed generally to vary with molecular weight, so that both a C16 alkyldisulphonate and a C10 dialkyldisulphonate gave detergency similar to a C11 LAS.

The effect of alkyl chain length on solubility is as might be expected, following a similar trend to other surfactant types. A C16 monoalkyl disulphonate has an aqueous solubility of 40%, while the C10 analogue is soluble up to 70%.

Applications. The short chain ADPODS products, usually C6, are effective hydrotropes. They are able to significantly reduce the viscosity of LAS-based formulations and to solubilise up to 25% octanol in water. The ADPODS hydrotropes are more efficient than short chain alkylbenzenes (such as toluene and xylene sulphonates) but are significantly more expensive.

The combination of high solubility, low viscosity, high tolerance of electrolytes and good detergency makes ADPODS effective ingredients in laundry liquids. It is possible to formulate a liquid with a high level of soluble builders (sodium citrate or disilicate) which would

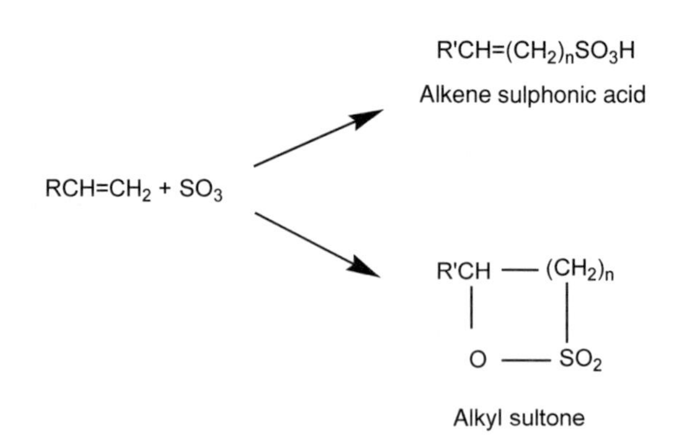

Figure 4.8 Sulphonation of α-olefin.

not be possible with LAS. ADPODS will also perform much better than LAS in unbuilt liquids.

4.1.2 α-Olefin sulphonates

Although used very little in Europe, α-olefin sulphonates are a very important class of surfactants with widespread use in Asia, India and many developing regions.

Chemistry and general properties. The chemistry of the olefin sulphonates is more complex than the sulphonates previously considered. They are manufactured in a three-stage process: sulphonation, neutralisation and hydrolysis. Sulphonation is carried out in standard sulphonation reactors and yields a mixture of sulphonates (Figure 4.8).

Note that the olefin can rearrange to an internal olefin, giving a product containing a mixture of 1-, 2-, 3- and 4-alkene sulphonates [3]. The alkene sulphonic acid is simply neutralised and is unaffected by the hydrolysis process which is intended to open the sultone ring to a sulphonate group but the alkaline hydrolysis also generates a hydroxyl group, normally at C3 or C4 (see Figure 4.9).

The properties of the two species are very different in terms of their solubility, viscosity and detergency. The alkene sulphonate is the better performing surfactant and the manufacturing process is developed to maximise this more desirable product. The sulphonation process is significantly more exothermic than LAB sulphonation, requiring careful control of reactor temperatures and more dilute SO_3 which often results in lower reactor loadings and hence lower productivity, compared to LAB. Low product colours can be achieved but require very careful control of sulphonation, neutralisation and hydrolysis. Earlier processes relied heavily

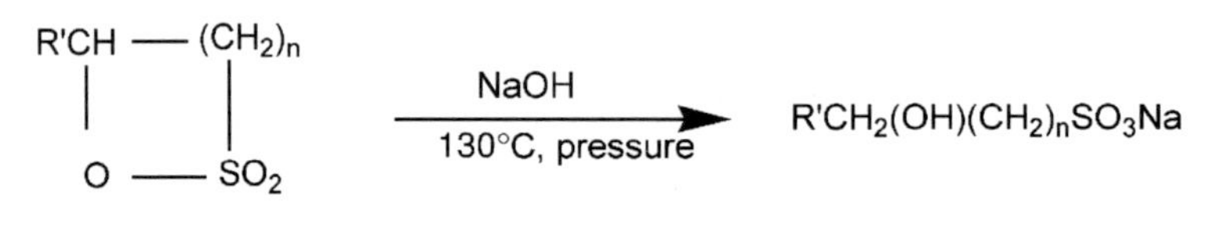

Figure 4.9 Hydrolysis of sultones.

on bleaching to give an acceptable product colour but the use of hypochlorite bleach during the hydrolysis stage leads to the formation of highly irritant and sensitising impurities – chlorosultones.

In the 1960s, there were instances of chlorosultones from poor manufacturing procedures being present in consumer products which caused a number of cases of severe irritation and sensitisation in consumers. The memory of these unfortunate cases is, in the opinion of the author, still a reason for the low level of use of olefin sulphonate in Europe. In fact, the risk of sensitisation from olefin sulphonates made using current technology is negligible. Improved control of sulphonation, highly efficient hydrolysis, and the use of hydrogen peroxide as the bleaching agent mean that there are no chlorosultones, and only ppm level of residual sultones exists.

Commercially, olefin sulphonates are supplied as 40% active solutions which are of low viscosity and stable. The ratio of alkene sulphonates to hydroxy alkane sulphonate is 3:1 or better. Products with higher level of the hydroxyl alkane sulphonate could lead to much higher solution viscosities and reduced foaming and detergency in application. Olefin sulphonates can be dried without the use of builders to give powders at concentrations of $>$90%. The powders are still mildly hygroscopic but are easier to use than an LAS powder of a similar active. There are some examples of commercial use of AOS slurries of $\sim$60%, but these are not common, since heated storage is essential. Olefin sulphonates are more soluble than LAS and do not thicken as readily in formulation. This allows a formulator to use less (or no) hydrotrope in a formulation, but can result in some difficulties where a product needs to be thickened, since salt response is poor. One of the main attributes of olefin sulphonates is their ability to foam and clean well both in cold water and in the presence of electrolytes and sebum. This makes them suitable for regions where the washing temperature is low such as Japan, or in developing regions where river washing is still practiced.

Raw materials. The linear olefins required for sulphonation are petrochemicals, derived from a number of sources. Less common, and generally of lowest quality, are olefins from refinery or crackers which are higher C chain fractions from oil and also from the cleavage of higher carbon number waxes. The predominant olefin source is from polymerisation of ethylene, using the Ziegler process. These olefins can contain significant levels of branching in the carbon chain and a level of branching of up to 50% is common. Olefins are also supplied by Shell Chemicals, manufactured using their Shell higher olefin process (SHOP). While also based on ethylene, this process uses a single reaction step, with specific catalysts to yield predominantly linear olefins, the key manufacturers of which include Shell, Chevron Phillips and Ethyl Corporation.

The manufacturing processes produce a distribution of molecular weights which can be 'peaked' to give a dominant product. The ranges of olefins produced are from C_6 to C_{24}, the former being of interest for production of short chain alcohols, while the higher molecular weights are used in other processes. The molecular weights of interest for surfactant production fall in the range C_{12}–C_{18}.

Composition vs. performance. α-Olefin sulphonates are produced from a number of olefin cuts, with differing performance characteristics. Typical products are C_{14-16}, C_{16-18} and C_{14-18}. These differing cuts give different performance attributes with the more soluble, lower molecular weight products being better foamers and the higher molecular weight products having higher detergency, but reduced solubility. While one can generalise about performance trends related to gross composition, studies have been carried out highlighting the performance characteristics of the individual chemical species [4]. A large difference in the composition of e.g. a 12–14 AOS from two different suppliers could lead to performance differences in apparently similar products. Foam volume, density and viscosity vary depending on the type and degree of branching found in the parent olefin. The ratio of the alkene sulphonate to the hydroxyalkyl will also influence the solution properties, formulation viscosity and detergency of the product. As manufacturing processes become better controlled and more consistent, products from different sources have also become more consistent. Formulators should be aware of these potential performance differences between similarly specified products and check the suitability of a new source in their formulation.

Applications. AOS has excellent detergency properties and has a greater tolerance to hard water and sebum than LAS. Coupled with its relative ease of drying and good powder properties, AOS is highly suitable for detergent powder formulations, particularly at low temperatures. Many Japanese laundry products use AOS as the primary surfactant and use of AOS has also grown in the United States with C_{14-16} and higher used in laundry applications.

C_{12-14} olefin sulphonates are also used in personal care formulations since they are less aggressive than LAS and will not over-strip (i.e. degrease to leave an excessively dry or 'squeaky' feel) the skin or hair although some care may be required in formulating to compensate for a dry feel to the foam. AOS-based formulations are also more difficult to thicken than products based on alkyl sulphates or alkyl ether sulphates but use of alkanolamides or sarcosinates as secondary surfactants can overcome both problems and give a product more acceptable to the consumer.

The chemical stability of AOS and the ability to maintain surfactancy under extremes of pH, temperature and electrolyte concentration generate some industrial applications. AOS is used in enhanced oil recovery (EOR) applications where surfactants must withstand high shear forces, high temperatures and brine concentrations. AOS is commonly used as the foaming agent in salt water foam drilling where a foam is pumped into a drill hole to assist in the flushing of drilling waste from the bore.

4.1.3 Paraffin sulphonates

These surfactants are also called alkane sulphonates or secondary alkane sulphonates. They are a versatile class of surfactants with interesting chemistry and are significantly less commercialised than LAS or alkyl sulphates, due to the more complex manufacturing route and the lack of flexibility of a paraffin sulphonate manufacturing unit which can effectively produce only paraffin sulphonates. Compare this to a falling film SO_3 plant which can be used for the manufacture of a wide range of sulphonated and sulphated surfactants.

Chemistry and general properties. The main challenge in the manufacture of paraffin sulphonates is the lack of a reactive function in the paraffin (alkane) hydrophobe but this is overcome by using the Strecker reaction or a variant of it. The classical Strecker reaction (shown in Figure 4.10) uses a chloroalkane (which can be made by the reaction of a chlorinating agent with an alcohol) and sodium sulphite.

$$RCl + Na_2SO_3 \longrightarrow RSO_3Na + NaCl$$

Figure 4.10 Sulphonation using the Strecker reaction.

This process is scarcely used in industry and the processes used to produce paraffin sulphonates on a large commercial scale are a variation of the above reaction. This can be either a sulphoxidation reaction, using a mixture of air (oxygen) and SO_2, or a sulphochlorination, using a mixture of chlorine and SO_2 (an improved process using liquid SO_2Cl and a catalyst has also been reported [5]). In both cases, the gaseous reagents are mixed with the liquid organic substrate and the mixture is exposed to UV light (or other exciting radiation) to generate free radicals. As is common with radical reactions, the product is a complex mixture of primary and secondary alkylsulphonates, with sulphonation occurring at any position on the chain. The yield of this reaction is low and unreacted paraffin must be recovered and recycled for the process to be economic. Very few paraffin sulphonates are available commercially, one example being Hostapur SAS, manufactured by Clariant. This is a sodium salt of a C13–17 alkane, available as a 30% solution or 60% paste.

Raw materials. The paraffins used in the manufacture of paraffin sulphonates are essentially the same as those used in the production of LAB but favouring the higher end of the molecular weight range. For paraffin sulphonate manufacture, it is essential to use a normal paraffin, free of any aromatics because branched alkane (especially tertiary) and aromatic species will act as radical traps and reduce the reaction yield.

Composition vs. performance. Paraffin sulphonates have excellent solubility and surfactancy with detergent performance equivalent to LAS and solubility significantly better. As with olefin sulphonates, optimum detergency is found at a chain length of ~C15.

Applications. The very high solubility of alkane sulphonates makes them ideally suited for high concentration, liquid formulations. Their high tolerance of electrolytes also means that they can be used in highly acidic or alkaline formulations, such as industrial hard surface cleansers. The major commercial use of alkane sulphonates has been in manual dishwashing liquids where they bring detergency similar to LAS, but their improved solubility (which is synergistic with ether sulphates) allows concentrated formulations to be produced without the use of hydrotropes or additional solvent. Paraffin sulphonates are also tolerant of high concentrations of amine oxides which are commonly used foam and detergency boosters. This is in contrast to LAS, which can be precipitated from solution by amine oxide.

Paraffin sulphonates are becoming less popular in this application at the time of writing since their disadvantages include limited availability (one manufacturer in Europe) and higher cost than LAS. The products also tend to be more coloured and have a stronger base odour than LAS or SLS which is a disadvantage in the pale, lightly fragranced formulations popular in 2004.

There are a number of industrial applications for paraffin sulphonates, such as EOR, leather processing and metal cleaning, which exploit their high solubility, chemical stability and electrolyte tolerance.

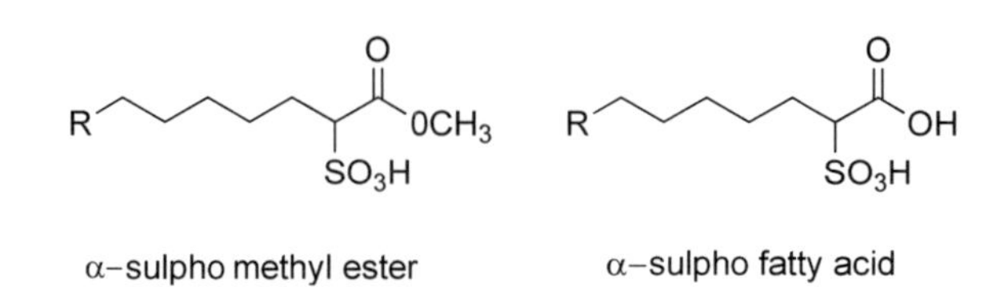

Figure 4.11 Products of methyl ester sulphonation.

4.1.4 Sulphonated methyl esters

As a class of surfactant, sulphonated methyl esters (SMEs) have been known since the 1980s, but have not been widely commercialised. Through the late 1990s into 2000, there were signs of increased use in Asia and the United States across a variety of applications. They share many similarities with olefin sulphonates but, importantly, they are made from renewable oleochemical feedstocks. This is preferred by many formulators, particularly in cosmetic and personal care applications.

Chemistry and general properties. SME surfactants are produced by the reaction of a strong sulphonating agent (normally SO_3) with the methyl ester of a fatty acid. As with AOS, the reaction path is not simple and a mixture of products is obtained as shown in Figure 4.11. Reaction with SO_3 generates a complex acid, forming initially the sulphonated ester and then a disulphonate. This second slow step adds the sulphonate to the carboxylate group in the α-position. The acid is 'digested' in a system in a manner similar to the hydrolysis step in AOS manufacture and bleaching agents and additional methanol are often added at this stage.

The product post-digestion contains a mixture of α-sulpho methyl ester and α-sulpho fatty acid. When the sulpho fatty acid is produced, a small amount of methanol is liberated which can be removed from the product by stripping under vacuum. The acid is neutralised with sodium hydroxide to give the final product, a mixture of sodium methyl ester sulphonate and the disodium α-sulpho fatty acid. The nature and application of the SME are determined by the relative proportions of these two species, as will be shown. The manufacture of low coloured surfactant requires a well designed process and a high level of control over reaction conditions. Recent improvements in process design have allowed the production of much lower colours but a bleaching step is usually required, especially for personal care and dishwash formulations. Modern processes also give a high degree of control over the ratio of sulphonated methyl ester to sulphonated fatty acid (SFA) which allows a range of products to be made, with attributes optimised for specific applications.

As esters, SMEs can be hydrolysed under certain conditions. In the pH range 5–9, SMEs are very stable, even at temperatures close to their boiling points but, as pH moves outside the optimum, hydrolysis rates increase. In the great majority of personal care and household formulations, hydrolysis is not an issue.

Products are normally supplied as a 30–40% solution, although products with a high level of sulphofatty acid will be viscous pastes at these concentrations, so secondary surfactants are often blended into such products to improve their storage and handling properties.

SMEs can also be dried and their powder properties are better than LAS, making SME an attractive ingredient in laundry powders. Products containing high sulphonated fatty acid as the disodium salt also have good solid forms and are used in personal care applications as ingredients in bar products.

Raw materials. The hydrophobe for SME is currently derived exclusively from oleochemical sources, rather than from petrochemicals, as in the case of LAS and AOS. While these two sources can often provide surfactants of equivalent performance, oleochemcials are frequently preferred (especially in personal care applications) because they are derived from 'natural' ingredients. The use of renewable resources is also cited as an additional benefit of oleochemical-based surfactants and this is discussed more fully in Section 4.2.1.

SMEs are often based on the methyl ester of coconut oil or palm kernel oil, both of which give a carbon distribution predominantly of C_{12-14}. Products based on palm stearine, a lower cost oil with mainly C_{16-18} carbon chains, are more difficult to process, and additional care is needed to avoid producing a dark coloured surfactant. The sources and processing of oleochemcials used in surfactant manufacture will be discussed in more detail later.

Composition vs. performance. SME is a highly versatile system, in that the performance can be tailored by varying not only the carbon chain distribution but also the ratio of the two principal components, SME and SFA. The nature and surfactant properties of these two species are significantly different and maximising the concentration of either in a surfactant product can produce a variety of performance attributes. The SME, which is normally the major component, shows good foaming and detergency properties with excellent performance in hard water. SMEs also have some hydrotropic effect and can reduce the overall viscosity of a formulation. The SFA (as the disodium salt) shows reduced detergency and foaming compared to the SME and markedly reduced solubility. It can be said therefore that a surfactant consisting of predominantly SME with low levels of SFA will have good solubility and give excellent foaming and detergency. When SFA levels increase, the surfactant becomes less soluble (often presenting as a paste or slurry) with lower detergency. Formulators using high-SFA products have reported that excellent skin feel can be obtained from personal wash products.

SMEs also follow the general trends of performance vs carbon chain length, i.e. detergency increases with increasing chain length, while solubility and foaming decrease. This means that the optimum product for a liquid personal wash product would be a high foaming, mild C_{12-14}, while C_{16-18} might be more suited to machine laundry.

Applications. SMEs are currently used in three main areas of application: laundry powders, manual dishwash and syndet/combi-bars. SMEs have been used in laundry powders in Japan for many years, their performance attributes being well suited to the Japanese washing habit of using cold, soft water. In studies of the comparative performance of LAS, AS and SME in laundry applications [6, 7] SMEs were shown to perform extremely well, with the same detergency performance as LAS and AS being obtained with only half the concentration of SME. These studies also showed that SME had less effect on enzyme stability than LAS and AS. When looking at European wash conditions, the overall performance is affected by the level of the SFA in the product, from which we can conclude that, to produce an SME for European laundry products, it will be important to minimise SFA and maximise SME. The use of SME in laundry is increasing in the United States as an alternative to LAS and one large U. S. laundry powder manufacturer is now producing, drying and formulating its own SMEs for its laundry powders.

In manual dishwash, a high-SME product (such as Alpha-Step® MC48 from Stepan, SME:SFA ~5:1) based on a distilled coconut methyl ester is used as a partial replacement for LAS and alkyl ether sulphate. Substitution of SME for LAS/AES can give enhanced performance at similar surfactant levels and SME has been shown to be synergistic with ether sulphate with a mixture of SME and AES giving greater foam volumes than either surfactant alone. The combination also gives better foaming performance in hard water than soft water.

The performance of SME-based dishwash formulations can be further enhanced by adding Mg ions to the formulation. In a comparison of formulations of similar overall composition, those using SME as a total or partial replacement for AES were shown to have their plate-wash performance improved by 8–10%. Many European dishwash formulations are based on alkyl sulphate instead of LAS but using SME instead of AS in these formulations can also give enhanced performance. Formulations using SME/AES instead of AS/AES have been shown to give superior foaming and plate washing at the same overall surfactant concentration.

The use of SMEs in synthetic detergent (syndet) and combi-bars (containing both soap and syndet) is another growth area where a different set of performance attributes is required from both laundry and dishwash. Here high solubility is a disadvantage and a less aggressive cleansing is required. These attributes can be found in the SFA component of SMEs. As already seen, SFAs are markedly less soluble and are milder surfactants than the sulphonated ester, making them well suited to this application. Modern processing allows SME products with SME:SFA ratios approaching 1:1 to be manufactured and such products are viscous slurries at 40% active matter. They often incorporate an additional surfactant to make processing easier and bring additional performance attributes, such as lubricity and enhanced foam density or stability. In the United States, so-called combo bars form a popular market sector between standard soaps and more expensive soap-free syndet bars. Combo bars seek to provide the consumer with a product that is noticeably milder than soap alone but at a lower price-point than a syndet. Typically, these products are 50– 85% soap, the balance being synthetic (non-soap) surfactants. Using 15% of an SME-based product gives a measurable decrease in skin irritancy and improved skin feel after washing compared to soap. The SME is commonly incorporated into the soap base prior to drying where it can reduce the viscosity of the soap slurry and allow the drying of higher solid slurry, bringing an additional benefit to the manufacturer.

4.1.5 Sulphonated fatty acids

This is an unusual and interesting class of surfactants that, although limited to some specialist applications, has properties not found in other surfactant types.

Chemistry and general properties. The product is prepared by reacting a fatty acid, typically oleic acid (a C18:1 acid), with oleum, or preferably sulphur trioxide. If a saturated fatty acid is used, the product is an α-sulphofatty acid, $R(SO_3H)COOH$ and the reaction mechanism is thought to be similar to that previously suggested for the sulphonation of methyl esters. With the use of an unsaturated acid, such as oleic, the picture becomes more complex. The reaction chemistry is not fully understood, but the product is a mixture of γ-hydroxy sulpho fatty acid and α-sulphonated oleic acid.

In the former product, the SO_3 has added across the unsaturation in the carbon chain, resulting in a saturated alkyl chain, with an internal hydroxyl and sulphonate and a terminal carboxylate. This mechanism probably proceeds via a sultone intermediate, with the final product being formed during the digestion and neutralisation of the intermediate acid. Both products have two acid groups to be neutralised and, in theory, it should be possible to prepare both the mono and di-sodium salts. The sodium salts of the α-sulphonated fatty acids have been studied in some detail and, generally, the disodium salts are between three and ten times more soluble than the mono-sodium salts (depending on C chain), but general solubility is very poor. For example, the solubility of disodium α-sulpho lauric acid is ~7% at 40°C. These compounds also give very low foam and have poor detergent properties.

The properties of the γ-sulphonated product appear to be markedly different, although they have not been studied in as much detail. Sulphonated oleic acid, in which the γ-sulphonate is the major component, is fully miscible with water at room temperature as is the di-potassium salt.

The potassium salt of sulphonated oleic acid is one of very few commercial examples of fatty acid sulphonates and is sold as a ~50% solid low viscosity (~200 cPs) solution.

Raw materials. It is possible to use any fatty acid as a feed material for sulphonation but economic considerations dictate that oleochemical material be preferred. Fatty acids are readily obtained from vegetable and animal oils and fats which are fatty acid triglycerides. These are transesterified to generate glycerol and three moles of a fatty acid ester, normally a methyl ester. The methyl ester can be distilled to give a specific cut and the fatty acid finally isolated by hydrolysis or hydrogenation of the ester. It is common to use animal fats (tallow) in which case the dominant C chains are 16 and 18.

Composition vs. performance. With this product group, little work has been done in this area. As already seen, the α-sulpho fatty acids show poor solubility, even at higher temperatures and they do not foam like a usual anionic surfactant. It is well known that many sulphonates do not foam well in the presence of soaps and it may be that, since these surfactants contain both the sulphonate and 'soap' function, they effectively have internal foam control.

The γ-sulphonates perform quite differently and their enhanced solubility and detergency may be due to the increased separation between the main functional groups and the additional hydroxyl.

The sulphonated oleic acid sodium salt is a mixture of α- and γ-sulphonates and shows the unique combination of good anionic detergency and very low foaming. The γ-sulphonate provides detergency and solubility, while the α component controls foaming. This is analogous to laundry powders, where it is usual to incorporate a low percentage of soap to reduce the foaming of the primary surfactant.

Applications. At present there are very few known applications, although the surfactants have significant potential due to their unique properties. Sulphonated fatty acids are used in some hard-surface cleaning formulations where their low foam is a benefit and in emulsion polymerisation, where they perform similarly to LAS but with greatly reduced tendency to foam. Future applications for these products may include machine dishwash, extended use in detergent products and industrial applications such as pigment dispersants. For these to be realised, further process development will be required to give a more consistent and better defined product.

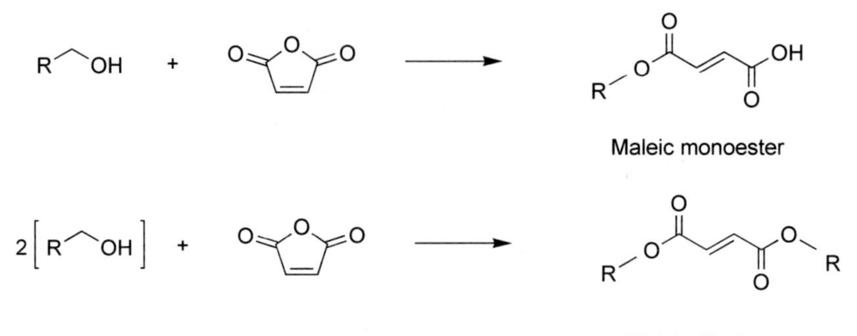

Figure 4.12 Preparation of maleic esters.

4.1.6 *Sulphosuccinates*

These are sulphonates with some atypical properties and are more complex in that they are polyfunctional, containing sulphonate, ester and, in the case of the half-ester, carboxylate groups. Sulphosuccinates are valued as very mild surfactants in personal care applications and, as the diester, also as highly effective oil-soluble wetting agents.

Chemistry and general properties. Sulphosuccinates are made in a two-stage synthesis. Firstly an ester is made by reacting maleic anhydride with an alcohol or an ethoxylated alcohol (Figure 4.12). If the molar ratio of alcohol to maleic anhydride is 1:1, the product is called the half-ester, but where 2 mol of alcohol is used per mole of anhydride, the product is a diester.

The ring opening is moderately exothermic and the reaction is initially controlled by limiting the addition rate of the alcohol to the molten maleic anhydride. The reaction proceeds similarly for both mono and diester products.

The second stage is sulphonation with sodium sulphite (Figure 4.13) which adds the sulphonate group across the double bond. This, too, is an exothermic reaction and care must be taken to limit the temperature rise to avoid discolouration of the sulphosuccinate.

Both stages of the reaction are ideally carried out in nitrogen atmosphere as oxidation can lead to yellowing of the final product, which is undesirable, particularly in personal care

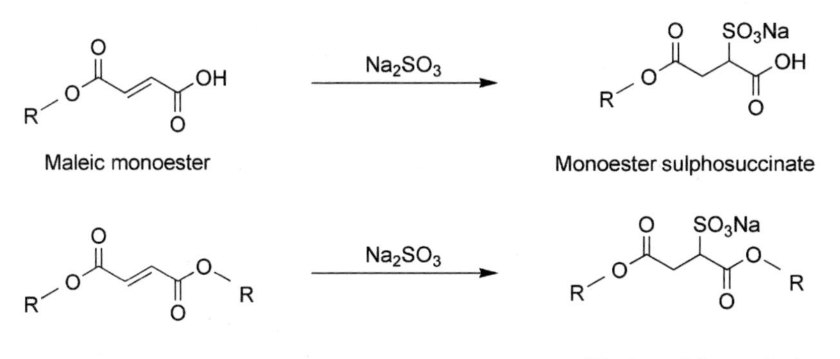

Figure 4.13 Sulphonation of maleic esters.

applications. Sulphosuccinates of ethoxylated alcohols seem to be particularly sensitive to oxygen during manufacture. The sulphite is normally added as an aqueous solution, the concentration of which can be chosen to give the final product concentration. With diester sulphosuccinate, additional care is needed when adding the sulphite solution since the diester has very low water solubility, making the initial reaction very slow but, once some of the surfactant has been formed, it emulsifies the ester, greatly increasing the rate of reaction. The majority of diester sulphosuccinates sold commercially contain some additional solvent (usually isopropanol) to maintain the clarity of the solution.

Half-ester sulphosuccinates are highly water soluble and are usually sold as 40–50% solutions. With care, alkylsulphosuccinates may also be spray dried to give a free flowing powder of >90% active matter. Diesters have low aqueous solubility but are highly soluble in organic solvents. A typical product, di-isooctyl sulphosuccinate, is sold as a 70% solution in a mixture of water and isopropanol.

As a class of surfactants, sulphosuccinates differ from most other sulphonates in their chemical stability and, due to the presence of the ester linkages, sulphosuccinates will hydrolyse at extremes of pH and with elevated temperature. Monoesters are more sensitive than diesters, with optimal stability of pH 6–8, whilst diesters are more stable and will tolerate pH of 1–10 at room temperature. This allows the use of diesters in a much wider range of environments, particularly under moderately acidic conditions.

Sulphosuccinates are not particularly effective detergents but they are good wetting agents and the monoesters are favoured in personal care formulations because of their very low irritancy.

Raw materials. Maleic acid is a petrochemical prepared by catalytic oxidation of either benzene or, preferably, butane. It is a commodity product (approx 900 000 Te global production) used in many chemical syntheses and polymers.

Sulphosuccinates are prepared using a wide variety of alcohols and the choice of alcohol is a major determinant of the properties of the surfactant. In some instances, the consumers' view of what materials are acceptable limits the choice of alcohol source with one example of this being the preference for oleochemical alcohols for personal care applications.

Monoester sulphosuccinates use a wider variety of alcohols than diesters and tend to use longer carbon chain alcohols to obtain the required HLB value from a single alkyl group. A typical product for cleansing applications would use a C12–14 alcohol derived from coconut or palm kernel oil. These materials are easily available due to their use as raw materials for sulphation (see later). Effective sulphosuccinates can also be prepared from petrochemical alcohols but these seem to be less popular. Ethoxylated alcohols (typically 3 mol of EO) are also used and can provide additional benefits in personal care applications. In some cases, alkanolamides or ethoxylated alkanolamides are used as the alcohol, such as ethoxylated cocomonoethanolamide, but they are relatively uncommon, since they are difficult to manufacture and are prone to colouration.

Diester sulphosuccinates are prepared from short chain alcohols, typically in the range C6–10, with some branching. The most used diester sulphosuccinate is di-isooctyl sulphosuccinate, or DOSS, which is commonly made using 2-ethyl hexanol. This is a Guerbet alcohol and is, therefore, monobranched. True isooctanol may contain a much wider range of isomers, leading to variability in the sulphosuccinate properties. Other popular variants use cyclohexanol, hexanol or longer chain alcohols.

Composition vs. performance. The greatest effect of the properties of a sulphosuccinate is undoubtedly the degree of esterification. All monoesters are water soluble, with diesters being very much less so. Variation of the alcohol used in the sulphosuccinate has the expected effect – higher molecular weights give reduced solubility and this is particularly so in diesters, where the dihexyl product is up to 30% soluble in water, but the dioctyl is only 1% soluble.

Applications. As mentioned previously, mono and diesters will be considered separately. The former are used in water-based cleansing applications where they provide good foaming properties with low irritancy, although they exhibit relatively poor detergency. Products based on ethoxylated alcohols (e.g. laureth-3) are favoured since they foam better and are milder than the non-ethoxylated equivalents. Use of sulphosuccinates in shampoo can bring some desirable attributes to a formulation but is not common at the present time. Use of a monoester sulphosuccinate in combination with an ether sulphate will reduce the irritancy of the ether sulphate and give a very high foaming, mild product. Ethoxylated alkanolamide sulphosuccinates are claimed to perform particularly well in this application, with some variants also reportedly having additional biocidal activity [8]. Monoester sulphosuccinates are useful ingredients in personal wash formulations where modest detergency and low irritancy potential are important. Alkylsulphosuccinates are commonly used, although surfactants based on ethoxylated alcohols are milder. Sulphosuccinates in liquid formulations are difficult to thicken and so are normally used in combination with another surfactant, such as an ether sulphate, to help give some salt response to the formulation and provide additional detergency. Alkylsulphosuccinates can also be incorporated into soap-free cleansing bars (syndet bars). Alkylsulphosuccinate is often favoured here since it can be spray-dried and incorporated as a dry solid, without increasing the moisture content of the base. ICI developed a range of syndet formulations in the 1980s and 1990s, using varying ratios of sodium lauryl sulphate, sodium lauryl sulphosuccinate and sodium cocoyl isethionate (SCI), with starch used as a filler and to improve the wear properties of the bar. By varying the ratio of these key surfactants, products with a range of cleansing, mildness and skin feel (and cost) attributes were produced.

The properties of alkylsulphosuccinate as a dry surfactant also lead to a detergent application in carpet shampoo. A formulation based on alkylsulphosuccinate with alkyl sulphate gives good foaming and detergency and the residue (containing the soils removed during the washing process) dries to a crisp solid which can be removed with a vacuum cleaner. These formulations are less popular with the advent of wet and dry vacuum cleaners which require lower foaming detergents.

Diester sulphosuccinates are used almost exclusively in industrial applications where their efficiency of wetting (Draves wetting time <25 s at 0.025% active, cf. 25 s at 0.25% active for LAS) and low surface tension make them excellent dispersants and wetters. DOSS is used as a wetting agent in agrochemical formulations and also a dispersion and wetting aid in dispersible powder formulations (where the product is supplied as a powder to be mixed into water at the point of use). Sulphosuccinates can also be used in a similar way to aid dispersion of pigments into non-polar media, though care has to be taken to avoid foaming in the final product. Perhaps the use of diester sulphosuccinate with which consumers are most familiar is in water repellent lubricant sprays such as WD-40. Diester sulphosuccinates are also used in large volumes in emulsion polymerisation processes.

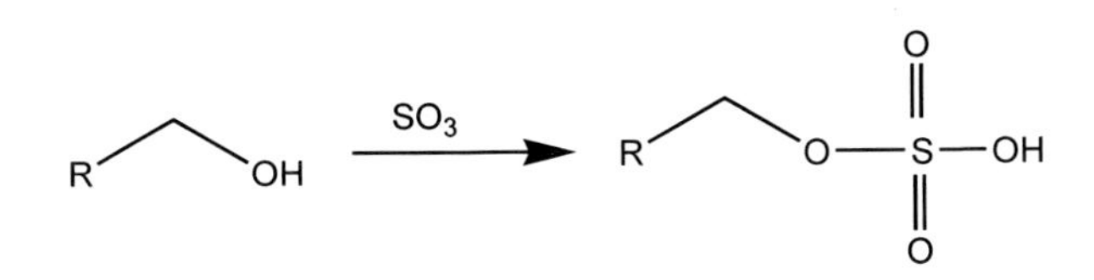

Figure 4.14 Sulphation of an alcohol.

4.2 Sulphates

Sulphates are the second most important class of anionic surfactants in terms of volumes and range of application and share many features with sulphonates in that they are manufactured in the same way. However, there is one very important difference between the two that is the chemical stability of the sulphate group compared to the sulphonate.

4.2.1 Alkyl sulphates

Alkyl sulphates are a versatile and economic class of surfactants with applications in such diverse areas as polymerisation and toothpaste.

Chemistry and general properties. Alkyl sulphates are prepared by reacting an alcohol with a sulphonating agent, normally sulphur trioxide (as illustrated in Figure 4.14).

The product is a sulphuric acid ester or 'sulpho acid' which is susceptible to acid hydrolysis, reverting to the alcohol and free sulphuric acid. Since the rate of hydrolysis is dependent on the concentration of the acid, the hydrolysis reaction is effectively self-catalysing so that the sulpho acid must be neutralised as quickly as possible after manufacture to prevent reversion. This simple chemistry has a profound effect on the manufacture, use and economics of alkyl sulphates.

Since the reaction is much more sensitive than LAS sulphonation, milder reaction conditions are used which normally means a lower concentration of SO_3 in the air/SO_3 mixture – typically 5%, compared to up to 10% for LAS. This, in turn, can limit the output from a reactor, since the operating limit is normally determined by the maximum gas flow through the tubes and, unlike LAS, the intermediate acid cannot be stored for any significant period of time. Good design of a sulphation plant normally minimises the time between sulphonation and neutralisation by achieving minimum possible inventory in the falling film reactor and the shortest possible distance (hence minimum time and inventory) between the falling film reactor and the neutralisation plant.

The need to neutralise immediately after sulphation also changes the mode of use and the economics of alkyl sulphates compared to LAS which is commonly traded as the free acid at typically 97% active matter. This reduces the price (fewer unit operations, hence lower manufacturing costs) and transport costs per unit of active matter and gives the formulator freedom of choice in counter ion and concentration of the final surfactant. Since this is not possible with AS, the manufacturer must neutralise the sulpho acid, often needing to add a buffer and preservative. (This increases the price paid to the supplier, but not necessarily the total cost of the surfactant in formulation.) Most alkyl sulphates are supplied as 20–30% active solutions, so the transport cost per unit of active matter is 3–5 times higher for LAS.

As discussed above, the typical commercial form of an alkyl sulphate is as a 25–30% solution. As the C_{12-14} sodium salt, normally referred to as SLS, this solution is stable at temperatures above 20°C, but will solidify at temperatures between 10°C and 15°C. Higher chain lengths will solidify at higher temperatures, while the use of ammonia or amine salts will allow use of higher concentrations and lower storage temperatures. Sodium, potassium and lithium salts of alkyl sulphates are available as dry powders. The physical properties of these powders are superior to those of LAS, being non-hygroscopic and free flowing, without the use of builders. The highest grades of powders have >95% active matter and less than 1% moisture and a typical dry product for use in laundry powders would have ~90% active matter and 3–5% residual moisture. Commercial supply of alkyl sulphates, especially the sodium salts at concentrations between 40% and 90%, is rare but not unknown. The influence of the micellar structure can be noted in solutions of alkyl sulphates but is not as pronounced as in ether sulphates. There is a viscosity minimum at approximately 70% active, which does allow products of this concentration to be offered commercially. The specialised transport, handling and storage equipment required to deal with a product with very high viscosity which must be maintained at above 65°C put products of this type out of reach of all but extremely high volume users, where the savings on transport might offset the additional equipment costs.

As surfactants, alkyl sulphates are good detergents and foamers and the foam from an alkyl sulphate is dense and stable (often described as creamy).

Raw materials. Alkyl sulphates are the products of the reaction of SO_3 with fatty alcohols which are key materials in surfactants, being used in a wide range of anionics (and non-ionics). Since the source and composition of alcohols are so important to such a wide range of surfactants, it is worth examining them in some detail.

Fatty alcohols, by which the author means those in the range C_8 and above, are split into two classes, petrochemical and oleochemical, or, as they are more usually referred to, synthetic and natural. The discussion of the relative merits of synthetic vs natural products has been at the forefront of surfactant technology for many years and has produced a wealth of literature. It is beyond the scope of this work to discuss whether oleochemicals have an inherent environmental benefit over petrochemicals. A good deal of scientific study on life cycle analysis and macro environmental impact is available but social and ethical arguments, as well as the perceptions of the end consumer, also play a part. On a strictly scientific basis, the author sees no inherent advantage in either source. The performance of a surfactant based on synthetic materials may differ from a naturally derived one but neither is intrinsically better than the other. In terms of impact on humans and the environment, there is also no clear evidence to suggest a difference between the two sources of hydrophobe.

The majority of surfactants for detergents have optimal carbon chains between C12 and C18, depending on the attributes required. From an examination of the composition of natural oils shown in Figure 4.15, it is apparent that the choice of oils is limited.

To obtain substantial yields of surfactant hydrophobes, especially in the most useful C12–14 range, the choice is restricted to coconut, palm and palm kernel oils. In the C16–18 range, the same oils are also used but animal (normally beef) tallow can be used. The use of animal fats raises some ethical issues but these are not commonly used to produce anionic surfactants. The oils may be converted to methyl esters by transesterification which allows easier distillation to remove heavy/light fractions and the esters are finally hydrogenated to fatty alcohols. Alternatively, the fat or oil can be hydrolysed to fatty acid prior to esterification

C chain	8	10	12	14	16	18:0	18:1	Others
Coconut	8.0	6.5	47.6	17.3	8.5	2.7	6.4	**2.1**
Palm				2.5	40.8	3.6	45.2	**7.9**
Palm kernel	4.0	3.9	50.4	17.3	7.9	2.3	26.6	**58.7**
Olive				0.7	10.3	2.3	78.1	**7.2**
Sunflower			0.1	0.3	5.9	4.7	26.4	**61.5**
Soya			0.1	0.4	10.6	2.4	23.5	**51.2**
Rape				1.1	5.3	1.8	54.5	**24.4**
Tallow (beef)		**0.1**	**0.2**	**2.5**	**22.8**	**17.1**	**45.5**	**1.7**

Figure 4.15 Composition of common oils.

and hydrogenation but, in both routes, glycerol is the by-product from the initial treatment of the fat or oil. The commercial value of glycerine can play a key role in the economics of alcohol production.

Petrochemical-derived alcohols use a wider range of chemistry but, in each case, an olefin is the starting point. The olefins may be derived from n-paraffins which give internal olefins, or more linear α-olefins from ethylene oligomerisation. The olefins are converted to alcohols using the oxo process (see Figure 4.16).

The first step of the reaction is the production of an aldehyde from the olefin, followed by hydrogenation to alcohol. In most embodiments of this chemistry, the stages of the process are separated, the intermediate aldehydes being isolated and purified before being hydrogenated in a second, separate process. The alcohols produced are a mixture of branched and linear, a typical alcohol from α-olefins being ca 50% branched. Shell has developed a variation of the oxo process, which combines the hydroformylation step with the hydrogenation in a single reactor. This process depends on a catalyst developed specifically for this application and is used to produce the Shell Neodol series of alcohols. It has an added benefit of producing a much higher yield of linear alcohols than a standard process, from both internal and α-olefins. Neodol alcohols are up to 80% linear and can show performance

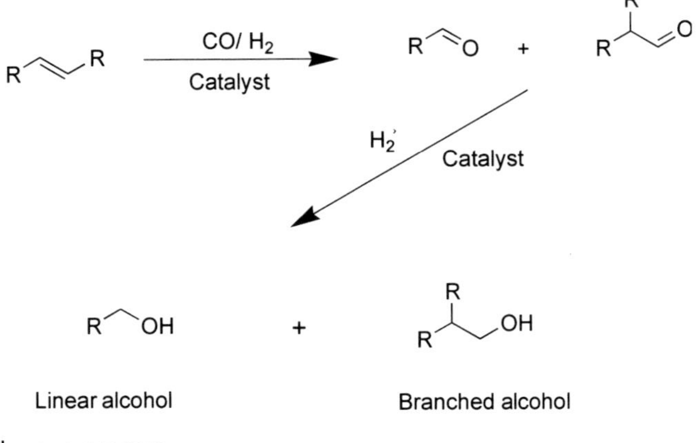

Figure 4.16 The oxo process.

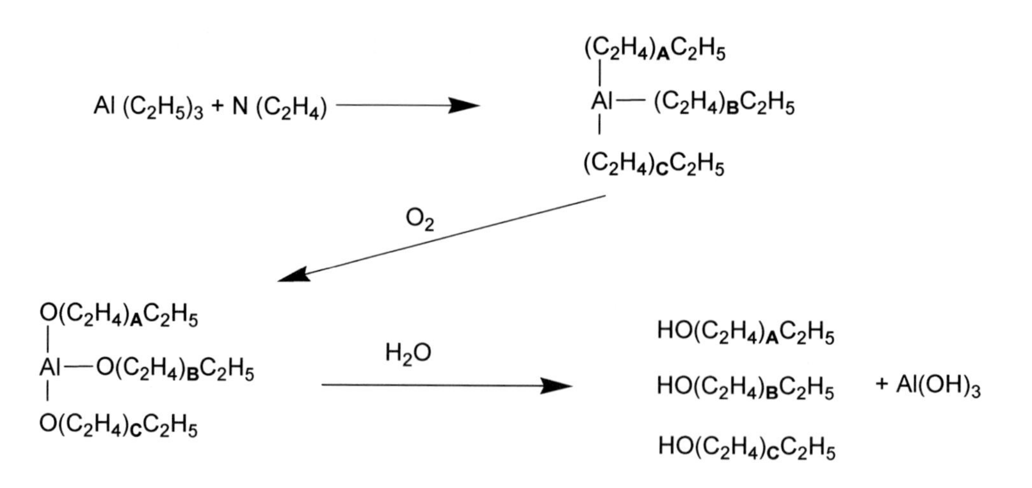

Figure 4.17 Production of alcohols from ethylene growth products. Note: A+B+C=N.

similar to oleochemical-derived products. A third process gives alcohols commonly known as Ziegler alcohols, although they are more correctly known as ethylene growth products.

This process (shown in Figure 4.17) develops alkyl chains of the required molecular weight by adding ethylene onto triethylaluminium. The resulting 'growth product' is oxidised to aluminium trialkoxide in the presence of air which is then hydrolysed to liberate the alcohol.

A newer development is the technology of the production of alcohols from gas to liquid where high molecular weight olefins can be made from natural gas or from coal gasification. This is done using the Fisher–Tropsh reaction.

Alcohols for sulphation should have a defined and consistent carbon chain distribution, which gives a consistent molecular weight (this is important to ensure an accurate SO_3: alcohol ratio) and consistent performance. The level of carbonyls, which can arise from oxidation of an alcohol or from residual aldehyde from manufacture, should be as low as possible but <150 ppm (as C=O) since higher levels of carbonyls will lead to the development of yellow colouration on sulphation. Other impurities include unreacted paraffins but, in most commercial products, these are present at very low levels.

Composition vs. performance. There are a number of key trends that apply to alkyl sulphates, related to the influence of the hydrophobe on surfactant performance, and understanding these can help a formulator select the optimum surfactant for a product.

Increasing the molecular weight of an alkyl sulphate has a significant effect on solubility, as might be expected. Across the series of linear hydrophobes, the solubility decreases with increasing molecular weight and, perhaps more importantly, the Kraft point increases. A 30% solution of a C_{12-14} alkyl sulphate will be a low viscosity liquid at 25°C, whereas the corresponding C_{16-18} sulphate will be solid at 25°C, becoming a mobile paste at 50–55°C. Detergency increases somewhat with increasing molecular weight but, in practice, this must always be balanced against the reduced solubility. The foaming power decreases with increasing molecular weight, optimum foaming being obtained at a carbon chain length of 12 and the C_{16-18} sulphate producing significantly lower levels of foam.

Branching of the hydrophobe can produce some useful performance benefits and a branched hydrophobe will give slightly reduced detergency and foaming but will be more soluble and more tolerant of hard water than a linear equivalent.

Applications. Alkyl sulphates have been widely used in personal care products for many years. SLS is the surfactant most commonly used in toothpaste formulations where the use of ~1% of SLS provides the foaming that consumers perceive as a sign of cleaning, as well as helping to remove particles from the mouth by entraining them in the foam. As a product taken orally, the SLS must have a good toxicity profile (persons who swallow 10 mg of toothpaste per brushing will consume ~5 g of surfactant in their lifetime) and must not adversely affect the taste of the toothpaste formulation. The low taste profile is normally achieved by making a very simple solution, with low excess alkalinity, no buffer system and the lowest possible levels of unsulphated alcohol – typically 0.5% free alcohol in a 30% solution. It is believed that the unsulphated alcohol, or rather oxidation products of the alcohol, can cause raised flavour profiles in SLS. To ensure chemical stability and even lower free alcohol, the slurry is spray dried. The steam stripping effect of the spray drier removes most of the remaining free alcohol, such that a feed solution of 30% active matter and 0.5% free fatty alcohol dries to a powder of up to 97% active matter and 0.5% free alcohol – a 70% reduction in the level of unsulphated alcohol. Toothpaste manufacturers set very demanding specifications on their surfactant suppliers and some include a 'taste test' by an expert panel to ensure that the surfactant has no effect on the flavour of their product.

Alkyl sulphates are used in combination with ether sulphates as the surfactants in most shampoo formulations generally as sodium, ammonium and amine salts. The use of counter ions other than sodium is favoured where an increase in solubility (with slight reduction in irritancy) is required. Shampoos produced in Europe by Procter and Gamble Company use a blend of ammonium lauryl and laureth sulphates but most manufacturers now favour sodium salts. The use of alcohol sulphates generates a dense, long lasting foam, well suited to this application. As a reasonably aggressive detergent, it removes and emulsifies sebum from the hair and scalp but can leave the hair 'stripped' if not moderated by, e.g., a betaine. Alkyl sulphates are commonly formulated with an alkanolamide, such as cocomonoethanolamide (CMEA). These non-ionic surfactants give increased viscosity to the formulation, increased foam stability and can give improved after-feel to the hair.

The use of alkyl sulphates in household products grew significantly during the 1990s. This was partially due to performance requirements of newly developed formulations, such as compact detergent powders, and partly because of the good environmental profile of alkyl sulphates and corresponding concerns over the environmental impact of LAS. One of the main areas where the physical and surfactant properties of AS surfactants helped to drive the development of enhanced consumer products was compact laundry powders and, later, laundry tablets. It has been demonstrated earlier that AS surfactants, particularly as sodium salts, have good powder properties even as (effectively) the pure surfactant. This is in contrast to LAS, which does not form good powders at high concentrations due to its hygroscopicity and amorphous nature, even at relatively low moisture contents. In traditional laundry powders, this was of little consequence, due to the low surfactant and high filler content (see earlier). In the early to mid 1990s a trend to increasingly compact laundry powders emerged and these products saw surfactant concentrations increase from 10% or less to 20% or above whilst the powder dose per wash fell from >200 g to 70 g, leaving less room in the formulation for builders and fillers. In formulations such as these, the sticky nature of solid LAS (and increased concentrations of non-ionic surfactants) can create difficulties in manufacturing a free-flowing laundry powder. At the same time, there was also a trend (mainly in Northern Europe) away from products built with phosphates and towards zeolite builders. Alkyl sulphates built with zeolites were shown to give better washing performance than LAS-based products, particularly in low dose, hard water washing. As powder densities continued to rise, the use of pre-dried alkyl sulphates also grew because the good powder properties allowed solid alkyl sulphates to be added to powders at the agglomeration stage to boost the surfactant content, without reducing the density. Initially, C_{16-18} alkyl sulphates were widely used, due to their quality as solid products, good detergency and low foaming. As typical washing temperatures fell from >60°C to 40°C, solubility became an issue, so use of C_{12-14} sulphates or C_{12-18} blends became more popular. The rise in the use of alkyl sulphates has slowed as the environmental profile of LAS has been thoroughly evaluated through the 1990s and found to have no significant adverse impact. This position could change dramatically, should a future revision of the 2005 Detergent Regulation (648/2004/EC) impose a requirement for anaerobic biodegradation on surfactants for laundry detergents. In this scenario, alkyl sulphates are one of the few commodity anionics capable of meeting this requirement.

The use of alkyl sulphates has grown significantly in manual dishwash formulations which have been predominantly based on LAS and ether sulphate, where the LAS contributed good detergency (especially on greasy soils). As the trend for light coloured products in clear bottles developed (mainly in northern Europe, in the early 21st century) the base colour of formulations using LAS could affect the appearance of the final product. One solution to this was to substitute an alkyl sulphate for LAS, since alkyl sulphates tend to be practically colourless at the concentrations used in dishwash formulations. The performance of the AS based products is very similar to a LAS based formulation but, whilst the cost can be slightly higher, improvements in the quality of LAS, especially to colour, may see a reversal of this trend.

Alkyl sulphates are used industrially as wetting and dispersing agents, and also in emulsion polymerisation.

4.2.2 Alkyl ether sulphates

This class of surfactants has possibly the widest range of use of any anionic surfactant. It is found in almost every product where foaming is desirable, in industrial, household and personal care applications. Alkyl ether sulphates are described in terms of their parent alcohol and the degree of ethoxylation. Thus, sodium laureth–2 is the sodium salt of a sulphated (predominantly) C_{12} alcohol, with an average of 2 mol of ethylene oxide added. Often, the alcohol is assumed to be the typical C_{12-14} and the surfactant simply called a 2- or 3-mol ether sulphate.

Chemistry and general properties. The chemistry of ether sulphates is very similar to that of alkyl sulphates. The 'backbone' of the molecule is a fatty alcohol and often the same alcohols are used as feedstocks for alkyl sulphates, and alkyl ether sulphates and, with higher degrees of ethoxylation, as non-ionic surfactants. The ethoxylation process is more fully described in Chapter 5.

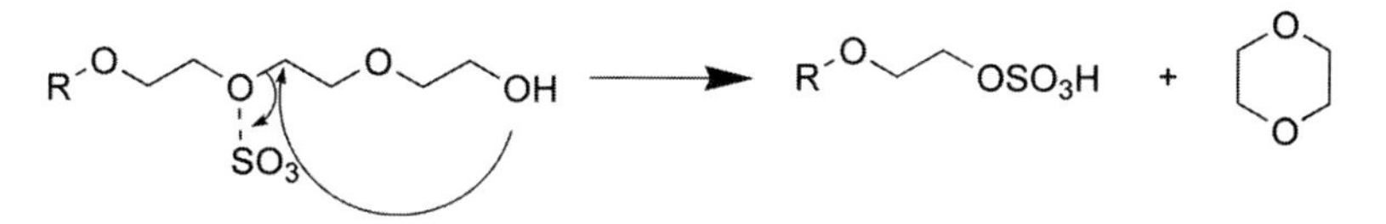

Figure 4.18 Formation of 1,4 dioxane.

The manufacture of the sulphate is again very similar to the process used to prepare alkyl sulphates but milder conditions are required, so the concentration of SO_3 is reduced to as low as 2.5% (compared to 5% for AS, 10% for LAS). To achieve this concentration, large volumes of process air are required to dilute the SO_3 and the limiting airflow that can be passed down the reactor often sets the maximum production rate. The mild conditions are required both to avoid hydrolysis and preserve colour, as with alkyl sulphates, and also to minimise the formation of 1,4 dioxane. Figure 4.18 shows how this impurity is formed by cleavage of two ethylene oxide groups by acid hydrolysis, with subsequent ring closure to form the dioxane ring.

Formation of 1,4 dioxane during sulphation is minimised by slightly undersulphating the feedstock, using the lowest practical concentration of SO_3 in air and minimising the time between sulphation and neutralisation. It is increasingly common to disregard these principles during manufacture and rely instead on the use of a dioxane stripper to remove the dioxane after it has been formed. These devices are wiped film evaporators operating at moderate temperature and reduced pressure. The surfactant is fed to the stripper at a low active matter content (typically 68%). The stripper removes ~2% moisture which also removes most of the dioxane by steam distillation. If a stripper is not used, a well controlled sulphation process can give 1,4 dioxane levels of 30–50 ppm with a 2-mol ether sulphate (note that it is normal to express dioxane as concentration at 100% active matter – the actual level in a 70% ether sulphate would therefore be 20–35 ppm). By using a stripper, the level can be reduced from >100 ppm to 10 ppm. The low dioxane specification demanded by many formulators is driven by a mixture of toxicology (1,4 dioxane is moderately toxic, with an acute LD_{50} of >2000 mg kg^{-1}, and some evidence of carcinogenicity in animals), regulatory control and consumer preference.

It is common to use a buffer system with ether sulphates to maintain an alkaline pH and, whilst phosphate is the most effective, it may not be acceptable in all applications. Alkyl ether sulphates show very strong phase changes with increasing concentration. The low active products (~30%) are low viscosity free flowing liquids (spherical micelles). Between 30% and 70% viscosity increases dramatically, forming a solid, rubber-like gel at ~50% active (rod-like micelles), falling to a viscosity minimum at ~70% (lamellar phase). At this concentration, most alkyl ether sulphates are soft gels of ca 10 000 cPs which are highly shear thinning and can be stored and pumped at temperatures of 20°C–10°C. The use of high active ether sulphate by formulators is very popular in Europe, but less so in North America. Transport and formulation costs are lower (the high active product does not require a preservative to be added) but the formulator needs to exercise care in storage, since the temperature is best maintained at 20–25°C for ease of handling; yet temperatures in excess of 60°C may lead to hydrolysis. Special high-shear equipment is also needed to dilute the surfactant to useable concentrations, since the viscosity peak must be overcome and, once diluted, the surfactant also must be preserved to prevent microbial attack. The majority of

European users whose demand exceeds 500 Te of surfactant per year will use high active material.

The addition of an ethylene oxide chain to what is essentially an alkyl sulphate changes its properties in several important ways. Firstly the Kraft point is very significantly reduced. Low active solutions of ether sulphates are clear are fluid at temperatures close to 0°C, and the Kraft point reduces with increasing levels of ethoxylation. Secondly, the nature of foam changes, from the dense stable foam of an alkyl sulphate to a much more open foam structure. The tolerance of the surfactant to water hardness is also improved, with ether sulphates showing better foaming in the presence of moderate hardness.

Raw materials. Feedstocks for ethoxylated alcohols are made from a large number of alcohols and practically every fatty alcohol used to make alkyl sulphates is also ethoxylated to make non-ionic surfactants, or feedstock for ether sulphates.

Ethoxylation is carried out in batches, or occasionally in a continuous process. A catalyst, typically KOH is used and must be removed at the end of the process. Oxygen must be excluded from the reaction for safety reasons and also to avoid the formation of carbonyls, which lead to yellowing of the surfactants. Typical feedstocks for ether sulphate have degrees of ethoxylation up to 3. The composition of the feedstock (and the resulting surfactant) is complex, with the EO number representing an average, rather than an absolute value. Most ethoxylates for detergents contain significant quantities of both un-ethoxylated alcohol, and higher ethoxylates and the absolute composition of an ethoxylate, as well as the average degree of ethoxylation, can affect surfactant performance. Significant work has been carried out into improving the distribution of oligomers produced by the ethoxylation process and these narrow-range products have generated some clear improvement in non-ionic surfactants but in the low mole product use for sulphates, they appear to generate little benefit. A very wide spread of oligomers can be detrimental, since higher levels of high-mole ethoxylates can lead to increased formation of 1,4 dioxane.

A possible by-product of the ethoxylation process is polyethylene glycol (PEG) caused by polymerisation of the ethylene oxide. PEG levels in the ethoxylate can give performance issues in the sulphated product. PEG is readily sulphated and as the sulphate it acts as a hydrotrope, interfering with the salt response of the surfactant. Very high levels of PEG can cause high levels of 1,4 dioxane to be formed due to the high availability of cleavable EO groups. The majority of detergent ethoxylates will contain $<1\%$ PEG; levels $>5\%$ will create serious problems for the surfactant producer.

Composition vs. performance. Formulators often do not fully appreciate the complex composition of ether sulphates. Without taking into account formulating aids such as buffers, etc., a typical 3 mol ether sulphate will be a 'blend' of over 50 chemical species, all contributing to the overall performance of the surfactant. Formulators often treat all ether sulphates as equal and will switch suppliers only to find that some of their formulations are no longer stable or that the performance changes.

It is possible to draw some general trends and it is certainly true to say that the differences between ether sulphates are less than the difference between ether sulphate and its corresponding alkyl sulphate. Figure 4.19 shows the foaming properties of a series of alkyl sulphates and their corresponding 2 and 3 mol ether sulphates.

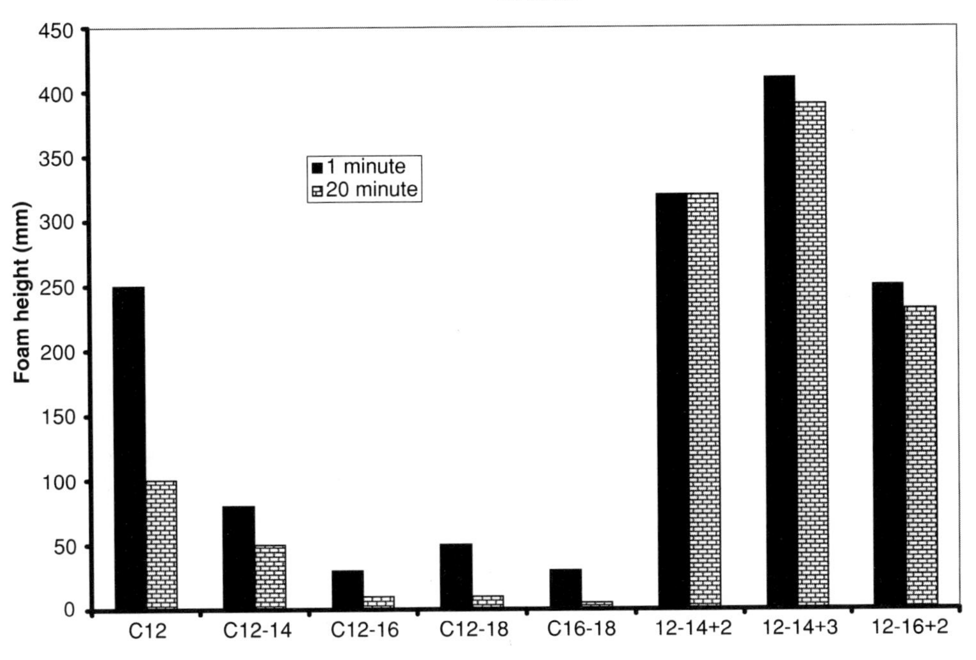

Figure 4.19 Foaming properties of alkyl and alkylether sulphates.

All ether sulphates perform similarly but do also reflect the overall trend of the alkyl sulphates. This may be understood better by examining the composition of the surfactants more closely. A 1-mol ether sulphate will contain ~50% alkyl sulphate but this reduces to typically 25% for a 2-mol and 20% for a 3-mol ether sulphate. This significant proportion of alkyl sulphate influences the surfactant properties and in the author's experience, surfactants varying only in the level of alkyl sulphate (same degree of ethoxylation, same base alcohol) can perform differently. The ethoxylate is the major determinant of surfactant performance and foam is one of the most obvious effects. The most common attribute brought to a formulation by use of ether sulphates is foaming. As demonstrated above, the choice of base alcohol has some influence on foaming. The degree of ethoxylation also affects foaming with higher degrees of ethoxylation giving reduced foam volumes and density.

Ether sulphates show a strong salt effect – that is an increase in viscosity on addition of salt (or other electrolyte). The response to electrolyte (the 'salt curve') can be very different between ether sulphates, even from different suppliers of the same product. Generally, the more soluble the surfactant, the lower the salt response but higher degrees of ethoxylation reduce salt response, as does branching in the alcohol as shown in Figure 4.20.

In practice, the salt response is determined by multiple factors, including the level of unsulphated matter which can have a significant impact on viscosity. Ether sulphates are moderately good detergents, being less effective than the corresponding alkyl sulphate – this moderation of detergency is favoured in personal care applications as the product is less aggressive than an alkyl sulphate and avoids stripping hair and skin. Increasing the degree of ethoxylation reduces detergency and increases mildness.

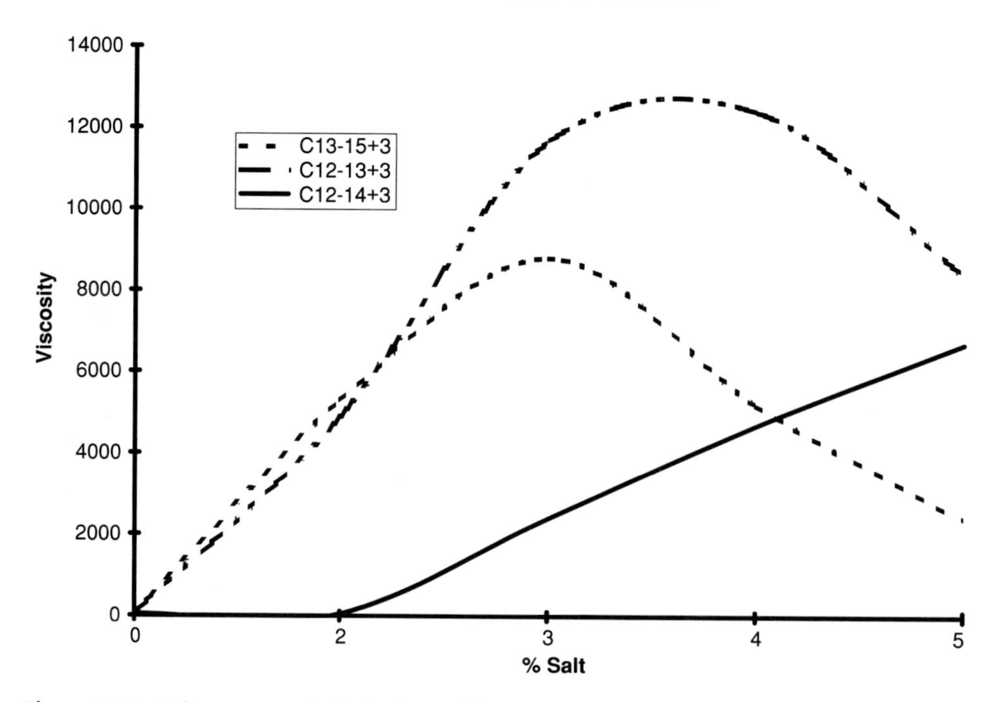

Figure 4.20 Salt response of alkyl ether sulphates.

Applications. Alkyl ether sulphates are almost ubiquitous in cleaning products but are rarely found as the primary surfactant as they bring only some detergency but mainly contribute to foaming and viscosity building.

Alkyl ether sulphates are used relatively little in laundry products, especially in solid ones. The surfactant has a very poor solid form, the pure product being amorphous, somewhat hygroscopic and sticky. Commercial dry ether sulphates are unknown other than as blends of ~50% with carrier solids, such as sodium carbonate. In liquid formulations, 3-mol ether sulphates can be used to give improved solubility and hard water tolerance in unbuilt liquids and amine salts are occasionally used.

Manual dishwash liquids are a major area of application. Whether the primary surfactant is LAS or SLS, ether sulphate is usually present as the secondary surfactant. The foaming generated by the ether sulphate is often the visual cue used by the consumer to judge the effectiveness of a product. This is reflected by the use of the loss of foam as the endpoint in many protocols used to evaluate dishwashing performance.

4.3 Phosphate esters

They are a versatile surfactant type, with some properties analogous to those of ether sulphates. Unlike sulphate (which is a sulphuric acid mono alkyl ester), phosphate can form di- and triester, giving a wider range of structures and, with them, the ability to tailor the product to a greater number of application areas.

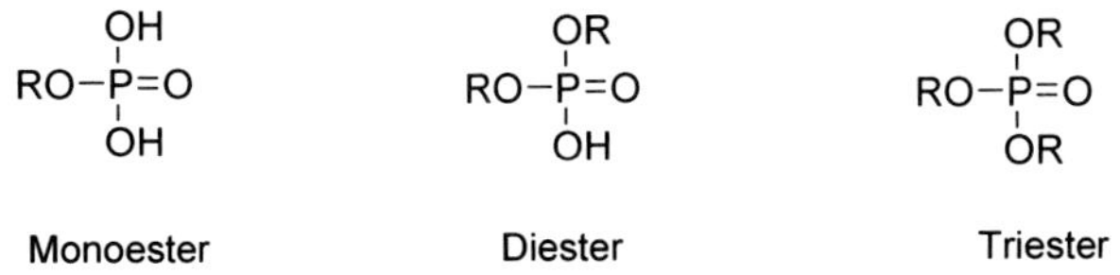

$$\begin{array}{ccc} RO-P(=O)(OH)-OH & RO-P(=O)(OR)-OH & RO-P(=O)(OR)-OR \\ \text{Monoester} & \text{Diester} & \text{Triester} \end{array}$$

Figure 4.21 Structure of phosphate esters.

Chemistry and general properties. As with all esters, phosphate esters are the reaction products of an acid and an alcohol. It is also possible to use phosphorous pentoxide (P_2O_5), which is a more aggressive reagent than the acid and yields a different ester product. Examples of their structures are given in Figure 4.21.

In Figure 4.21, R is alkyl or ethoxylated alkyl and the ester can be mono-, di- or tri-alkyl. If phosphoric acid is used as the phosphating agent, then the product is predominantly monoester. The reaction is milder and more easily controlled but formation of the di- and triesters is more difficult. P_2O_5 is a more aggressive (solid) reagent, making the reaction a two-phase one; this requires careful control over reaction conditions, particularly addition of the P_2O_5, to avoid colour formation. The dominant product here is the diester, although there are still significant levels of monoester present.

The esters can be used as acids, but are also often used as sodium or potassium salts. With such a wide potential for structure and properties, it is difficult to draw general properties for the whole class. They are more stable to hydrolysis than are sulphates and can be used across a wide range of pH and temperatures, but other properties are highly dependent on the degree of esterification and the alcohol used.

Raw materials. The alcohols used in PE manufacture are typically 'detergent' alcohols but shorter chains may also be used. Ethoxylated alcohols, used as non-ionic surfactants in their own right, can also be phosphated to give a surfactant with properties intermediate between non-ionic and sulphated anionic. The provenance of the alcohols has already been covered in detail in the Section 4.2.

Composition vs. performance. Given the variety of alcohols, degree of esterification and salts that can be prepared, the surfactant properties that can be obtained from this group of surfactants are very wide. The predominant types use feed materials which are also used as non-ionics or as sulphation feedstocks for improved economics and these would typically be in the range C_{12-16} with 0–15 mol of ethylene oxide. Increasing the degree of ethoxylation will, as might be expected, increase solubility, and tolerance to hard water and electrolytes. Foam generation will also increase with EO addition, though not as much as with sulphate. Generally, the foaming of phosphate esters is intermediate between the parent alcohol/ethoxylate (often regarded as low foaming) and the sulphated alcohol/ethoxylate. Wetting is better with lower molecular weight alcohols and the shorter chains (e.g. C_8) have hydrotroping powers.

Applications. As medium foaming/hydrotropic surfactants, PEs can be used in detergent cleansing applications with the short chain alkyl esters being effective hydrotropes for non-ionic surfactants. Application of PEs in detergents is relatively limited, due to their cost compared to a sulphate/sulphonate, or non-ionic but long chain diesters can be used as effective de-foamers in anionic systems.

Phosphate esters are widely used in metalworking and lubricants. A C_{12-14} with 6 mol of ethylene oxide (diester) can be used as an emulsifier but also as an 'extreme pressure' additive – it can reduce wear where there is high pressure metal to metal contact. PEs can also show corrosion inhibiting properties, as with petroleum sulphonates and the emulsifying power of PEs with low foam is used in agrochemical formulations. PEs can act as dispersants or hydrotropes in plant protection formulations, allowing the development of easy-to-handle and dilute formulations of both poorly miscible and insoluble herbicides.

The surfactant and anti-corrosion properties of PEs also find use in textile auxiliaries – low foaming gives further benefits and C_8 diesters are also reported to give antistatic effects on synthetic fibres.

4.4 Carboxylates

The final section of this chapter looks at anionic surfactants which derive their functionality from a carboxylate group. These include one of the earliest surfactants made by man (soap) to more complex 'interrupted soaps' where these structures give mild, hard-water-tolerant surfactants.

4.4.1 Soap

The alkali-metal salts of fatty acids – soaps – are the oldest synthetic surfactants and they have been prepared in various forms and in varying purity since pre-historic times. The technology has progressed from boiling animal fats with wood ash to an ultra-efficient high volume process, with a very extensive knowledge and literature base.

Chemistry and general properties. As mentioned above, soap is normally the alkali metal salt of a fatty acid, normally the soluble sodium or potassium salt. Amine soaps are also used. The manufacturing process (see Figure 4.22) normally starts with a triglyceride (oil or fat) which is hydrolysed with a strong base to give the soap and glycerine, or hydrolysed at elevated temperature to give the free fatty acid which can be neutralised after further refining.

Soap is stable under alkaline conditions but, at acidic pH, the fatty acid is liberated and is precipitated. Soap is also very sensitive to the presence of electrolyte and is readily precipitated by salt. The phase behaviour of soap is well defined, but solubility is generally low.

Raw materials. Soap is made commercially from a number of oils and fats, which are typically animal tallow, palm and coconut. The by-products of edible oil refining are often rich in fatty acids and are also used in soap production (in fact, the by-products are sometimes referred to as soapstock). The properties of soap are highly dependent on the quality of the oil or fat used and, to produce a good quality soap, the oil is often put through a series of processes aimed at improving the attributes of the finished product. A simple washing process with hot water can remove many of the soluble components of the raw material that may cause quality issues in subsequent processing. Colour is one attribute that is often targeted and unsaturations in the oil will oxidise during subsequent processing giving a coloured soap. The colour and properties of the oil may be improved by hydrogenation ('hardening') but this is a relatively expensive process. A great deal of technology has been developed to

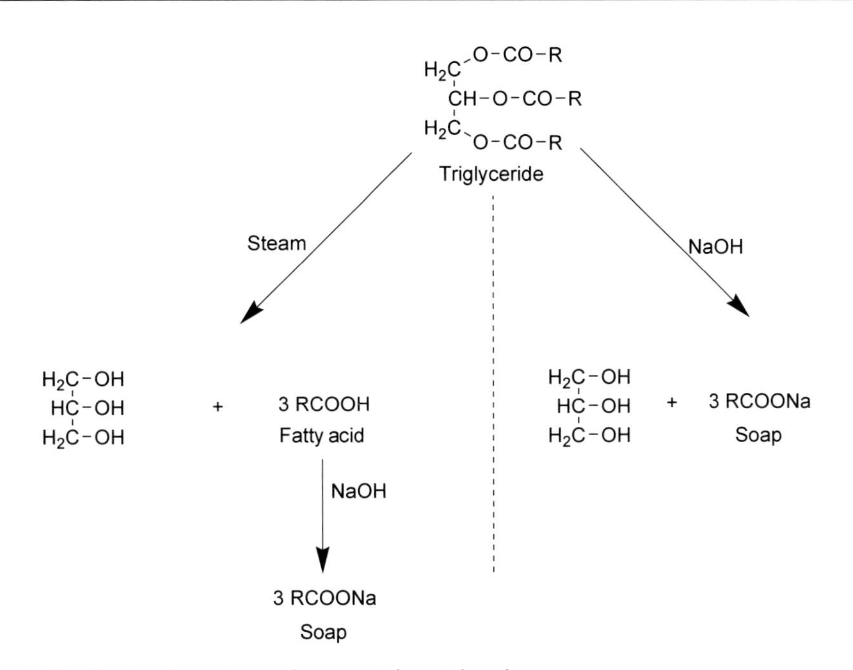

Figure 4.22 Production of soap from a triglyceride oil.

bleach the oil or fat but a detailed discussion of oil bleaching is beyond the scope of this chapter. The processes used employ a variety of bleaches, including acid activated earths which absorb colour forming impurities. The fat is also often deodorised which involves purging with steam to remove volatile compounds that can cause unpleasant odours in the finished soap. This deodorising process can also reduce colour forming bodies.

If soap is to be produced from fatty acid rather than triglyceride, the oil is 'split', i.e. hydrolysed to give 3 moles of fatty acid plus glycerol and this can be achieved by treating the oil with steam at temperatures of ca 300°C. Fatty acids are often further distilled, to 'top and tail' the carbon distribution, to focus on the 12–18 range.

Composition vs. properties. One of the key properties of a soap, key to determining applications, is solubility. As with other surfactants, the solubility of the soap is dependent upon the carbon chain distribution, which is, in turn, determined by the choice of raw material oils. C_{12-14} gives a more soluble soap with very high foam generation whereas C_{18+} soaps have much reduced solubility. The use of unsaturated acids, such as oleic, gives improved solubility compared to the saturated equivalents and, where high solubility is required, potassium salt or an amine salt may be used instead of sodium salt.

Foaming is a further key property controlled by carbon chain and maximum foaming is obtained from C_{12} saturated soaps. Generally, higher molecular weights will give lower foaming properties but introduction of unsaturations, or use of different counterions can affect both the volume and the nature of the foam.

Applications. The most familiar application is as a bar found in the kitchen or bathroom. This application requires a combination of good foaming and also good bar qualities. The bar should be resistant to mushing – formation of a gel layer on the bar surface – and be long lasting, with good resistance to mechanical wear. This combination of properties may be obtained by using a combination of a C12-rich soap for foaming and a C16–18-rich soap for improved bar properties. A common combination is a mixture of palm kernel or coconut oil with animal tallow (or a vegetable equivalent, such as palm stearine) and a mixture of sodium and potassium salts may be used to control solubility. Bar soap formulation is more complex than this section can cover, with many additives available to the soap formulator. These additives can moderate the less desirable properties of soap, such as irritancy, drying and 'scum' formation in hard water. In extremis, this leads to the development of combi- and syndet bar products where the soap is only a part of a more complex surfactant and additive package.

Soaps are also used in the formation of shaving foams and gels. Dense foam, solubility and reduced irritancy are key attributes for this application and amine salts are used almost exclusively for foams and gels. A typical shaving soap formulation uses the triethanolamine salt of palmitic acid (from palm oil, predominantly C_{16}). Other ingredients such as emollient esters are used to give the required lubricity and skin feel. Industrial uses for soaps tend to employ more soluble and electrolyte resistant unsaturated acids, with oleic acid (C18 unsaturated) finding a number of applications. Potassium oleate, as a 20% solution or a 35% gel, is used as a foaming agent in foamed latex carpet backing and the same soap may also be used as an emulsifier in emulsion polymerisation.

4.4.2 *Ether carboxylates*

Ether carboxylates are a very versatile class of surfactants, used in diverse applications from mild personal care formulations to lubricants and cutting fluids. They are interrupted soaps, with the addition of a number of ethylene oxide groups between the alkyl chain and the carboxylate group. The additional solubility imparted by the EO groups gives much greater resistance to hardness and reduced irritancy compared to soap.

Chemistry and general properties. Figure 4.23 shows how a carboxylate group is added to an ethoxylate by reaction with chloroacetic acid.

Shell Chemicals has also developed a process where the carboxylate is prepared by direct oxidation of an ethoxylated alcohol.

Ether carboxylic acids are much more stable than the corresponding ether sulphate acid, with the higher ethoxylates giving acids with sufficient stability to allow them to be stored almost indefinitely and some are commercially available in acid form. This improved stability compared to the sulphates allows ether carboxylates to be used in acidic formulations where the ether sulphate would hydrolyse. There are some similarities in the properties and application of ether carboxylates and ether sulphates and they can be combined to give useful synergistic effects. The carboxylates are generally milder that the sulphates and the higher ethoxylates are very mild with good surfactant properties.

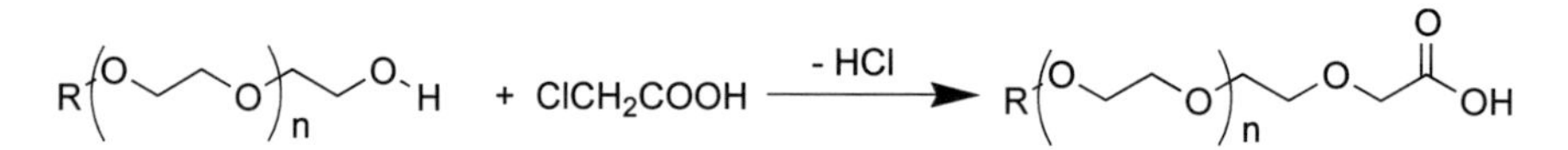

Figure 4.23 Preparation of ether carboxylate.

Raw materials. The base materials for ether carboxylates are typically ethoxylated alcohols, although ethoxylated aromatics or alkanolamides may also be used but a wider range of alkyl chains and degrees of ethoxylation are used in ether carboxylates than in ether sulphates. Carbon chains from C_4 to C_{20} and degrees of ethoxylation from 2 to 20 may be combined to give the required properties in the surfactant.

Sources and properties of alcohol ethoxylates are covered in more detail under alkyl sulphates and alkyl ether sulphates.

Composition vs. properties. With such a wide pallet of raw materials, it is possible to produce a very wide range of attributes in the surfactant and the HLB may be varied from 8 (low C number, low EO), to give surfactants soluble in organic media, to >20 (very good aqueous solubility). Ether sulphates (laureth-2 or laureth-3) would have HLB values of $\gg 20$.

A key attribute of ether carboxylates is mildness which increases with EO number but this can also reduce detergency. Comparing sodium salts with a predominantly C_{12} alkyl chain, the 3-mol carboxylate would have a Zein score of ~ 150, while the 13-mol one would score ~ 80 (cf. laureth-2 sulphate at 270 and sodium lauryl sulphate at 490) [9]. The sodium salts show phase behaviour similar to ether sulphates but the position and scale of the viscosity minimum can be varied with C chain, degree of ethoxylation and, unlike ether sulphates, by the degree of neutralisation [9].

Applications. Ether carboxylates, with their attributes of mildness and good foaming, are ideally suited for use in personal care products. They can be used alone to produce very mild formulations or in combination with sulphates where they have a detoxifying effect (they reduce the irritancy of the sulphate) and can give synergistic foaming, with enhanced stability. Laureth-4 to laureth-7 are well suited to this application to formulate high performance shampoos and foam baths. Carboxylates with lower degrees of ethoxylation, such as laureth-3 carboxylate, are useful for producing viscosity (in e.g. shampoos). Due to their stability at low pH, these products may also be used to formulate thickened acid cleaners. This application may grow in importance, since the amine ethoxylates commonly used to thicken acids have a poor environmental profile and may not meet the requirements of the 2005 Detergent Regulation. Carboxylates can also be used to thicken hypochlorite bleach.

Short chain (C_4) ether carboxylates are low foaming wetting agents which are used in metal working industrial degreasing and in bottle washing.

4.4.3 Acyl sarcosinates

These surfactants are interrupted soaps, in that they have additional functionality added to the carboxylate. In this case an amide function is created and the surfactant head group is a carboxylate salt, as in soap. The additional function produces a surfactant which shares some properties of soap but with generally enhanced performance.

Chemistry and general properties. Sarcosine is N-methylglycine, an amino acid, and acyl sarcosinates are prepared by reacting an acyl halide (normally a chloride) with sarcosine (Figure 4.24). The use of an acyl halide provides a good leaving group – the halogen.

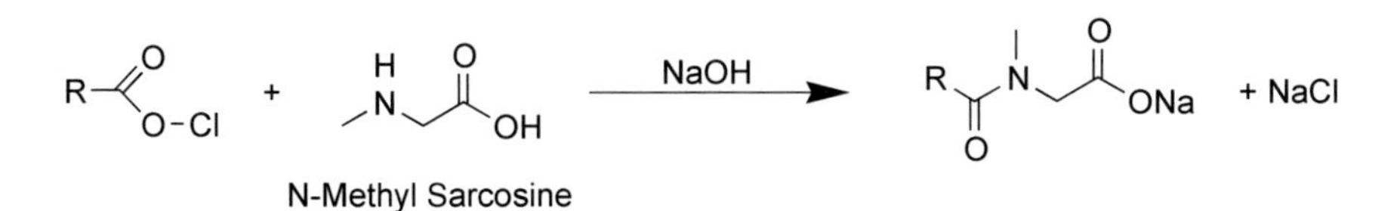

Figure 4.24 Preparation of an acyl sarcosinate.

Salt is a by-product. Due to the stability of the amide group, the free acid can be formed and separated from the reaction mixture to give a salt-free product. The stability of the amide group also allows sarcosinates to be used in a wider range of chemical environments than isethionates (see below). Sarcosinates are stable under moderately acidic conditions but will degrade at low pH or with elevated temperature. The surfactants are moderately soluble at high pH and the sodium salts are supplied as a 30% solution.

Raw materials. Acyl halides are prepared by treating fatty acid with a chlorinating agent, such as PCl_5, PCl_3, thionyl chloride or $SOCl_2$. Thionyl chloride has the advantage that the by-products are gases (SO_2, HCl) and may be more readily separated from the reaction mixture than phosphorous based reagents which have liquid by-products that must be removed by distillation.

Sarcosine is a naturally occurring amino acid but is made industrially by reacting methylamine with monochloroacetic acid (MCA), a common reagent also used in the manufacture of betaines.

Structure vs. properties. Few data exist on variants of sarcosinates, with the cocoyl variant being dominant. The lauryl (C_{12}) variant has been prepared and shows a higher CMC and higher surface tension at the CMC than the cocoyl and the surface tension also shows some dependency on pH.

Applications. Sarcosinates show low irritation potential and are good foamers. Due to these properties they find applications in personal care products where synergistic effects with other surfactants may also be exploited. In combination with other anionics, sarcosinates will often detoxify the formulation and give improved foaming and skin feel. Sarcosinates are also used for their hydrotropic properties – the addition of sarcosinate to other anionics often gives a reduced Kraft point or a raised cloud point if combined with non-ionic surfactants. Lauroyl sarcosinate is used to formulate SLS-free toothpastes which are claimed to have improved taste profile.

In household products, sarcosinates may be used to give the lower Kraft point/raised cloud point effects discussed above but their use is relatively uncommon in Europe.

4.4.4 *Alkyl phthalamates*

A minor, but interesting variant of a sarcosinate, developed by Stepan Company, with some unique and useful properties.

Chemistry and general properties. As shown in Figure 4.25, Phthalamates may be considered as variations of the sarcosinates, since they consist of an alkylamide, a 'spacer' and a terminal carboxylate and, in this case, the spacer is an aromatic ring.

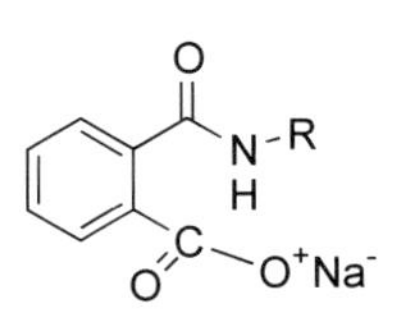

Figure 4.25 An alkyl phthalamate.

Synthesis is by the reaction of phthalic anhydride and a primary amine to give a cyclic imine, which can be ring opened by a strong base, such as NaOH or KOH.

Structure vs. properties. A series of alkyl phthalamates was synthesised by Stepan to determine the optimal composition for different applications [10]. The alkyl chain length was varied from C_8 to C_{18} and all variants were shown to give enhanced foaming and wetting performance in deionised water, with substantially reduced performance in the presence of calcium ions. The C_{12} variant gave the best overall foaming and wetting performance. The surfactants were also shown to be good emulsifiers, with the C_{12} again being the best performer.

Applications. N-octadecylphthalamate has been commercialised as an emulsifier for high HLB systems (water in oil) for cosmetic creams and lotions. The emulsions made using this product are very fine and it is claimed that a much richer skin feel can be obtained with lower oil content, compared to conventional emulsifiers.

The same phthalamate surfactant has also been used in the preparation of sun lotions based on titanium dioxide. The matrix formed by the phthalamate gives additional stability to the dispersed TiO_2, allowing formulators to obtain the highest possible SPF (sun protection factor) rating from a given concentration of TiO_2.

4.4.5 Isethionates

Acyl isethionate could also be classed as an interrupted soap but, unlike ether carboxylates, the additional functionality is added after the carboxylate and the labile metal ion are replaced with an ester, terminated in a sulphonate group.

Chemistry and general properties. Sodium isethionate is 2-hydroxyethane sulphonate, sodium salt and will form an ester with a fatty acid halide, normally an acyl chloride (Figure 4.26).

The chloride by-product may be removed by washing or neutralised and left in the product as salt. Due to the ester group, acyl isethionates are prone to hydrolysis at both acid and alkaline pH, their stable range (at elevated temperature) being 6–8. The aqueous solubility of isethionates is generally poor (0.01% at 25°C for the cocoyl derivative) and, due to this restricted solubility and tendency to hydrolyse, isethionates are generally sold as solids. In terms of surfactancy, isethionates perform similarly to alkyl sulphates in terms of

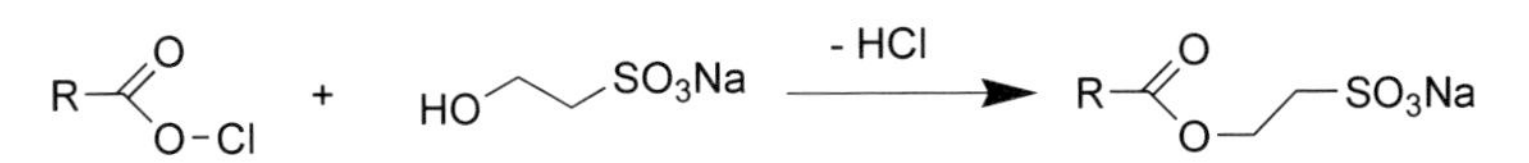

Figure 4.26 Preparation of an acyl isethionate.

their foaming, producing slightly lower foam volumes than a sulphate with the same carbon chain. Their detergency and wetting is good, particularly with greasy soils.

Raw materials. Sodium isethionate is produced by the reaction of ethylene oxide with sodium bisulphite. The most common commercially available isethionates are based on coconut fatty acid.

Composition vs. properties. Very few variants of isethionate are exploited commercially, so there is little information available on structure/property trends.

Applications. By far the largest application of acyl isethionate is as SCI in personal care products where it is used in most solid personal wash products which claim mildness and/or improved skin feel. The SCI is a good detergent in its own right and performs well in the presence of soap (unlike many other anionics). The additional benefit of the isethionate is the after-wash skin feel. Whereas soap alone leaves the feeling skin dry and tight (due to stripping of sebum), SCI leaves a smooth silky feel to the skin which leads to extensive use of SCI in combi- or syndet bars. Combi-bars use a high percentage of soap, with sufficient secondary surfactant to moderate the above-mentioned undesirable effects of soap on the skin. Combi-bars have a significant market presence in North America but, at the time of writing, are almost unknown in Europe. One major exception is the 'Dove' combi-bar, from Unilever which uses a high level of SCI to achieve a mild, moisturising effect. Unilever has a patented process to form the SCI in situ which also gives this product a significant economic advantage over similar products.

Syndet bars contain no soap and, again, often rely on SCI to give mild cleansing, often in combination with sulphosuccinate and alkyl sulphates. Syndets have a small share of the cleansing bar market and are often marketed as a 'care' product, rather than as a simple cleansing bar.

4.4.6 *Taurates*

Taurates can also be classed as modified soap, being similar to isethionates in structure and function. Taurates are useful secondary surfactants, used to modify the properties of primary surfactants.

Chemistry and general properties. Taurine is amino ethane sulphonic acid, and although it is possible to prepare true acyl taurates, superior performance is obtained by using N-methyl taurine (see Figure 4.27). 'Taurate' in this section is used to mean the N-methyl derivative, as is common in industry.

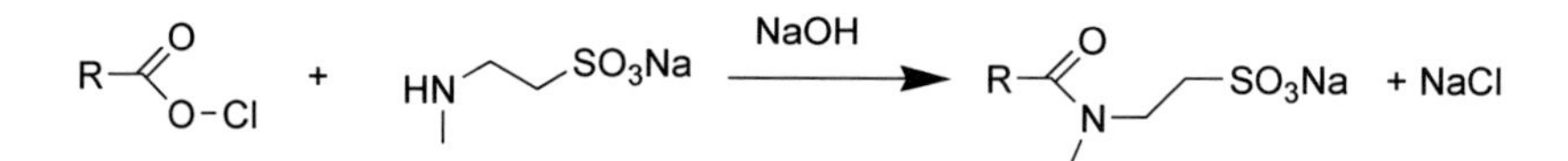

Figure 4.27 Preparation of an N-methyl taurate.

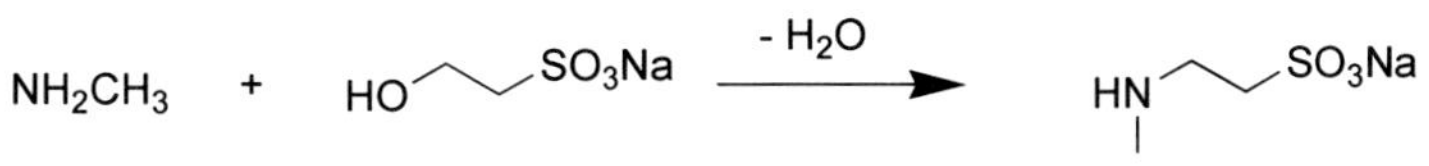

$$NH_2CH_3 + HO\text{-}CH_2CH_2\text{-}SO_3Na \xrightarrow{-H_2O} HN(CH_3)\text{-}CH_2CH_2\text{-}SO_3Na$$

Figure 4.28 Preparation of N-methyl taurine.

The chemistry of taurates is analogous to isethionates, with the ester link replaced by an amide. This gives taurate improved stability to temperature and pH compared to the equivalent isethionate.

The most common derivative, N-methyl cocoyl taurine has limited solubility and is commercially available as a 30% paste. Formulators should be aware that the figure is typically a solid content, with the surfactant content being ~25%, the balance of solids being mainly salt.

Raw materials. N-methyl taurine is the reaction product of sodium isethionate and methylamine (see Figure 4.28). Taurine can be made by using ammonia instead of methylamine, but has little use in surfactants.

Fatty acid chlorides have been discussed in Section 4.4.3. As with isethionates and sarcosinates, the cocoyl derivative is predominant.

Composition vs. properties. Taurates are mild (interrupted soap) high foaming surfactants, mainly used in personal care applications. As one might anticipate, the C_{12-14}-rich cocoyl derivative is used for optimum foaming and detergency. Few other derivatives are in common use. Taurates based on narrower cuts than cocoyl are available (such as lauroyl, myristyl, etc.). Palmitic and oleic acid derivatives are also manufactured, the former having some interesting properties and applications. It is claimed that N-cyclohexyl palmitoyl taurate has particularly low foaming properties. Generally, the trends observed are same as for alkyl sulphates.

Applications. At the time of their invention in the 1930s taurates were used as primary surfactants in a range of applications. They have good foaming and detergency, are stable in a wide range of formulation and use conditions, and can be used in combination with other surfactants. Importantly, they could be used in combination with soap, without reduction in the foaming of the soap – unlike many other anionics. This use declined as other sulph(on)ated surfactants became more readily available and more economic. Interestingly, there is some renewed interest in taurates as primary surfactants. Formulators who market sodium lauryl sulphate free products (mainly in the personal care sector) often use sodium cocoyl taurate as the replacement primary surfactant. (We should recognise that while this is a valid formulation decision, the reasons given for the elimination of SLS from personal care products have no foundation in fact.)

Taurates are mainly used as secondary surfactants as are isethionates and sarcosinates. Taurates have a positive effect on skin feel and are used in bodywash formulations for this effect. Taurates can be especially useful in combination with AOS, since the taurate can moderate the dry feel of the AOS foam, and gives a synergistic foaming performance [11]. The good foaming properties of taurates in hard water, or in the presence of high electrolyte levels, make them suitable for washing in e.g. river or even salt water. Taurates are used industrially in a number of applications. N-cyclohexyl palmitoyl taurate is used in agricultural formulations as a dispersant/wetting agent. It is also used to form a complex with iodine, and produces a biocidal product. The N-methyl derivative has applications as a wetting and dispersing agent, including the manufacture of water-based explosive gels.

References

1. Brent, J. (2004) Petrochemical based surfactants. *Proceedings 6th World Surfactants Congress.*
2. Loughney, T.J. (1992) Comparative performance properties of alkylated diphenyloxide compositions. *Proceedings 3rd CESIO International Surfactants Congress.* vol. D, pp. 255–64.
3. Gee, J.C. (2000). The origin of 1-alkene, 1-sulphonate in AOS manufacture. *Proceedings 5th World Surfactants Congress.* vol. 1, pp. 347–56.
4. Hu, P.C. (1992). Foaming characteristics of alpha olefin sulphonate and its components. *Proceedings 3rd CESIO International Surfactants Congress.* vol. D, pp. 334–47.
5. Canselier, J.P. (2004) A new process for the manufacture of alkanesulphonates. *Proceedings 6th World Surfactants Congress.*
6. Horie, K. (2004) New process of methyl ester sulphonate and its application. *Proceedings 6th World Surfactants Congress.*
7. Umehara, K. (1992) Washing behaviour of alpha-sulpho fatty acid methyl esters and their physico-chemical properties. *Proceedings 3rd CESIO International Surfactants Congress.* vol. D, pp. 234–42.
8. Hunting, A.L. (1983) *Encyclopaedia of Shampoo Ingredients.* Micelle Press, UK p. 382.
9. Jackson, S.W. et al. (1996) Ether carboxylic acids and their salts, advances in technology and applications. *Proceedings 4th World Surfactants Congress.*
10. Bernhardt, R.J. et al. (1992) Synthesis, characterisation, and properties of N-alkylphthalamate surfactants. *Proceedings 3rd CESIO International Surfactants Congress.* vol. C, pp. 301–08.
11. Hunting, A.L. (1983) *Encyclopaedia of Shampoo Ingredients.* Micelle Press, UK, p. 364.

Chapter 5
Non-ionic Surfactants

Paul Hepworth

5.1 Introduction

The term non-ionic surfactant usually refers to derivatives of ethylene oxide and/or propylene oxide with an alcohol containing an active hydrogen atom. However other types such as alkyl phenols, sugar esters, alkanolamides, amine oxides, fatty acids, fatty amines and polyols are all produced and used widely throughout the world in a multitude of industries. This chapter covers the production of these materials and how they can be modified to meet the desired end product use.

There are over 150 different producers and some 2 million tonnes of commercial non-ionic surfactants manufactured worldwide of which at least 50% are alkoxylated alcohols. Ethoxylated nonylphenol production is falling and accounts for 20% of the market while alkoxylated fatty acids account for some 15%. Fatty acid amides and sugar esters account for another 10% and there are a large number of specialities making up the balance. In general, non-ionic surfactants are easy to make, relatively inexpensive and derived from a variety of feedstocks.

5.2 General alkoxylation reactions

The nature of ethylene oxide and, to a lesser degree, the higher alkylene oxides, because of their high reactivity, flammability and explosion hazards mean that plants handling these reactants must be designed to eliminate all possible ignition sources. Reactions must be operated in inert conditions and have explosion pressure rated plant design [1–4].

Many plants have been designed and operated on a batch system with various stirring systems and recirculation loops since the early 1950s but the latest thinking for bulk production is probably the plants that Davy Process Technology has developed for alkoxylation. At least 10 plants of its design have been put into beneficial operation since they were first introduced in 1990 and are based on the Buss Loop Reactor technology.

There are three stages to the production of an ethoxylate or other alkoxylate. An amount of initiator or catalyst (normally NaOH or KOH) is accurately charged to the pre-treatment vessel and mixed with some of the feedstock which is to be ethoxylated. This is warmed to 120–130°C and dehydrated until the water level is less than 200 ppm. The balance of the material to be ethoxylated is added to the reactor together with the initiator and warmed

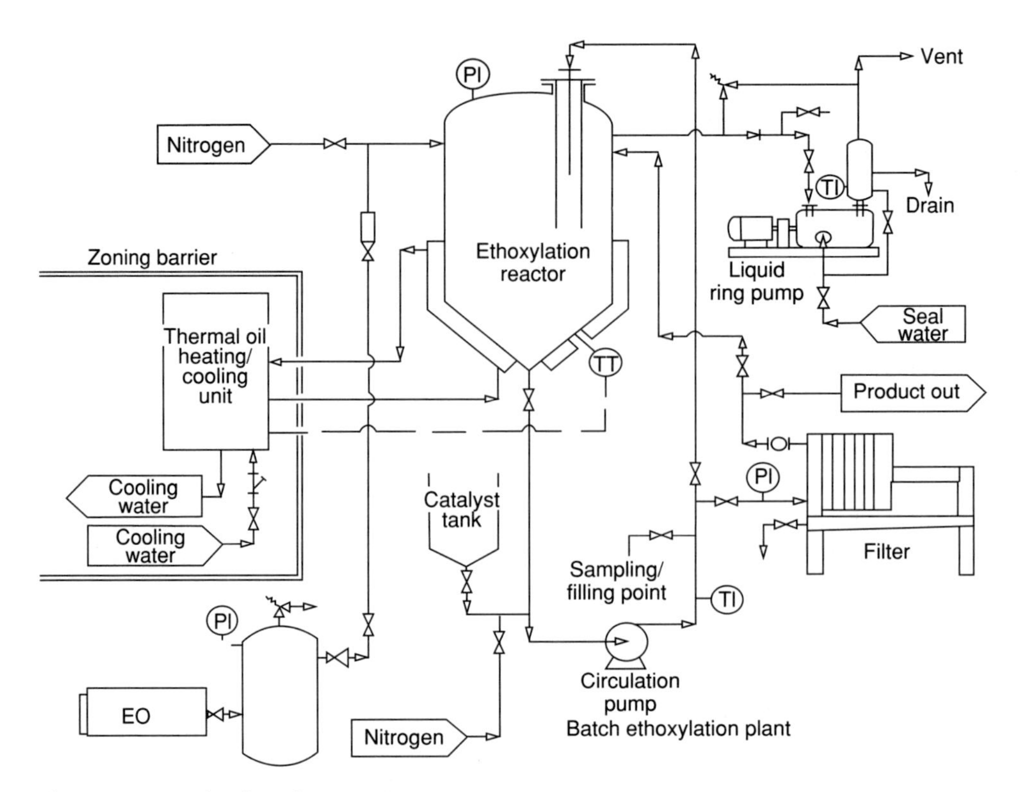

Figure 5.1 Batch ethoxylation unit.

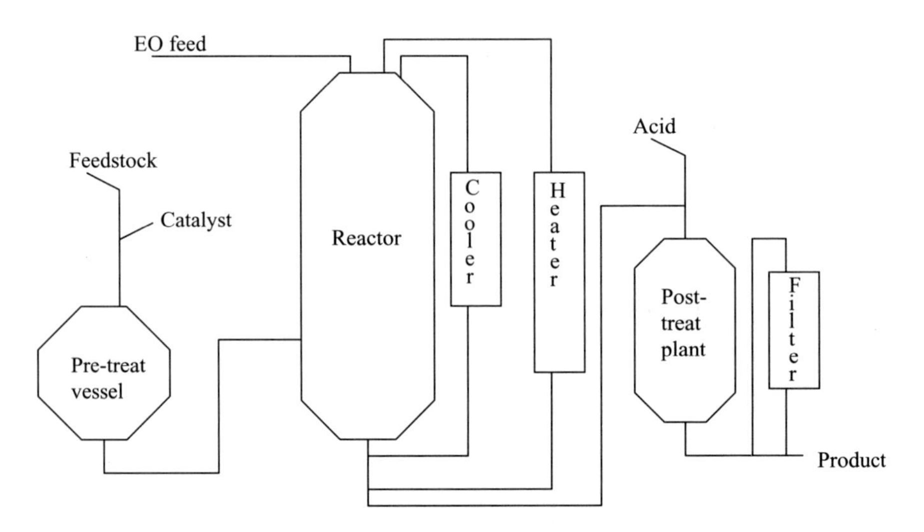

Figure 5.2 Schematic of batch/continuous process.

to 130–140°C and dried under vacuum. Once the reactants have been dried, the heating is switched off, the vacuum broken with nitrogen and ethylene oxide added to produce the desired degree of ethoxylation. Most modern reactors are based on load cells. The temperature is maintained at about 150°C by the controlled addition of the oxide until the specification is met. Reactions are usually carried out between 120 and 180°C. The product is nitrogen sparged to reduce free ethylene oxide to 1 ppm, cooled, neutralised and filtered to remove catalyst [5, 6]. Batch neutralisation, filtration, bleaching [7], etc., if required, takes place in the post-treatment vessel. While the batch is in the reactor the catalyst for the next product to be made can be prepared in the pre-treatment vessel so the system can be regarded as batch continuous.

Standard plant provides a build ratio of 1:8 but by special design can be 1:25 or even 1:50. A build ratio of 1:25 means 25 mol of ethylene oxide is added to 1 mol of alcohol or in round terms 4 tonnes of ethylene oxide is added to 1 tonne alcohol. If the plant had a build ratio of 1:12.5, it would mean the batch would have to be stopped half way, split in two, dried and recatalysed with attendant loss of production. It is a convenient measure of reactor size and recycle volume so that, at the start of the reaction, there is enough volume of alcohol to circulate round the plant and, at the end, enough volume to hold the finished product.

5.3 Alkyl phenol ethoxylates

These were probably the first ethoxylates produced in large quantities and were important from the mid-1940s. They were largely based on nonylphenol with much smaller quantities of octyl and dodecylphenol derivatives. The first mole of ethylene oxide adds with relative ease to the alkyl phenol and then additional moles of ethylene oxide add to produce a Poisson distribution certainly up to the 10-mol derivative [8, 9].

Production of a 9-mol nonylphenol ethoxylate is carried out at 130–150°C as described in Section 5.2. The importance of water removal from the feedstock can be seen in Table 5.1.

Derivatives of nonylphenol up to about the 12-mol ethoxylate are liquid at ambient temperature and do not require heated storage. They are used for reducing oil–water interfacial tension and are excellent for removing oily soils. The major drawback is the biodegradation resistance of the benzene ring, which limits the use to industrial applications in which waste can be treated before any discharge to waterways. However, their relative cheapness has maintained their use in some formulations destined for the household market in certain parts of the world.

Table 5.1 Effect of water on polyethylene glycol (PEG) content and molecular weight (MW) of nonylphenol 9 ethoxylate (NP9)

% water in feedstock	NP9 MW average	PEG MW	Wt% PEG
0.01	615.2	808.1	0.16
0.1	608.6	791.1	1.57
0.2	601.3	773.1	3.07
0.5	580.5	723.7	7.18

Table 5.2 Typical cloud points of nonylphenol ethoxylates

Product	Cloud point (°C)	Comments
NP4	Insoluble	Oil soluble detergents and emulsifiers
NP6	Insoluble	Oil soluble detergents and emulsifiers
NP8	30–34	General purpose emulsifier, textiles
NP9	51–56	General purpose detergent, textiles
NP10	62–67	General purpose detergent, textiles
NP12	87–92	Emulsifier agrochemicals

Nevertheless, some industrial markets with their own sewage treatment plants have been forced to change to suitable alternatives, usually alcohol ethoxylates, because there is a large body of evidence which says the biodegradation goes quickly until it reaches NP2 and then slows dramatically or even stops. Of course, at this stage, the product of biodegradation is no longer soluble in water, so one questions whether the product is just not being seen by the bacteria doing the degradation when it becomes entrapped in the sludge at the bottom of the rivers.

As can be seen from Table 5.2, nonylphenol ethoxylates have a steeply increasing cloud point for very little addition of ethylene oxide. Most industrial products have a rounded up/down value of ethylene oxide in their nomenclature. Thus, NP9 from one company could be actually NP9.25 and from another could be NP8.75. The cloud point for these two products could be 15°C different and in some applications, such as in solubilisation of a fragrance or flavouring, this could be crucial. This is almost certainly due to the sharp (compared to alcohol-based products) Poisson isomer distribution and also variable polyethylene glycol levels in different manufacturers' products. Therefore, it is suggested that product should always be purchased on a cloud point specification and not to an EO number.

Regulatory action is likely to affect greatly the use of nonylphenol ethoxylates and major detergent manufacturers have not used them for many years. The UK has had a voluntary agreement against the use of alkyl phenols in domestic cleaning products since 1976 and Switzerland has banned completely the use of all alkylphenol ethoxylates. The European Union Existing Substances Process has a document proposing bans on their use in textiles, the pulp and paper industry, metalworking, emulsion polymerisation, phenol/formaldehyde resin production and the plastic stabiliser industries. The USA is somewhat slower in coming to the same conclusions but it looks as though some sort of regulatory control is inevitable. However, there is a strong lobby for keeping them and they will probably be used in certain industries for several more years. In general, whilst it is bad news for this section of the non-ionics market it will open up many opportunities for development for the alcohol ethoxylate producers.

5.4 Fatty alcohol ethoxylates

With the slow demise of the nonylphenol ethoxylate market due to legislation, the fatty alcohol market has the chance to design alternatives by subtle changes to the hydrophobe chain lengths and alkoxylate levels. The effects must be achieved with biodegradability as a key parameter and fish toxicity as an up and coming extra requirement to keep in mind. There is no direct natural source of fatty alcohols: they all have to be synthesised and there are five major processes used. Historically the first alcohol ethoxylates were based on tallow or stearyl alcohol. The general processes for producing fatty alcohols are mentioned below.

1. Catalytic hydrogenation of fatty acids from natural fats and oils $RC00H \rightarrow RCH_2OH$
2. The OXO process from alpha olefins $RCH{=}CH_2 + CO + H_2 \rightarrow RCH_2CH_2CH_2OH + RCH(CH_3)\,CH2OH$
3. Ziegler process to give linear even numbered chains
4. From *n*-paraffins to give essentially linear even and odd numbered chains
5. From the Shell 'SHOP' process to give linear even and odd numbered chains

From these various processes it is possible to produce a selection of alcohols from C_6 to C_{20}.

From route 1, the 'natural products', one can have C_{10}, C_{12} and C_{14} from palm oil and C_{16} C_{18} and C_{20} from tallow. It is also possible to have $C_{18\cdot 1}$ from rapeseed. From roule 2, one gets the alcohols with the odd chain lengths C_9 through to C_{15} in various cuts determined by the alpha olefin used. From route 3 one can make both plasticiser C_6 to C_{10} and detergent alcohols C_{12} to C_{20}. Once again these are even chain numbers. From routes 4 and 5 one gets mixtures of odd and even chain numbers in roughly equal proportions.

It can be seen that it is possible to vary the alcohol chain greatly and the alkoxylate chain can be varied in the same way as with nonylphenol to produce both water and oil soluble products. The major difference between nonylphenol and alcohol ethoxylates is the distribution of ethoxylate chains. The rate constants for the addition of ethylene oxide to primary alcohols are comparable and are essentially the same as for the 1-mol or the 2-mol adduct. These addition products, of course, are still primary alcohols. Thus, if one were making a 2-mol adduct of the alcohol, there would be a fair proportion of free alcohol still present – of the order of 10–20%. Chain growth starts well before all the starting alcohol has reacted and alcohol ethoxylates have therefore a much broader ethoxylate chain distribution than the comparable nonylphenol ethoxylate. It has been shown that ethylene oxide consumption becomes constant after 8 or 9 mol of ethylene oxide per mole of alcohol has been added [10, 11].

Detergent alcohol ethoxylates have been used in the detergency industry for many years but interestingly have changed depending on the location of their use and also on the domestic laundry processes. For example, in hot climates $C_{18} + 11$ or even 18 mol of ethylene oxide tends to be used because there is less migration of the ethoxylate due to its higher melting point in the finished detergent powder and less bleed of the alkoxylate from the powder onto the packaging.

On the other hand, as wash temperatures reduced in Europe, better detergency was found in products with 6–7 mol of ethylene oxide in C_{12} to C_{15} alcohols. Then, as the builder was changed from phosphate to zeolite on environmental grounds there was a need to reduce the chain length down to C_9/C_{11} to get better oily soil removal whilst still employing the $C_{12/15} + 7$-mol ethylene oxide for detergency properties. Use of this chain length also coincided with the development of agglomeration processes instead of spray drying for making powder detergents so there is now no loss of the more volatile low ethoxylate components of the alcohol ethoxylates in the end use application. An interesting way of designing a molecule for a particular end use is to add a little propylene oxide to ethylene oxide. A blend of 5–10 wt% propylene oxide in ethylene oxide where propylene oxide is added to alcohol randomly leads to a slightly more liquid product (good for unheated storage

Table 5.3 Range of alcohol ethoxylates produced by Shell Chemicals

Product	End use	Chain length	EO addition moles
Neodol^{TM} 135–3	Sulphation	13/15	3
Neodol^{TM} 135–7	Detergency	13/15	7
Neodol^{TM} 23–1.1	Sulphation	12/13	1.1
Neodol^{TM} 23–2.2	Sulphation	12/13	2.2
Neodol^{TM} 23–2	Sulphation	12/13	2
Neodol^{TM} 23–3	Sulphation emulsifier	12/13	3
Neodol^{TM} 23–6.5	Detergency	12/13	6.5
Neodol^{TM} 25–2.5	Sulphation	12/15	2.5
Neodol^{TM} 25–3	Sulphation emulsifier	12/15	3
Neodol^{TM} 25–7	Detergency	12/15	7
Neodol^{TM} 25–9	Detergency	12/15	9
Neodol^{TM} 45–4	Sulphation	14/15	4
Neodol^{TM} 45–5	Detergency emulsifier	14/15	5
Neodol^{TM} 45–7	Detergency	14/15	7
Neodol^{TM} 91–2.5	Sulphation	9/11	2.5
Neodol^{TM} 91–5	Detergency emulsifier	9/11	5
Neodol^{TM} 91–6	Detergency emulsifier	9/11	6

as with nonylphenol ethoxylates), a slightly lower foam profile and equivalent detergency whilst still maintaining biodegradability to meet current legislation. The same product was used to replace nonylphenol ethoxylates in wool scouring where better cold temperature liquidity for storage and improved biodegradability were required. A typical range of alcohol ethoxylates produced by Shell Chemicals is given in Table 5.3.

Production of all these products is exactly the same as that for nonylphenol ethoxylates. The alcohol feedstock is dehydrated at around 130°C under vacuum, the relevant amount of catalyst, (NaOH or KOH) added, the reactor padded with nitrogen and ethylene oxide added, the feed rate being controlled by monitoring the reaction conditions.

There are many producers in Europe with similar ranges of products including some based on C_{13} alcohols e.g. Lansurf^{TM} AE35. Lankem also has a range based on C_{16-18} alcohols with 4, 19 and 35 mol of ethylene oxide added e.g. Lansurf^{TM} AE735. In addition Lankem produces random alcohol alkoxides such as Lansurf^{TM} AEP66, which are based on C_{12-15} alcohols with a random mix of ethylene and propylene oxides.

A further development in the 1980s/1990s was the introduction of some newer catalysts. Narrow range or peaked ethoxylates can be made using acid activated metal alkoxides, metal phosphates or activated metal oxides as catalyst. These catalysts are insoluble and therefore heterogeneous in nature and the major process difference is that catalyst slurry is added to the reactor after which the conditions are exactly as with normal alkaline catalysts. The reactions are slightly quicker and need less catalyst but it must be filtered out. Most producers [12–20] have patents on these systems, the advantages of which are seen in the finished products as:

1. Lower viscosity
2. More of the surface active oligomers
3. Lower free alcohol content

However, despite these advantages there is little evidence for these products taking off commercially.

It is worth mentioning two more types of alcohol that have relatively small and specialised markets. Secondary alcohols such as those used to make the Tergitol® 15-S range of ethoxylates produced by Dow Chemicals are ethoxylated in the same way as primary alcohols. However, secondary and tertiary alcohols are less reactive than primary alcohols when using alkali catalysts and give products which contain a much wider distribution of the ethylene oxide adducts. This is because the monoethoxylates are much more reactive than the starting alcohols and growth is preferred over initiation. The products are much more liquid for a given amount of added ethylene oxide, e.g. Tergitol 15-S-7 has a pour point of 1°C compared to linear 12–14 7-mol ethoxylate having a pour point of 20°C.

The second alcohol family is Guerbet alcohols which have been known for over 100 years when Marcel Guerbet synthesised these beta branched primary alcohols. The process is a modified Aldol reaction as shown below:

$$2CH_3(CH_2)_9OH \rightarrow CH_3(CH_2)_9C[(CH_2)_7CH_3]HCH_2OH$$

This is very simplified: there are potential side reactions to be minimised and complicated purification steps before the final alcohol is available for use. Therefore, they are expensive which probably accounts for their underutilisation but they have some very interesting physical properties. The alcohols are liquid up to C_{20}, whereas the saturated linear alcohols are all solids. The only liquid linear long chain alcohols are unsaturated and lack the oxidative stability of the Guerbet alcohols.

Ethoxylation is carried out in the same manner as for primary alcohols described earlier but, in general, only up to the 3-mol ethoxylate as a feedstock for some specialised ether sulphates. These products show some advantages in wetting and foaming applications compared to the straight alcohol sulphates. These 'twin tail' surfactants require less co-surfactant to make microemulsions and emulsify 3–5 times more oil than sulphates made from linear hydrophobes.

During the last 10 years, Sasol has introduced a complete range of Guerbet alcohols under the trade name Isofol^R with chain lengths from C_{12} to C_{20}, C_{24}, C_{28} and C_{32}. Other uses suggested are in the various ester products (mentioned later in this chapter) where they should give low irritation and be more effective products if they can be cost competitive.

5.5 Polyoxethylene esters of fatty acids

Probably the third largest group of ethoxylated products is the esters of fatty acids. The fatty acids used are almost entirely derived from natural products by fat splitting in which the triglyceride, (fat or oil) is reacted with water to form $CH_2OH–CHOH–CH_2OH$ (glycerol) plus 3 mol of fatty (e.g. stearic) acid $C_{17}H_{35}COOH$. These are homogeneous reactions taking place in the fat or oil phase because water is more soluble in fat than fat is in water. Continuous fat splitting plants usually have counter current oil and water phases and operate at high temperatures and pressures, which reduce reaction times. The acids quite often contain some unsaturation, e.g. oleic acid, and this, in particular, should be stored and transported under nitrogen to prevent oxidation if the later ethoxylation product is to be of good odour, colour and quality. Peroxide values of starting acids should in particular be measured. A

Table 5.4 Typical range of fatty acid ethoxylates

Product	Chemical description	HLB	Water solubility	Colour and form at 25°C
Myrj™ 45	POE (8) stearate	11.1	Separates	Cream solid
Myrj™ 49	POE (20) stearate	15	Clear	White solid
Myrj™ 59	POE (100) stearate	18.8	Clear	Tan solid

very cheap source of acid used for some products is tall oil fatty acid obtained from the pulp and paper processes.

Ethoxylation is carried out in the same plants and manner as all the alcohol ethoxylates described earlier. Initially, the ethylene oxide reacts with the acid to produce ethylene glycol monoester, $RCOO(CH_2CH_2O)H$, and then reacts rapidly with further ethylene oxide to produce the polyethoxylated product $RCOO(CH_2CH_2O)nH$. However, the reaction conditions are ideal for ester interchange and the final product contains free polyethylene glycol, the monoester and the diester $[RCOO(CH_2CH_2O)nOCR]$ in the ratio 1:2:1 [21, 22]. An alternative method of preparation of these products is to react polyethylene glycol of desired molecular weight and esterify it with acid in an ester kettle. Reaction temperatures and catalysts vary but are in the region 100–200°C. An equimolar ratio of fatty acid to polyethylene glycol results in a mixture similar to the product via ethoxylation, i.e. dominant in monoester [23]. If high excesses of polyethylene glycol are used, monoester dominates but even purified monoester products revert to the mixture on storage, with an adverse effect on wetting properties.

The esters formed in this process are hydrolysed in both acid and base conditions and are much less stable than alcohol ethoxylates. This limits the applications in detergents but they have many industrial uses. In the textile industry they have good emulsifying, lubricating, dispersive and antistatic properties. They are also used widely in personal care, institutional and industrial cleaning, crop protection, paints and coatings and adhesives.

Table 5.4 mentions some commercial products.

5.6 Methyl ester ethoxylates

Hoechst and Henkel first attempted ethoxylation of these materials in 1989 with alkali/alkali earth and aluminium hydroxycarbonates respectively but these catalyst activities were too low for commercial application [24, 25]. Vista, in 1990, patented [26] the use of activated calcium and aluminium alkoxides and Lion Corporation, in 1994, filed a patent using magnesium oxide [27]. There was a flurry of activity in the 1990s and Michael Cox and his co-workers have written most of the literature [28–30]. The proprietary catalysts are more expensive than those for standard alcohol ethoxylates and generally have to be removed from the final product. They are more reactive than the standard alkali catalysts with the result that the reaction proceeds faster and at lower temperature and uses less catalyst.

These materials, of course, contain no active hydrogen, so how does the reaction work? The mechanism is complex and not fully understood but is thought to involve transesterification. The actual distribution of the ethoxymers depends on the catalyst used but a calcium/aluminium alkoxide yields a distribution between the conventional alkali catalysed alcohol ethoxylates and the peaked or narrow range alcohol ethoxylates. The properties of the products look very similar to those of alcohol ethoxylates but as yet are not commercialised.

5.7 Polyalkylene oxide block co-polymers

These materials were introduced by Wyandotte Chemicals Corp and are made by the sequential addition of propylene oxide and ethylene oxide to a low molecular weight reactive hydrogen compound [31]. The polypropylene oxide mid-block is water insoluble and acts as the hydrophobic part of the molecule in the same way as fatty alcohol in conventional ethoxylates. The addition of ethylene oxide to polypropylene oxide mid-block gives water soluble polyols having surface-active properties and the structure:

$$HO(C_2H_4O)a(C_3H_6O)b(C_2H_4O)cH$$

where b is at least 15 and $a + c$ is between 20 and 90% by weight of the molecule. The commercial products made by BASF are shown on the Pluronic™ grid in Table 5.5.

The prefixes L, P and F represent the physical forms of the products as liquid, paste and flake respectively.

Manufacture of the polyols is usually carried out in the same reactors as for ethoxylates. The first step is to dissolve sodium hydroxide in propylene glycol and warm to 120°C. The required amount is charged to the reactor, dehydrated and padded with nitrogen. Once this is achieved propylene oxide is added as fast as it will react, maintaining the temperature at 120°C until the required molecular weight is reached. Then ethylene oxide is added at a rate, which maintains the temperature at 120°C. When all the ethylene oxide is added, the

Table 5.5 Pluronic™ grid of poly alkylene oxide block co-polymers

Typical molecular weight of polyoxypropylene hydrophobic base	First digit								
			L101		P103	P104	P105		F108
3250	(10)								
2750	(9)			L92		P94			F98
2250	(8)		L81			P84	P85	F87	F88
2050	(7)			L72			P75	F77	
1750	(6)		L61	L62	L63	L64	P65		F68
1450	(5)								
1200	(4)			L42	L43	L44			
950	(3)		L31				L35		F38
Second digit		(1)	(2)	(3)	(4)	(5)	(6)	(7)	(8)
0		10	20	30	40	50	60	70	80

% polyoxyethylene (hydrophilic units) in total moleculer espectively

reaction mixture is stripped of all low boiling material, neutralised, usually with phosphoric acid, filtered and cooled. These products are more demanding on reaction conditions to maintain batch-to-batch consistency. Therefore, the rate of addition of ethylene oxide must be controlled carefully. Too fast addition will lead to production of polyethylene glycol instead of the block polymer and temperature and stirrer speeds can be critical for production of high quality block co-polymers.

There are widespread uses in industrial applications for the complete range of block co-polymer surfactants. L61 is an exceptionally good defoamer while F68 finds applications in pharmaceutical products largely because of low foaming properties. Another application where defoaming is required is in refining sugar produced from sugar beet, and special products have been designed for this industry. However, most of these products do not biodegrade and, for this reason, are no longer widely used in the detergent industry. Many manufacturers are attempting to achieve the properties of Pluronics™ but with improved biodegradability. Thus, products are made from fatty alcohols with larger proportions of propylene oxide added to ethylene oxide reducing the foam and increasing the wetting properties but, so far, every product made is a compromise between foam, biodegradability and toxicity, particularly to fish. The Pluronics™ offer a unique balance of properties but clearly have some restrictions on their applications.

A further extension to the range is the Pluronic™ R surfactants in which the hydrophobic and hydrophilic blocks are reversed to give the structure below:

$$HO(C_3H_6O)x(CH_2CH_2O)y(C_3H_60)xH$$

The process is the same as for the normal block co-polymers; the hydrophilic block is first made by adding ethylene oxide to ethylene glycol in the normal conditions to produce a sufficiently long chain molecule which is then capped with propylene oxide to produce the hydrophobic blocks. A similar but less extensive series is available offering an even broader selection of surfactant properties from this type of chemistry.

5.8 Amine ethoxylates

They are a small class of surfactants with applications in the industrial sector rather than the detergent industry. They are produced using the same equipment as for alcohol ethoxylates but the first step is uncatalysed. Primary amine is dehydrated as normal under vacuum and the reactor nitrogen padded and the required ethylene oxide to produce diethanolamine is added. The addition of the second mole is more rapid than the first but then, even in the presence of excess ethylene oxide, there is little or no ethoxylation of the diethanolamine [32, 33]. This reaction is carried out at about 120°C:

$$RNH_2 + CH_2CH_2O \rightarrow RNHCH_2CH_2OH$$
$$CH_2CH_2O + RNHCH_2CH_2OH \rightarrow RN(CH_2CH_2OH)_2$$

Having achieved the first step the further ethoxylation is catalysed with base, sodium or potassium hydroxide, to give the degree of ethoxylation required. The reaction temperature is about 150°C. The two main materials produced commercially are the diethanolamines, which are used in plastics as antistatic or anti-fog agents and the 15-mol ethoxylate, which is used as an adjuvant in herbicide formulations.

Table 5.6 Poly oxyethylene fatty amines

Trade name	Alkyl radical	Ethylene oxide	Average mol wt.	Water solubility
Ethomeen^{TM} C/12	Coco amine	2	285	Insoluble
Ethomeen^{TM} C/15	Coco amine	5	422	Milky at 60°C
Ethomeen^{TM} C/20	Coco amine	10	645	Clear at B Pt
Ethomeen^{TM} C/25	Coco amine	15	860	Clear at B Pt
Ethomeen^{TM} S/12	Soya amine	2	350	Insoluble
Ethomeen^{TM} S/15	Soya amine	5	483	Separates at 40°C
Ethomeen^{TM} S/20	Soya amine	10	710	Clouds at 95°C
Ethomeen^{TM} S/25	Soya amine	15	930	Clear at B Pt
Ethomeen^{TM}T/12	Tallow amine	2	350	Slightly soluble
Ethomeen^{TM} T/15	Tallow amine	5	482	Milky at B Pt
Ethomeen^{TM}T/25	Tallow amine	15	925	Clear at B Pt
Ethomeen^{TM} 18/12	Stearyl amine	2	362	Insoluble
Ethomeen^{TM} 18/20	Stearyl amine	10	710	Clear at B Pt

The amines used are typically coco or tallow amines although others can be used for specialities. The major producer is probably Akzo Nobel with their Ethomeen^{TM} range, which is shown in Table 5.6.

More specialised products are made from secondary amines such as Ethomeen^{TM}2C/25 with the structure $R_2N(CH_2CH_2O)xH$ based on dicocoamine. Also, there is a small series of products based on n-alkyl-1,3-propanediamines where all three hydrogen atoms on the nitrogen atoms are available for ethoxylation. These are Ethoduomeen^{TM} T/13 and T/20 with 3- and 10-mol ethylene oxide added respectively.

Propylene oxide can also be introduced into all of these products for specialised applications.

5.9 Fatty alkanolamides

Alkanolamides are produced by condensation of fatty esters or acids with an alkanolamine. The monoalkanolamides account for 30% of market while diethanolamides account for the bulk of the rest with the balance being made up of a few specialised materials.

The monoethanolamides are not soluble in water and are hard waxy solids, which render them useful in laundry detergent powders since they give easier flowing powders than the sticky diethanolamides and improve the foam and foam stability properties. The simple chemistry is given below:

$$RCOOH + H_2NCH_2CH_2OH \rightarrow RCONHCH_2CH_2OH + H_2O$$

The simplest method of preparation involves heating equimolar quantities of acid and the monoalkanolamine and distilling off the water. However there are competing reactions [34, 35] leading to several potential co-products:

- The amine soap $RCOO^{-}\,{}^{+}H_3NCH_2CH_20H$
- An amino ester $RCOOCH_2CH_2NH_2$
- An ester amide $RCOOCH_2CH_2NHOCR$

These are minimised by avoiding excess acid and a longer reaction time with cook down temperature of typically 150°C.

Preparation from the methyl ester rather than free acid gives a product with much less of the unwanted side products. Commercial grades of monoethanolamides are relatively complex mixtures and with 60 trade names there is likely to be variation in properties depending on the exact manufacturing procedures employed. The dialkanolamides are prepared by similar chemistry which seems even more complicated:

$$RCOOH + HN(CH_2CH_2OH)_2 \rightarrow RCON(CH_2CH_2O)_2 + H_2O$$

The original work by Kritchevsky [36] involved heating equimolar quantities of dietha nolamine and fatty acid. This yielded, as expected, a water insoluble waxy solid. However, when the diethanolamine content was doubled the reaction was modified to produce a liquid, which was soluble in water and foamed and wetted well even in hard water. Its properties were quite unlike a physical blend of the 1:1 diethanolamide with diethanolamine. It became known as a 'low activity', a 2:1, Kritchevsky or Ninol type alkanolamide.

A typical composition [37] of the low active material is as given below.

Composition	(in %)
Diethanolamide $RCON(CH_2CH_2OH)_2$	55
Amino ester $RCOOCH_2CH_2NHCH_2CH_2OH$	10
Free diethanolamine	22
Diethanolamine fatty acid soap	10
Amide ester $RCOOCH_2CH_2N(COR)CH_2CH_2OH$	1
Water	2

Preparation from the methyl ester of the fatty acid using only a slight excess of alkanolamine gives what is known as the 1:1 or super amide with more than 90% diethanolamide and up to 5% unreacted ethanolamine, 4% amide ester and 1% unreacted methyl ester.

The chemical differences between low active and super amides dictate the fields of application. The low active products are liquids used where high purity is not required and where the amine soap helps disperse other ingredients. Super amides are produced in twice the quantity and used in solid products, shampoos and light duty detergents as foam stabilisers. They are generally waxy solids.

5.10 Amine oxides

Amine oxides are the reaction products of tertiary amines and hydrogen peroxide. In aqueous solutions fatty amine oxides exhibit non-ionic or cationic properties depending on pH, and under neutral or alkaline conditions they exist as non-ionised hydrates.

The oxides are produced from tertiary amines by adding the amine to 35% solutions of hydrogen peroxide over a period of 1 h at 60°C with very good mixing. During the addition period, small amounts of water must be added to prevent the formation of gel [38–40]. The

Table 5.7 Typical commercial range of amine oxide products

			Carbon chain distribution					
Trade name	Chemical structure	Active level	C8	C10	C12	C14	C16	C18
Aromox^R C/12-W	Cocobis (2-hydroxyethyl amine oxide)	30–32%	5%	6%	50%	19%	10%	10%
Aromox^R C/13-W	Cocopolyoxyethylene [3] amine oxide	28%	5%	6%	50%	19%	10%	10%
Aromox^R 14D-W 970	Tetradecyl dimethyl amine oxide	24–26%			2%	97%	1%	
Aromox^R MCD-W	Cocodimethyl amine oxide	30–32%			68%	29%	3%	
Aromox^R B-W 500	Alkyldimethyl amine oxide	29–31%			35%	14%	50% inc. C18	
Aromox^R T/12 HFP	Tallowbis (2-hydroxyethyl) amine oxide	39–41% in propylene glycol/H_2O			1%	4%	31%	64%
Aromox^R T/12	Tallowbis (2-hydroxyethyl) amine oxide	49–51% in diethylene glycol/H_2O			1%	4%	31%	64%

amount of water varies depending on which amine is being oxidised and should be only sufficient to prevent gel formation. A 10% molar excess of peroxide is used. After the amine addition has been completed, water is added to produce a 30–40% solution and the reaction temperature is raised to 75°C. The reaction continues with stirring for further 3 h and the reaction product is then cooled and any excess peroxide removed with sodium sulphite to give a specification of less than 0.1% H_2O_2. An interesting alternative source of tertiary amine is the 2-mol alkoxide of a fatty primary amine, which will undergo the same oxidation reaction as described above. The ethoxylated products are used in industrial and institutional cleaners as foam boosters/stabilisers and wetters and in personal care products such as foam baths, shampoos and aerosol mousses. The main use for dimethyl fatty amine oxides is in disinfection and food industry cleaning. The C_{14} dimethyl amine oxide is particularly good and stable in thickened sodium hypochlorite solutions. A typical range of amine oxides is given in Table 5.7 based on products from Akzo Nobel.

5.11 Esters of polyhydric alcohols and fatty acids

Partial esters of fatty acids with polyhydroxy compounds of the type $CH_2OH(CHOH)_nCH_2OH$ where $n = 0$–4 are the basis of a useful class of surfactants. The weak hydrophobic properties are provided by the unesterified hydroxyl group, which essentially limits the range to monoesters. In general, monoesters of lauric, palmitic, stearic and oleic acids are produced from ethylene glycol, propylene glycol, glycerol, sorbitol, mannitol and sucrose.

5.12 Glycol esters

These are usually prepared by direct esterification of the ethylene glycol with whichever fatty acid is required. Equimolar concentrations yield a mixture of mono and diesters and the reaction is usually carried out at 170–200°C. A multitude of base and acid catalysts can be used but, generally, sodium hydroxide or a metallic soap is used commercially. To obtain a high monoester content a 3-mol excess of glycol to fatty acid will give up to 70% yields of monoester.

5.13 Glycerol esters

Mono- and diglycerides are the most commercially important members of this series being used extensively as emulsifiers in the food and cosmetic industries. They can be prepared from individual fatty acids but, more commonly, directly from oils or fats by direct glycerolysis. Thus, oil or fat is heated directly with glycerol at 180–230°C in the presence of an alkaline catalyst. Ideally, 1 mol of coconut oil plus 2 mol of glycerol will yield 3 mol of monoglyceride. In practice, the reaction product is 45% monoester, 44% diester and 11% triester and any unreacted glycerol is removed by washing with water. The substitution on the monoester is 90% on position 1 and that on the diester is mostly in positions 1.3.

Commercially available grades of monoglycerides usually contain 40 or 60% monoglycerides since the solubility of glycerol in fats limits the conversion. It is about 20% soluble at 180°C and 40% at 250°C giving approximately 40% and 60% monoglyceride, respectively. However, the reaction is reversible and even heating distilled monoglyceride at 180°C for 3 h will convert 30% of the monoglyceride to glycerol, di- and triglycerides.

5.14 Polyglycerol esters

These are produced in a two-step process from glycerol and fatty acids [41]. The first step is a controlled polymerisation of glycerol into a polymeric form by heating the glycerol in the presence of an alkaline catalyst, such as 1% caustic soda, at a temperature of 260–270°C:

$$HOCH_2CH(OH)CH_2OH + HOCH_2CH(OH)CH_2OH \rightarrow$$
$$HOCH_2CH(OH)CH_2OCH_2CH(OH)CH_2OH$$

As the reaction proceeds, the material becomes more viscous such that most commercial products are only 2–10 units long.

Esters of oligomers can be made, with or without more catalyst addition, by reacting with any fatty acid. The addition of more hydroxyl groups with each additional glycerol means that a large range of polyglycerol esters can be made with various fatty acids from C10 to C18 and the hydrophile/lipophile balance (HLB) range of these products can vary from 3 to 16 making a very good series of emulsifiers.

5.15 Anhydrohexitol esters

The only hexitol-derived surfactants to achieve commercial importance are those where a portion of the polyol has been anhydrised. They are manufactured by the direct reaction of hexitols with fatty acids during which internal ether formation as well as esterification occurs.

The plant used for these reactions is a typical ester kettle which can have internal or external heating, recirculation of the reactants, low shear or high shear stirring of the immiscible reactants, condensation for return of free acids, water removal facilities, etc. The quality of ester obtained depends not only on reactor geometry and mixing abilities but, particularly, on how the heat input is achieved. Local 'hot spots' must be avoided and it is essential that high quality acids with low peroxide values, where applicable, are used. Being insoluble, the reactants need good mixing to achieve faster reaction times and an alkaline catalyst is used which, in effect, is the sodium salt of the acid [42–44]. Acidic catalysts such as phosphoric acid are also used [45].

Esterification takes place between 180°C and 240°C. Although esterification and dehydration occur simultaneously, esterification takes place faster and the reaction is cooked down at high temperatures to complete the anhydride ring formation and meet the hydroxyl specification. The total reaction time is about 6 h and, after completion of the reaction, the product is neutralised and filtered. The majority of the commercially available products are monoesters although one or two triesters are sold. The esterification occurs mainly on the primary hydroxyl group but small quantities of monoester do occur on the three secondary hydroxyl groups together with some di- and triesters. The anhydrisation can also be carried out further to yield isosorbides and their derivatives. A typical product list is given for the Uniqema range in Table 5.8.

Span™ surfactants are lipophilic and are generally soluble or dispersible in oil, forming water in oil emulsions. They are used for their excellent emulsification properties in personal care, industrial cleaning, fibre finish, crop protection, water treatment, paints and coatings, lubricant and other industrial applications.

Some more specialised products for the personal care industry are given in Table 5.9.

Arlacel™ 83 is widely used in eye makeup, face powders, makeup bases, nail care products and shaving preparations.

Table 5.8 Typical commercial range of sorbitan esters

Product	Chemical composition	HLB	Acid value max (mgKOH g^{-1})	Colour and form at 25°C
Span™ 20	Sorbitan monolaurate	8.6	7	Amber liquid
Span™ 40	Sorbitan monopalmitate	6.7	7	Tan solid
Span™ 60	Sorbitan monostearate	4.7	7	Tan solid
Span™ 65	Sorbitan tristearate	2.1	15	Cream solid
Span™ 80	Sorbitan monooleate	4.3	7	Amber liquid
Span™ 85	Sorbitan trioleate	1.8	15	Amber liquid

Table 5.9 Typical personal care range of sorbitan esters

Grade	Description	Acid value max (MgKOH g^{-1})	Viscosity at 25°C (cSt)	HLB
Arlacel™ 83	Sorbitan sesquioleate	7	1500	3.7
Arlacel™ 987	Sorbitan monoisostearate	12	4100	4.3
Arlacel™A	Mannide monooleate	1	300	4.3

5.16 Polyoxyalkylene polyol esters

The addition of ethylene oxide to sorbitan esters raises the HLB of the resultant ester products. Thus, ethoxylation of sorbitan monolaurate with 4 mol of ethylene oxide per mole of ester will raise the HLB value from 8.3 to 13.3. Values for different sorbitan esters are given in Table 5.10.

As with other ethoxylation reactions described earlier in the chapter, the reaction is carried out under base catalysis (sodium or potassium hydroxide) at temperatures between 130 and 170°C. However, ester interchange also takes place during the ethoxylation reaction. Thus, although the ester group was originally attached to the sorbitan ring system, this can easily rearrange to join any of the growing oxyethylene chains. The net effect of this ester interchange would appear to be as if the ethylene oxide inserts itself between the sorbitan ring and the fatty acid moiety. In this way, the overall shape of the final ester is not as simple as the earlier idealised version of a surfactant with a hydrophilic head and hydrophobic tail. It is closer to a core sorbitan ring, with radiating oxyethylene chains which are capped to a greater or lesser extent with the fatty acid function.

By analogy with the polysorbate ester reaction, similar surfactant ester feedstocks that feature a free hydroxyl function can undergo oxyethylation to materials of higher HLB and these include glycerol, polyglycol and sucrose esters. In addition, many naturally occurring fats and oils can react with ethylene oxide to raise their water solubility. Castor oil is a common base for oxyethylation and the obvious site is the hydroxyl group in the ricinoleic

Table 5.10 HLB values of sorbitan esters and polysorbate derivatives

		Poly sorbate product degree of ethoxylation	
	HLB value	4–5 mol EO HLB value	20 mol EO HLB value
Monoester laurate	8.6	13.3	16.7
Monoester palmitate	6.7		15.6
Monoester stearate	4.7	9.6	14.9
Monoester oleate	2.1	10.0	15.0
Triester stearate	4.3		10.5
Triester oleate	1.8		11.0

acid but the ester groups under the normal ethoxylation conditions lead to the insertion of a polyether chain into the glyceride linkage e.g.:

$$RC{:}O(OC_2H_4)nOCH2\text{-}C(-O-)HO\text{-}O(-O-)CH2$$

Common products available commercially are castor oil plus 30- or 45 mol ethylene oxide, which are very effective emulsifiers.

5.17 Alkyl poly glucosides

Alkyl glucosides are probably one of the oldest synthetic surfactants and were first made by Fischer in 1893 [46] but, until the 1980s, they were not produced industrially. Even from these early days, they have always been of interest because of the concept of being made from natural materials and renewable resources, i.e. glucose and palm- or coconut-oil-based alcohols [47, 48]. They have proved to be highly effective surfactants in washing and cleansing preparations. Commercial use of these products began in 1992 following the building of a 25 000-tonne-per-annum plant in the USA by Henkel Corporation which was followed by another plant of similar size in Germany in 1995.

The industrial process is based on the Fischer synthesis and commercial development work started some 25 years ago. Once a route using coconut-based alcohols had been established, full-scale plants became viable and alcohol blends are now used to control hydrophobicity of the molecules produced. Thus, products based on C8 alcohols are water soluble and those based on C18 are virtually insoluble. The critical micelle concentration (CMC) values of pure alkyl glucosides and C_{12}/C_{14} alkyl polyglucoside (APG) are comparable with those of a typical non-ionic surfactant and decrease with increasing alkyl chain length.

The hydrophilic part of the alkyl polyglucoside is derived from a carbohydrate source, and raw material costs increase in the order starch/glucose syrup/glucose monohydrate/water free glucose while plant equipment requirements and hence cost decrease in the same order.

Direct synthesis is simple and covered by patents with all major producers (Henkel, Huls, Akzo, BASF) having patents in the field from 1988 onwards [49–51]. Here, the carbohydrate is suspended in the alcohol as very fine particles and, in some cases, crystal water is removed to minimise side reactions.

Highly degraded glucose syrup (dextrose equivalent >96) can be used in a modified direct process where a second solvent or emulsifier is used to provide stable fine droplet dispersions [52, 53]. The fine droplets/particles are important as the reaction is heterogeneous and good contact between the reactants must be maximised at as low temperature as possible to avoid formation of polymers and charring of the sugars.

The trans acetylation process is commercially less attractive with higher plant costs and alcohol recycling but several major producers have patents on this process [54–57].

The reaction proceeds as given below using sulphonic acid catalyst and a reaction temperature below 140°C commensurate with residence time and speed of reaction:

Glucose 1 mol + fatty alcohol 3–10 mol → mono alkyl glucoside + water (continually removed)

The products are a complex mixture of species mainly differing in the degree of polymerisation (DP) and in the length of the alkyl chains. The carbohydrate chain gives a mixture

Table 5.11 Examples of commercial poly alkyl glucosides

Trade name	Chemical structure	Active level	Comments
Glucopon^R 215 CS UP	Capryl glucoside	62–65%	Hard surface cleaning Good wetting and dispersing properties
Glucopon^R 225 DK/HH	Decyl glucoside	68–72%	Excellent caustic stability
Glucopon^R 425N/NH	Coco glucoside		Good wetting, hydrotroping and detergency properties
Glucopon^R 600 CS UP L	Lauryl glucoside	50–53%	Good foaming, manual dish wash and laundry products
Glucopon^R 650 EC	Coco glucoside	50–53%	Dish wash and institutional cleaners

of oligomers largely dependent on the ratio of glucose to alcohol (mono, di, tri, etc.). The DP is important in determining the physical chemistry and applications of APGs. Alkyl-monoglucosides are the main group of components, usually more than 50%, followed by di-, tri- and higher glucosides but most commercial products have a DP averaging between 1 and 2.

The work-up is probably as important as the reaction stage to maintain the colour and viscosity of product. Firstly the acid is neutralised with magnesium oxide. The water soluble APGs must essentially be free of fatty alcohol which is distilled at much reduced pressure at around 140°C in a falling film evaporator keeping contact times in the hot zone to an absolute minimum. There are patents to BASF on this topic [58], which suggest the first distillation uses glycol that has a similar boiling point to the residual alcohol and the second distillation uses dipropylene glycol. Addition of hypophosphorous acid reduces formation of dark coloured product and bleaching, if required, can be achieved with hydrogen peroxide in the presence of magnesium ions in alkaline conditions. Tallow-based polyglucosides are produced in the same manner but the work-up stage is different because of its insolubility in water. The reaction is also terminated at 70% conversion to minimise side products and colour because it is impossible to bleach as an aqueous paste since it is insoluble in water. After the neutralisation stage, the product is filtered to remove glucose and polydextrose: then the excess tallow alcohol is distilled under vacuum as for the water soluble grades, but, in this case, leaving the APG as a 50% solution in tallow alcohol.

5.18 Gemini surfactants

The technology supporting gemini surfactants has been in existence for more than 20 years and they are so-called because they have two hydrophobic head groups and two hydrophilic groups in the same molecule. The two portions of the molecule are linked by a 'spacer' which can be attached directly to the two ionic groups or can be positioned down the hydrocarbon chain as shown in Figure 5.3, positions 1 and 2 respectively.

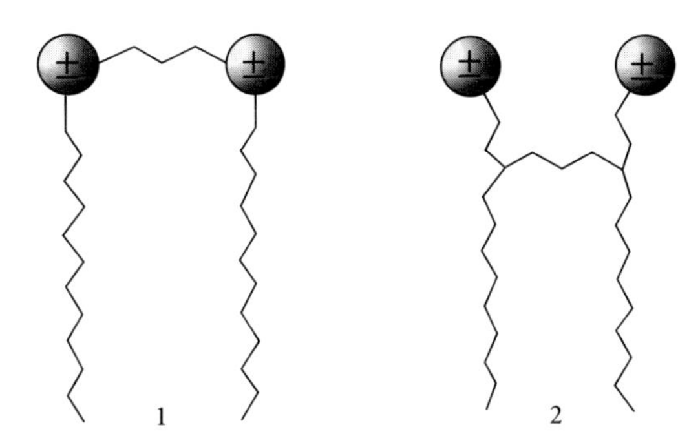

Figure 5.3 Possible structures of gemini surfactants.

These structures can impart properties somewhat different from their single hydrophobe/hydrophile analogues being, in general, more surface active. In particular, their lower surface tension and better wetting properties lead to their uses in coatings where they impart improved spreadability with a marked reduction in surface defects. Gemini surfactants usually have a cmc value one or two orders of magnitude lower than that of corresponding monomeric surfactants [59]. Another very interesting property of some gemini surfactants is the very different rheological properties (viscosity, gel and shear thickening) they exhibit at low concentrations.

A concise account of the structure, properties and uses of gemini surfactants is given by Rosen [60].

References

1. Dow Chemical Co. (1960) Alkylene oxides. Tech Bull. 125-273-60, Midland, MI.
2. Union Carbide Chemicals Co. (1961) Alkylene oxides. Tech Bull., New York.
3. Hess, L.G. and Tilton, V.V. (1950) *Ind. Eng. Chem.*, **42**, 1251.
4. (1999) Ethyleneoxide 2nd edition, Users Guide www.ethyleneoxide.com
5. Krause, W.P. (1961) (to Jefferson Chemical Co.). U.S. Patent 2,983,763.
6. Beauchamp, W.D. Booth, R.E. and Degginger, E.R (1962) (to Allied Chemical Corp.). U.S. Patent 3,016,404.
7. Stolz, E.M. (1957) (to Olin Mathieson Chemical Corp.). U.S. Patent 2,778,854.
8. Brusson, H.A. and Stein, O. (1939) (to Rohm & Hass Co.). U.S. Patent 2,143,759.
9. Flory, P.J. (1940) *Am. Chem. Soc.*, **62**, 1561.
10. Satkowski, W.B. and Hsu, C.G. (1957) *Ind. Eng. Chem.* **49**, 1875.
11. Wrigley, A.N., Stirton, A.J. and Howard, E., Jr. (1960) *J. Org. Chem.*, **25**, 439.
12. Yong, K., Hield, G.L. and Washecheck (1981) (to Conoco). U.S. Patent 4,306,693.
13. McCain, J.H. and Thuling, L.F. (1984) (to Union Carbide). U.S. Patent 4,453,022/3.
14. McCain, J.H., King, S.W., Knogf, R.J., Smith, C.A. and Hauser, C.F. (1989) (to Union Carbide). EP Patent 361,619A2.
15. Edwards, C.L. (1988) (to Shell Oil Co.). U.S. Patent 4,721816/7.
16. Edwards, C.L. (1995) (to Shell Oil Co.). EP Patent 665206A1.
17. Nakaya, H., Adachi, I., Aoki, N. and Kanao, H. (to Lion Corp.) (1989) Jap Patent 03185095A2.
18. BASF (1995) DE 4325237. Application.
19. Henkel (1990) DE 4010606. Application.

20. Sandoval, T.S. and Schwab, P.A. (to Vista Corp.) (1993) U.S. Patent 5,220,077.
21. Wrigley, A.N. Smith, F.D. and Stirton, A.J. (1959) *J. Am. Oil Chemists' Soc.*, **36**, 34.
22. Wetterau, E.P., Olanski, V.L., Smullin, C.F. and Brandner, J.D. (1964) *J. Am. Oil Chemists' Soc.*, **41**, 383.
23. Malkemus, J.D. and Swan, J.I. (1957) *J. Am. Oil Chemists' Soc.*, **34**, 342.
24. Hoechst, (1989) EP Patent 89105357.1.
25. Henkel (1990) DE 3914131.
26. Vista (1993) U.S. Patent 5,220,046.
27. Lion Corp. (1994) U.S. Patent 5,374,750.
28. Cox, M.F. et al. (1997) JAOCS, **74**, 847–859.
29. Cox, M.F. et al. (1998) *J. Surf Det.*, **1**, 11–21.
30. Cox, M.F. et al. (1998) *J. Surf. Det.*, **1**, 167–75.
31. Lundsted, L.G. (1954) (to Wyandotte Chemicals Corp.) U.S. Patent 2,674,619.
32. Santacesaria, E., Diserio, M., Garaffa, R. and Adino G. (1992) *Ind. Eng. Chem. Res.*, **31**, 2413.
33. Bartha, B., Faekes, L., Morgos, J., Sallay, P., Rusznak, I. and Veress. G. (1981) *J. Am. Oil Chemists' Soc.*, **58**, 650.
34. Jungermann, E. and Tabor, D. (1967) In M. Schick, (ed.), *Nonionic Surfactants.* Marcel Dekker, New York.
35. Cahn, A. (1979) *J. Am. Oil Chemists' Soc.*, **56**, 809A.
36. Kritchevsky, J. (1957) *J. Am. Oil Chemists' Soc.*, **34**, 178.
37. Farris, R.D. (1979) *J. Am. Oil Chemists' Soc.*, **56**, 770A.
38. Lake, D.B. and Hoh, G.L.K. (1963) *J. Am. Oil Chemists Soc.*, **40**, 628.
39. Pilcher, W. and Eton, S.L. (1961) (to Procter and Gamble Co.) U.S. Patent 2.999.068.
40. Priestley, H. and Wilson, J. (1961) (to Unilever Ltd.) South African Patent 61–1798.
41. McIntyre, R.T. (1979) *J. Am. Oil Chemists' Soc.*, **56**, 835A.
42. Brown, K.R. (1943) (to Atlas Powder Co.). U.S. Patent 2,322,822.
43. Griffin, W.C. (1945) (to Atlas Powder Co.). U.S. Patent 2,374,931.
44. Kubie, W.L., O'Donnell, J.L., Tester, H.M. and Cowan, J.C. (1963) *J. Am. Oil Chemists' Soc.*, **40**, 105.
45. Brown, K.R. (1943) (to Atlas Powder Co.). U.S. Patent 2,322,820.
46. Fischer, E. (1893) Ber., **26**, 2400.
47. Bertsch, H. and Rauchalles, G. (1934) (to Th. Bohme AG.). U.S. Patent 2,049,758.
48. Th. Bohme, A.G. (1935) DRP Patent 611055.
49. Henkel, EP (1988) 0437460B1.
50. Schmidt, S. (1991) (to Huls) EP Patent 0495174.
51. Akzo (1994) EP 0617045 A2.
52. Huls (1990) EP 0448799.
53. BASF (1992) WP 94/04544.
54. Henkel (1987) EP 0301298.
55. Henkel (1988) EP 0357969.
56. Huls (1990) EP 0482325.
57. Huls (1991) EP 0514627.
58. BASF (1981) DE 3001064.
59. Zana et al. (1991) *Langmuir*, 7, 1072.
60. Rosen, M.L. (1999) In D.R. Karsa (ed.), *Industrial Uses of Surfactants IV.* Royal Society of Chemistry, Cambridge, UK, pp. 151–61.

Chapter 6
Other Types of Surfactants

6.1 Cationics

J. Fred Gadberry

6.1.1 Introduction and background

Cationic surfactants represent one of the smaller classes of surfactants when compared to anionic and nonionic surfactants. Annual volume estimates for worldwide production of cationics are 500 000 metric tons [1]. However, the uniqueness of the positively charged hydrophile provides specific properties which for many applications makes these materials indispensable. The surfactants in the class are dominated by a positively charged nitrogen as the core hydrophile. While other positively charged hydrophiles are possible, such as sulfonium and phosphonium, virtually no commercial products of this type exist [2].

Nitrogen-based cationics were developed by the Armour company in the 1940s as a means to utilize the tallow from its Chicago stockyards [3]. The first significant product was dihydrogenated tallow dimethyl ammonium chloride which today still finds utility in a variety of applications where a positive charge and hydrophobation are required. The single largest market for cationic surfactants is as the active ingredient in fabric softeners [4]. The market for fabric softeners was established as a consequence of the move to the higher performing synthetic anionic detergents such as branched linear alkyl benzene sulfonates [5]. The trade-off for the high performing surfactants was fabrics which possessed an unacceptably rough hand-feel. Fabric softeners made from dihydrogenated tallow dimethyl ammonium chloride compensated for this, allowing the consumer to have clean and soft clothes. A second significant market for cationics is their use as hydrophobation agent in organoclays.

6.1.2 Manufacturing processes

6.1.2.1 Amine preparation

The manufacturing process for cationic surfactants can be divided into two parts. The first part is the creation of an alkylated amine. Several processes can achieve this endpoint and are briefly reviewed below. The largest volume process which was developed by Armour starts with tallow triglyceride which is split to yield fatty acid and glycerine. The fatty acid is reacted with ammonia and converted to fatty nitrile under high pressure and temperature conditions

[3]. The nitrile is then hydrogenated to a primary amine from which amine derivatives can be made by utilization of a variety of reagents [6, 7]. The example in eqns 6.1.1–6.1.3 shows the conversion of fatty acid to dialkyl secondary amine via catalytic deammoniafication (for this and all subsequent structures $n = 80$–20 unless otherwise noted):

$$CH_3(CH_2)_n\overset{O}{\overset{\|}{C}}OH + NH_3 \xrightarrow[\text{heat}]{\text{cat}} CH_3(CH_2)_nC{\equiv}N + H_2O \tag{6.1}$$

$$CH_3(CH_2)_nC{\equiv}N \xrightarrow[H_2]{\text{cat}} CH_3(CH_2)_nCH_2NH_2 \tag{6.2}$$

$$CH_3(CH_2)_nCH_2NH_2 \xrightarrow[-NH_3]{\text{cat}/H_2} (CH_3(CH_2)_nCH_2)_2NH \tag{6.3}$$

Alternate routes to amine derivatives have been developed in the intervening years. The production of dimethylamines can be accomplished by the routes shown in eqs 6.1.4–6.1.7. [8]. Both routes involve the reaction of an alkyl halide with dimethylamine. The first route is the conversion of a fatty alcohol to fatty chloride using phosphorous trichloride. The alkyl chloride is reacted with dimethylamine giving the alkyl dimethylamine [9, 10]:

$$CH_3(CH_2)_nOH + PCl_3 \longrightarrow CH_3(CH_2)_nCl + H_3PO_3 \tag{6.4}$$

$$CH_3(CH_2)_nCl + (CH_3)_2NH \longrightarrow CH_3(CH_2)_nN(CH_3)_2 + (CH_3)_2NH_2{}^+Cl^- \tag{6.5}$$

In the second route an alpha olefin derived from ethylene reacts with hydrogen bromide and a free radical initiator resulting in an alkyl bromide. The alkyl bromide is reacted with dimethylamine providing the desired product [11, 12]:

$$H_2C{=}CH(CH_2)_nCH_3 + HBr \longrightarrow BrCH_2CH_2(CH_2)_nCH_3 \tag{6.6}$$

$$BrCH_2CH_2(CH_2)_nCH_3 + 2(CH_3)_2NH \longrightarrow (CH_3)_2NCH_2CH_2(CH_2)_nCH_3 + Br^-\ {}^+H\,N(CH_3)_2 \tag{6.7}$$

Alcohols and aldehydes are also suitable materials for the creation of an alkyl amine. In addition to the aforementioned formation of alkyl chloride as an intermediate, alcohols can be directly converted to amines under hydrogenation conditions in the presence of ammonia while aldehydes are prereacted to form imine followed by hydrogenation [13]. Selectivity of the primary amine with these techniques is difficult and this process is more typically utilized for the preparation of tertiary amines where the reaction can be driven to completion. In certain cases, alcohols and aldehydes provide structural elements which are not attainable from natural sources. An example is the formation of a hydrogenated tallow 2-ethyl hexyl amine. The amine is prepared as shown below in eqn 6.1.8 using a hydrogenated tallow amine reacted with 2-ethyl hexanal [14, 15]:

$$CH_3(CH_2)_nNH_2 + H\overset{O}{\overset{\|}{C}}\underset{CH_2CH_3}{\underset{|}{C}}H(CH_2)_5CH_3 \xrightarrow[(2)H_2/\text{cat}]{(1)\text{imine}} CH_3(CH_2)_n\overset{H}{\overset{|}{N}}CH_2\overset{CH_2CH_3}{\overset{|}{C}}H(CH_2)_5CH_3 \tag{6.8}$$

The intermediate imine is hydrogenated giving the secondary amine. Formation of tertiary is suppressed due to the steric hindrance of the branched chained substituent. The cationic derived from this amine is highly water soluble due to suppression of lamellar phase formation by the branched alkyl group.

Nitrogen cationic surfactants can also be created by the use of difunctional small molecule amines which, after formation of an amide or ester bond, leave an amine residue which is suitable for quaternization as shown in eqs 6.1.9–6.1.11. The amine residue is then reacted with a suitable alkylating agent to form the cationic. Similarly, reaction of a triglyceride with diethylene triamine gives initially the diamide which, under appropriate conditions, can be cyclized to imidazoline [16]:

$$H_2N(CH_2)_2\overset{H}{\underset{}{N}}(CH_2)_2NH_2 + 2CH_3(CH_2)_n\overset{O}{\overset{\|}{C}}OH \longrightarrow R\overset{O}{\overset{\|}{C}}N(CH_2)_2\overset{H}{N}(CH_2)_2N\underset{\underset{O}{\|}}{C}R \tag{6.9}$$

$$R\overset{O}{\overset{\|}{C}}N(CH_2)_2\overset{H}{N}(CH_2)_2N\underset{\underset{O}{\|}}{C}R + nH_2\overset{O}{C-C}H_2 \longrightarrow R\overset{O}{\overset{\|}{C}}N(CH_2)_2\overset{(CH_2CH_2O)_nH}{N}(CH_2)_2N\underset{\underset{O}{\|}}{C}R \tag{6.10}$$

$$R\overset{O}{\overset{\|}{C}}N(CH_2)_2\overset{(CH_2\overset{R}{C}HO)_nH}{N}(CH_2)_2N\underset{\underset{O}{\|}}{C}R + CH_3O\overset{O}{\underset{O}{S}}OCH_3 \longrightarrow R\overset{O}{\overset{\|}{C}}N(CH_2)_2\overset{(CH_2CHO)_nH}{\underset{CH_3}{\overset{+}{N}}}(CH_2)_2N\underset{\underset{O}{\|}}{C}R \quad CH_3O\overset{O}{\underset{O}{S}}O^- \tag{6.11}$$

6.1.2.2 Quaternization

Cationic surfactants are prepared by the reaction of a tertiary amine with an alkylating agent resulting in the formation of the positively charged nitrogen center. This reaction can be accomplished with a variety of agents as shown in Table 6.1, where each R represents a hydrocarbon residue.

Table 6.1 Typical quaternization agents

Reagent	CAS number	Product
Dimethyl sulfate	[77-78-1]	$RR'R''N^+CH_3$ $^-O\overset{O}{\underset{O}{S}}OCH_3$
Methyl chloride	[74-87-3]	$RR'R''N^+CH_3\ Cl^-$
Benzyl chloride	[100-44-7]	$RR'R''N^+CH_2C_6H_5\ Cl^-$
Ethyl chloride	[75-00-3]	$RR'R''N^+CH_2CH_3\ Cl^-$
Alkyl chloride	Various	$RR'R''N^+R'''\ Cl^-$

The most commonly used alkylating agents are dimethyl sulfate and methyl chloride [16]. Methyl chloride is typically reacted with a suitable amine at 90–120°C in a pressure vessel rated to 200 psi [16]. The methyl chloride is added over the course of the reaction frequently on a pressure demand basis. Excess methyl chloride is added to drive the reaction to a low free amine content of 1% or 2%. The excess methyl chloride is then vented from the vessel and scavenged with an appropriate scrubber. Sparging with an inert gas can further lower the final methyl chloride level to less than 1000 ppm. Dimethyl sulfate is reacted with an amine at 40–60°C in a closed vessel but this can be done at higher temperatures to increase the fluidity of the reaction medium [17]. The reaction can be completed under atmospheric conditions. It is exothermic, yielding 141 kJ kg^{-1} in the case of the reaction of dimethyl sulfate with a triethanol amine tallow fatty acid ester [18]. As a consequence of the exothermic reaction, it is imperative to add the dimethyl sulfate slowly over the course of the reaction. Most quaternizations are accomplished in the presence of a solvent due to the high melting point of most cationic surfactants. The solvent can range in concentration from 85% to as low as 10% and, in some cases, the quaternizations can be run neat [17]. Alcohols are the solvents of choice due to their hydrotropic properties and the most popular choices include ethanol and 2-propanol although glycols are becoming increasingly popular due to concerns over VOCs. Water is a viable solvent for many cationic surfactants but only in a few cases can the concentration of the cationic surfactant exceed 40% [18].

6.1.3 Applications of cationic surfactants

6.1.3.1 Fabric softeners

Fabric softeners represent the single largest outlet for cationic surfactants consuming 200 000 metric tons per year in 1992 [19]. The basic structural requirement for a surfactant useful for fabric softening is the presence of two alkyl groups each with 12–18 carbon atoms and at least one positively charged hydrophile which, in commercial fabric softeners, is exclusively nitrogen (Figure 6.1).

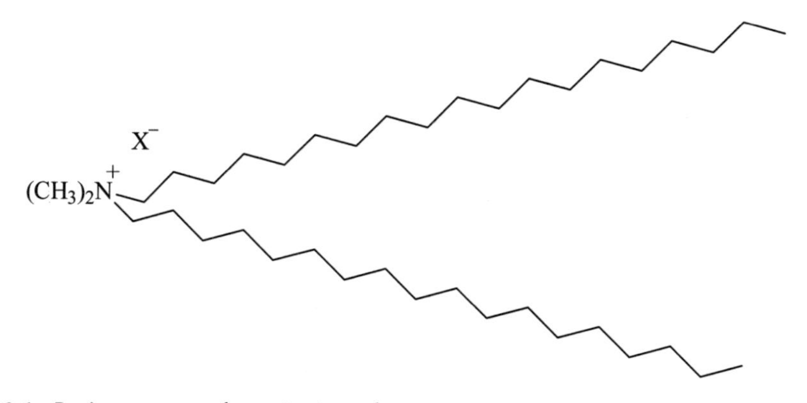

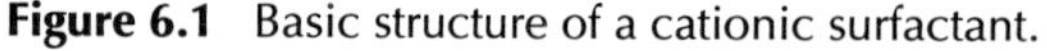
Figure 6.1 Basic structure of a cationic surfactant.

The positively charged nitrogen aids in the deposition of the surfactant onto the fabric and the 'twin tails' provide the lubricity and hand-feel desired by consumers [20]. From this basic structure a medley of materials arise each with its particular exigency being satisfied.

Structures. The driving force for development and existence of a large range of cationic softeners is the biodegradability requirement of the EU [5]. This Directive, which required a minimum of 60% biodegradability based on a 28-day closed bottle test with nonacclimated bacteria, doomed, in many countries within Europe, the use of nitrile-based quaternaries. Subsequent studies on the environmental fate of these nitrile-based products have questioned whether the closed bottle test is an accurate assessment of their ultimate environmental fate [21, 22]. Nevertheless, the requirements stand, driving a new generation of cationic softeners based on structures with enhanced biodegradability.

The new structures place a cleavable functionality within the 'twin tails' of the cationic surfactant which aids in biodegradation by separating the positively charged nitrogen hydrophile from the hydrophobic portion as shown in Figure 6.2.

The hydrophobic portion is typically a fatty alcohol or fatty acid, both of which are materials found in nature and exhibit excellent biodegradability [23–29].

Hydrophobe structural features. The twin hydrophobes or 'tails' of a typical softener have four primary elements which can be varied to meet the requirements of the application. The primary elements are the number of carbon atoms, the total degree of saturation, the quantity of polyunsaturates and the *cis* to *trans* ratio of the points of unsaturation [24–28, 30, 31]. Additional elements include substitution with noncarbon, hydrogen or oxygen substituents. To date, these have been of little commercial consequence and will not be discussed further.

The carbon atom number for fabric softener structures usually follows the distribution of common oleochemical feedstocks. The average number is 17.5 for animal-based products such as tallow [32] and this carbon atom number also appears to be optimal for many

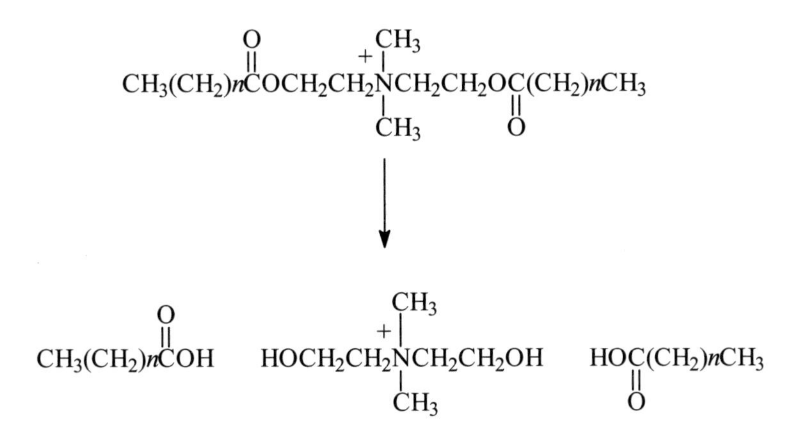

Figure 6.2 Biodegradation of a cationic surfactant.

Table 6.2 Iodine values and C16/C18 content of feedstocks for ester-based cationics

Oil source	C16 content	C18 content	Iodine value
Palm	34–47	45–70	38–56
Tallow	23–46	53–77	50–60
Rapeseed (low erucic)	3–6	77–95	100–120
Linseed	11–16	76–95	170–200
Soybean	7–11	73–95	125–140

structural variances. The physical manifestation of this is not understood but may be linked to the structural order on the surface of the fiber. Decreasing the carbon number decreases softening and lowers the deposition onto the fabric. Increasing the carbon number increases the difficulty of formulation due to formation of gels at concentrations higher than ~5%. Carbon atom numbers closer to an average of 16 are common for hair conditioner [33].

The total degree of unsaturation of any alkylene functionality is measured by its iodine value [34]. The iodine value measures the equivalents of iodine added across all the unsaturation points in a molecule. Table 6.2 shows the iodine value and 16 and 18 carbon atom percentages of typical feedstocks used in cationic surfactant manufacture [35].

The iodine value can be manipulated during manufacture by hydrogenating either the feedstock or its derivatives [24, 25, 36]. The total degree of unsaturation of cationic surfactants for softening of textiles has increased substantially due to the introduction of concentrated softeners in both the United States and Europe. Efforts to minimize solid waste and to improve economics led to the development of concentrated softener products. The new concentrated versions of fabric softeners are formulated at 17–26% active cationic surfactant [19]. Increasing the degree of unsaturation of the tallow backbone increases the concentration at which these materials can be formulated. The higher concentrations achievable with unsaturated materials is a consequence of increased liquidity of the higher iodine value material and the lowered tendency to form lamellar phases, due to the steric resistance against compacting unsaturated chains [24, 25].

The level of polyunsaturates in the hydrophobe of a cationic surfactant influences its liquidity and also its resistance to oxidative degradation and color formation [24, 37]. The higher the polyunsaturate level and consequently the iodine value, the higher the liquidity and the higher the aqueous concentration of a softener dispersion that can be achieved. Products with high or even modest degrees of unsaturation frequently require the addition of an antioxidant such as the hindered phenol derivatives, butylated hydroxy toluene and butylated hydroxy anisole [24, 25, 38].

The final element of the hydrophobe which can be manipulated is the *cis/trans* ratio of the unsaturated hydrocarbon fragments. Natural tallow has a *cis/trans* ratio of about 8–20 [39]. Metal catalyzed hydrogenation of fats and oils results in the reduction of the *cis/trans* ratio and an increase in the melting point of the oil when compared to a material of similar iodine value and a higher *cis/trans* ratio [40]. For concentrated fabric softeners, high *cis/trans* ratios are preferred to reduce the likelihood of gel formation in the final product or during processing [24, 40–42].

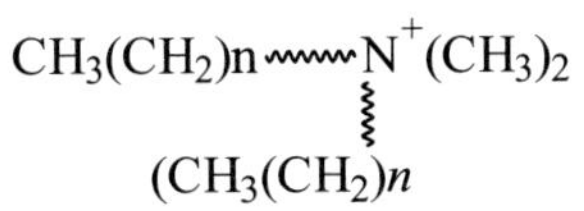

$$CH_3(CH_2)n \sim\sim \overset{+}{N}(CH_3)_2$$
$$\wr$$
$$(CH_3(CH_2)n$$

Figure 6.3 Classic hydrophile for the traditional cationic softener.

The manipulation of each of these elements will influence the final product properties and can be used to customize the molecule to the requirements of the softener system. Typically, a combination of these elements can best satisfy the needs of the softener composition. This combination can be best optimized by the use of experimental design techniques where each of the elements can be varied independently but the influence of each on the other can be evaluated [43].

Hydrophile structural elements. The hydrophile structure is the structural feature of the softener molecule which is most often manipulated in the development of new fabric softeners. The classic hydrophile for the traditional cationic softener was a positively charged nitrogen directly connected to the hydrophobe as shown in Figure 6.3 [44, 45].

The new, more biodegradable, cleavable structures have the hydrophobe connected to the positively charged nitrogen hydrophile via an ester linkage as in Figure 6.4. The structure of the hydrophobe in these materials can be more varied than with the traditional quaternaries.

Table 6.3 shows the common hydrophiles employed in ester-based cationic softeners. These structures are derived, for example, from an ethanolamine or from the reaction product of a chloro-substituted acid with a methyl amine.

The hydrophile influences the behavior of the softener primarily by its size which can alter the packing in the vesicular structure of a formulated material and also by influencing the hydrophilicity of the fabric upon which it is deposited.

Formulations. In the typical fabric softening formula, the surfactant exists in a vesicular structure and the number of bilayers is a function of the concentration of the cationic surfactant in the solution [48]. The quality of dispersion influences the softening observed on the fabric. A poor dispersion leads to an uneven coating of softener on the fabric and insufficient deposition [46]. Cationic surfactants which do not readily form stable vesicular structures can be made stable by the inclusion of appropriate stabilizers which include ethoxylated

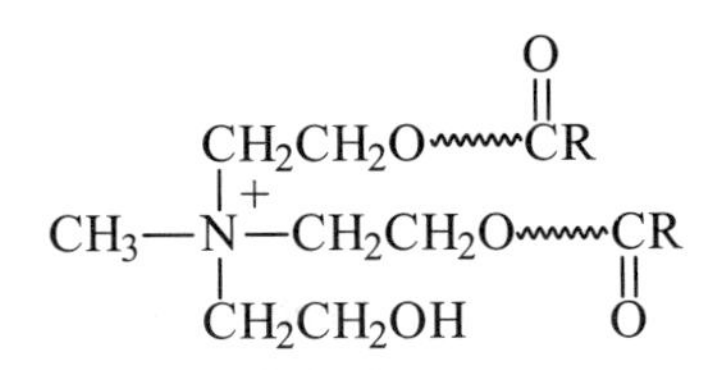

Figure 6.4 Structure of a biodegradable cationic softener.

Table 6.3 Structures of esterquaternary hydrophiles

Hydrophile structure	References
$-OCH_2CH_2\overset{+}{N}(CH_3)_2CH_2CH_2O-$	[25]
$-OCH_2CH_2\overset{+}{N}(CH_3)(CH_2CH_2OH)CH_2CH_2O-$	[19, 50, 51]
$-OCH_2CH(O-)CH_2\overset{+}{N}(CH_3)_3$	[52]
$-OCH_2CH(O-)CH_2OC(=O)CH_2\overset{+}{N}(CH_3)_3$	[30]
$-C(=O)CH_2\overset{+}{N}(CH_3)_2CH_2C(=O)-$	[24]
$-C(=O)CH_2\overset{+}{N}(CH_3)_2-$	[53]

alcohols [49]. Recently, microemulsion formulas of fabric softeners have appeared [50]. Additionally, clear fabric softener formulas have been prepared from esterquaternaries, a C8–C22 monoalkyl cationic surfactant and 17–75% of a nonaqueous organic solvent [51].

6.1.3.2 Dryer softeners

Dryer softeners are employed primarily in North America to control the buildup of static charge in the tumble clothes dryer. The product consists of a 7″ × 12″ sheet made of polyester and a mixture of a cationic surfactant and a 'release' or 'distribution agent' is applied to the sheet [52–54]. The level of the agent is typically 1.0–2.0 g per sheet and is in a solid form on the sheet [55, 56]. The sheet is tossed into the dryer with the wet clothes at the beginning of the cycle and controls the development of static charge throughout the dryer cycle.

The cationic surfactants utilized in dryer sheets are similar to those used in liquid softener compositions except that the counterion must be methyl sulfate. This counterion is necessary due to the corrosivity of chloride ion to the internal parts of the dryer and both

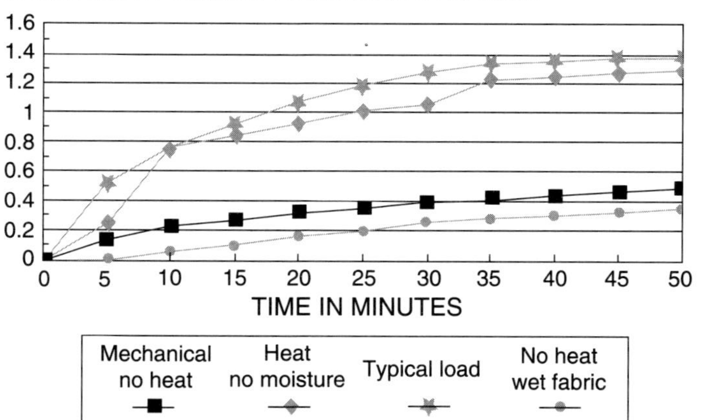

Figure 6.5 Determination of removal of cationic surfactant from a tumble dryer.

traditional-based nitrile cationic surfactants and ester-based cationics are suggested for use in this application.

Figure 6.5 shows that the removal of quaternary from the sheet is a result of the melting of the active from the sheet instead of removal as a consequence of mechanical action or dissolution by the water contained in the wet fabric [18, 55].

As a consequence, the melt characteristics of the softener active are important in determining its function. The melt characteristic is determined from the cationic surfactant employed and also from the selection of the release or distribution agent.

Additional components of the dryer sheet formula can include perfumes for marketing and soil release agents to prevent soiling of fabric during the use period prior to the next washing [43].

6.1.3.3 Softergents

Cationic surfactants are sometimes included in detergent formulas as softening agents. These 'softergents' provide ease of use but, typically, do not provide the cleaning level of detergent formulas without these materials. A recently exampled heavy duty laundry detergent contains lauryl trimethyl ammonium chloride as the softening agent where C_{12} alkyl benzene sulfonate is specifically excluded or limited to prevent phase separation and formation of a translucent viscous formula [57, 58]. Inclusion of a C8–C20 fatty acid into these formulas provides enhanced softening and cleaning.

Relatively few advances in new softener actives have been made in recent years. Typical efforts are modification of existing actives. The product shown in Figure 6.4 does not exist in a pure state as manufactured but is composed of mono-, di- and triester species. Best softening is found using products high in diester and relatively low in mono- and triester. A process to manufacture high diester product has been patented [59]. For the most part

Table 6.4 Cationic surfactant useful in hair conditioning products

Cationic surfactant	References
Dodecyl-trimethyl ammonium chloride	[60–63]
Di(palmitoyloxyethyl)-2-hydroxyethyl-methyl ammonium chloride	[34]
Di(hydrogenated-soyoyloxyethyl)-2-hydroxyethyl-methyl ammonium chloride	[19]
Dicetyl dimethyl ammonium chloride	[67]
Di(hydrogenated tallow) dimethyl ammonium chloride	[68]
Tricetyl methyl ammonium chloride	[71]

efforts have focused on formulation of fabric softeners as high load concentrates [60] or as clear formulations [61].

6.1.3.4 Hair conditioning

The washing of hair with synthetic detergents has a similar effect to the washing of clothes. In this case, the natural oils which normally coat hair are washed away and leave the hair with a lack of shine, tangles and a condition known as 'fly away'. Fly away, as the name implies, is due to hair fibers repelling one another due to the buildup of static charge and the hair then has an unkempt appearance. Conditioners restore the hair to its original condition by providing a synthetic oil to repair the damaged hair fibers. Cream rinse hair conditioners contain approximately 2% of a hydrophobic cationic surfactant and those which find utility in hair conditioners are shown in Table 6.4.

The cationic surfactant is selected on the basis of the amount of conditioning desired, the more hydrophobic cationic such as a dihydrogenated tallow dimethyl ammonium chloride provides a high level of conditioning, while the same molecular architecture based on di-palmityl-dimethyl ammonium chloride provides a lower level of conditioning. Still milder conditioning for oily hair can be provided by mono alkyl chain cationics such as dodecyl-trimethyl ammonium chloride.

Developments in this application of cationic surfactants follow the developments in the fabric softener field since traditionally fabric softener surfactants have been employed in hair conditioners. The use of ester-based cationics is drawing attention [68]. Again, the selection of the hydrophobe is seen to control the performance of the cationic surfactant. The conditioning performance of the ester-based cationic is excellent and, additionally, improved static control is demonstrated. This is potentially due to the improved hydration of the hair follicle as a consequence of the more polar nature of the ester-based cationics.

6.1.3.5 Detergents

Cationic surfactants are not used as the primary surfactant in laundry detergents, as this responsibility falls on the commodity surfactants such as linear alkyl benzene sulfonates, ethoxylated alcohols and alkyl sulfates. Nevertheless, over the past 20 years there has been extensive art claiming the use of cationics to improve the performance of their anionic counterparts [67, 69, 70]. A combination of an alkyl phosphate (AP) and (note: APE = alkyl

Table 6.5 Interfacial tension of anionic/cationic surfactant mixtures (dyne cm^{-1})

Surfactant	APE	AP/TTAB 0.3 g l^{-1} total	TTAB
Oil	1 g l^{-1}	4:1 ratio	1 g l^{-1}
Hexadecane	3.0 ± 0.3	1.1 ± 0.2	4.9 ± 0.2
Nujol	2.6 ± 0.3	1.1 ± 0.2	2.3 ± 0.1
Dirty motor oil	1.3 ± 0.7	0.8 ± 0.1	1.4 ± 0.2
Wesson oil	3.9 ± 0.6	1.4 ± 0.2	2.8 ± 0.6
Oleic acid	4.9 ± 0.9	4.7 ± 0.6	4.6 ± 0.6

phenol ethoxylate) tetradecyl trimethyl ammonium bromide (TTAB) substantially lowers the interfacial surface tension between the surfactant solution and a variety of oils compared to either surfactant itself as shown in Table 6.5 [69].

TTAB, in combination with an alkyl ethoxylated sulfate, also lowers the surface tension observed at CMC by approximately 10 dyne cm^{-1} compared to either of the surfactants alone. It is postulated that the combination of the anionic and cationic results in charge neutralization providing a 'net' surfactant which is similar to a nonionic surfactant. This is confirmed by the lowering of the cloud point of the surfactant combination as the mole fraction of anionic and cationic are adjusted to near the equivalency point.

Alkoxylated quaternary ammonium surfactants have been patented recently as performance boosters for detergent formulations [71].

6.1.3.6 Thickeners

As a consequence of their tendency to form lamellar phases or rod shaped micelles at low concentration, cationic surfactants are frequently employed as the primary surfactants to thicken high salt formulas [72, 73]. The viscoelastic nature of certain cationic surfactant solutions has been employed in a novel way to allow for a solution of sodium hypochlorite not to be easily diluted and therefore to remain at a higher concentration for the purpose of oxidizing clogs of human hair which form in drains [73]. Low concentrations of cetyl trimethyl ammonium chloride in combination with two hydrotropes form viscoelastic solutions with the values of viscosity and Tau/Go shown in Table 6.6.

Table 6.6 Properties of cetyl-trimethyl ammonium chloride solutions

Cetyl trimethyl ammonium chloride (wt. %)	Sodium xylene sulfonate (wt. %)	Chlorobenzoic acid (wt. %)	Viscosity (cP)	Tau/Go (s Pa^{-1})
0.370	0.260	0.080	47	0.35
0.5000	0.143	0.071	247	0.45
0.625	0.125	0.063	716	0.89
0.625	0.250	0.063	140	0.06

Table 6.7 Effect of cationic surfactants on effort required for cleaning

Composition	1	2	3	4
Nonionic(ethoxylated alcohol)	10%	10%	10%	10%
2-amino 2-methyl 1-propanol	4%	4%	4%	4%
K_2CO_3	1.2%	1.2%	1.2%	1.2%
Digol	8%	8%	8%	8%
Propylene glycol			0.6%	3.0%
Cetyl trimethyl ammonium bromide		1%		
$RC(=O)N(CH_2)_2\overset{+}{N}(CH_3)((CH_2CHO)_nH)(CH_2)_2NC(=O)R$ $CH_3OSO_2O^-$			0.2%	1.0%
R=saturated alkyl				
Effort Ns	930	469	189	137

6.1.3.7 *Hard surface cleaning*

Cationic surfactants, due to their positive charge, are attracted to many surfaces altering the surface and changing the force required for cleaning. Table 6.7 shows the effect of two cationic surfactant formulas when preapplied to a tile prior to soiling [74].

The author believes that the cationic surfactants modify the surface energy which raises the contact angle of the soil subsequently deposited on the surface. This allows for more effective cleaning with less effort compared to an untreated tile or one treated with a composition without a cationic surfactant.

When a low foaming cleaner is desired, inclusion of alkoxylated mono alkyl quaternary salts is useful. Such quaternaries, containing both ethoxyl and propoxyl moieties, are effective [75].

6.1.3.8 *Organoclays*

Production of organoclays is a large market for quaternary ammonium salts [16]. These products, prepared by ion exchange between the clay platelets and the quaternary ammonium salt, are useful for modifying and controlling the rheology of nonaqueous fluids such as lubricating oils, linseed oil and toluene [76]. This provides greases, solvent-based paints and drilling fluids. For these applications, dimethyl di(hydrogenated tallow alkyl) ammonium chlorides or dimethyl benzyl(hydrogenated tallow alkyl) ammonium chlorides are the quaternary salts most commonly used in combination with smectite-type clays. For more polar systems, i.e. containing polar organic solvents or water, poly-ethoxylated quaternary ammonium salts are needed. These products have been used to thicken paints [77] and to gel nail polish [78]. An interesting recent development is the preparation of a novel nanocomposite of clay/polymer by exchanging cationic surfactant monomer onto the clay platelets and then polymerizing these organoclay platelets with acrylamide [79].

An application of organoclays with great potential is the removal of contaminants from water by adsorption. This is the subject of studies of adsorption of phenolics onto hexadecyltrimethylammonium-modified montmorillonite [80] and organics from coal gas washing onto hexadecyltrimethylammonium-modified bentonite [81].

A detailed study of the interaction of hydrocarbons with cetyl trimethyl ammonium bromide modified montmorillonite has been done using Raman spectroscopy [82]. The quaternary salt was observed to be in a liquid-like state and it was concluded that interaction of organic compounds in this system is best classified as absorption.

Release of the quaternary salt, itself potentially harmful during sorption of the organic pollutant, was shown to be minimal in an earlier study [83].

6.1.4 Industrial applications of cationic surfactants

6.1.4.1 Drag reduction

The ability of cationic surfactant molecules to form long cylindrical micelles in water-based systems makes them attractive for drag reduction applications in heating and cooling systems. Unlike polymeric drag reducing additives which eventually degrade when subjected to shear, drag reducing surfactants can re-form the long cylindrical micelles and continue to function effectively. However, commercial applications of cationic surfactants for this purpose are limited due to the well known poor ecotoxicity of cationic surfactants. Amphoteric and anionic surfactant combinations [84] have supplanted cationic surfactants for most drag reduction applications in heating or cooling systems although cationic surfactants continue to receive significant academic attention [85].

One other industrial application of cationic drag reducing additives is found in oil and gas production where cetyltrimethylammonium salicylate and cetyl pyridinium salicylate have been found simultaneously to reduce drag and inhibit corrosion of tubular steel injection lines [86].

6.1.4.2 Oilfield applications

Gas hydrate inhibitors. Gas hydrates, solid water clathrates containing small hydrocarbons, are problematic for oil and gas production because they can precipitate and cause line blockage. Simple cationic surfactants containing at least two butyl groups were previously developed to inhibit formation of gas hydrate precipitates in gas production lines [87]. However, similar to the situation with cationic drag reduction additives, poor toxicity profiles prevent widespread commercial acceptance. Ester quaternaries with structures somewhat similar to those used in fabric care have been claimed as hydrate inhibitors [88]. Additionally, certain alkylether quaternary compounds, e.g. C_{12}–C_{14} alkyl polyethoxy oxypropyl tributyl ammonium bromide, were shown to have hydrate inhibition properties [89].

Hydraulic fracturing fluids. The ability of certain cationic surfactants to form long cylindrical micelles, also referred to as worm-like micelles, has led to the development of surfactant-based fracturing fluids. Hydraulic fracturing is the high pressure creation of a fissure underground to enable greater flow of oil. During the fracturing, solid packing material called proppant is placed in the fissure to prevent closure when the pressure is released. The fracturing fluids must be able to suspend the proppant until the fissure is packed. Long chain alkyl quaternary ammonium salts have been found to prove useful for fracturing fluid applications. The first effect product was N-rapeseed alkyl-N-methyl-bis(2-hydroxyethyl)ammonium chloride [90]. This product was superseded by a more pure version containing erucyl alkyl instead of rapeseed alkyl but, to be most effective, the erucyl derivative must contain minimal amounts of impurities such as traces of intermediates [91].

Fuel applications. Bitumen, the residuum of petroleum distillation, is gaining interest as a low cost fuel. The main problem with bitumen as a fuel is handling the viscous, almost solid product. This issue has been addressed by emulsifying molten bitumen in water using cationic surfactants such as tallow alkyl propanediamine [92] and salts of similar amines with fatty acids [93]. The emulsions thus prepared are pumpable and useful as fuels for stationary burning such as in power generation facilities.

A microemulsion fuel suitable for use in diesel engines has been prepared from diesel fuel, ethanol, traces of water and cationic surfactants as emulsifiers, plus other additives [94]. Suitable cationic surfactants are alkyl polyamines and their alkoxylates. The fuels benefit from improved lubricity.

Biocidal applications. The use of quaternary ammonium salts in disinfecting systems for household and industrial cleaners has been known for many years [95, 96]. Alkylbenzyldimethyl quaternaries, alkyltrimethyl quaternaries, and dialkyldimethyl quaternaries are the more commonly used biocidal quaternary ammonium salts [16]. Recently, dialkyldimethyl quaternary ammonium salts have received renewed attention as potential wood preservatives to replace the heavy metal types [97]. Metal-free wood preservative formulations containing dialkyldimethyl ammonium salts with non-halide anions, such as carboxylates, borates, and carbonates, have been developed [98, 99].

References

1. Chemical Market Reporter, September 13, 2004 p. 18.
2. Gadberry, James F. (1990) Non-nitrogen-containing cationic surfactants. In James Richmond (ed.), *Cationic Surfactants Organic Chemistry*, Chapter 7. Dekker, New York.
3. Ralston, A.W. (1948) *Fatty Acids and their Derivatives.* Wiley, New York.
4. Puchta, R., Krings, P. and Sandkuhler, P. (1993) *Tenside, Surfactant Determinant*, **30**, 186–91.
5. Cahn, Arno (ed.) (1994) *Proceedings of the 3rd World Conference on Detergents: Global Perspectives.* AOCS, Champaign, IL.
6. Fruth, A., Strauss, J. and Stuhler, H. (1992) U.S. Patent 5,175,370, assigned to Hoescht.
7. Fruth, A., Strauss, J. and Stuhler, H. (1993) U.S. Patent 5,231,228, assigned to Hoescht.
8. Wyness, G. and Jensen, D. (1968) U.S. Patent 3,379,764, assigned to Reilly Chemicals.
9. Eriksson, J.B. and Dadekian, Z.A. (1969) U.S. Patent 4,100,210, assigned to Lonza A.G.
10. Dadekian, Z.A. and Wilbourn, D.S. (1969) U.S. Patent 3,432,561, assigned to Baird Chemical.
11. McKinnie, B.G. and Harrod, W.B. (1994) U.S. Patent 5,347,053, assigned to Albemarle.
12. Davis, W.T. (1977) U.S. Patent 4,024,189, Ethyl Corporation.
13. Goe, G.L., Keay, J., Scriven, E., Prunier, M. and Quimby, S. (1995) U.S. Patent 5,424,436, assigned to Reilly Industries.
14. Stanley, K.D and White, K.B. (1986) U.S. Patent 4,569,800, assigned to Akzo Nobel Inc.

15. Stanley, K.D. and White, K.B. (1987) U.S. Patent 4,675,118, assigned to Akzo Nobel.
16. Kery, M. (1997) Quaternary ammonium compounds. In Mary Howe-Grant (ed.), *Kirk Othmer Encyclopedia of Chemical Technology*, 4th edn, vol. 20. Wiley, New York.
17. Brown, D.M., Gatter, E. and Littau, C. (1995) U.S. Patent 5,463,094, assigned to Hoescht Celanese.
18. Technical Data, Akzo Nobel Chemicals, Dobbs Ferry, NY.
19. McConnell, R.B. (1994) *Inform*, 5(1), 76.
20. Berenbold, H. (1995) *Inform*, 5(1), 82.
21. McAvoy, D.C., White, C.E., Moore, B.L. and Rappaport, R.A. (1994) chemical fate and transport in a domestic septic system: Sorption and transport of anionic and cationic surfactants. *Environ. Toxicol. Chem.*, **13**(2), 213–21.
22. ECETOX Technical Report No. 53 (1993) DHTDMAC: Aquatic and Terrestrial Hazard Assessment, CAS No. 61789-80-8, February, Brussels, Belgium.
23. Weuste, B. and Weissen, H.J. (1997) U.S. Patent 5,596,125, assigned to Akzo Nobel Inc.
24. Wahl, E., Bacon, D., Baker, Bodet, J.-F., Demeyere, J., Severns, J., Siklosi, M., Vogel, A. and Watson, J. (1995) U.S. Patent 5,474,690, Procter and Gamble.
25. Bacon, D.R. and Trinh, T. (1996) U.S. Patent 5,500,138, assigned to Procter and Gamble.
26. Content, J.-P., Courdavault Duprat, S., Godefroy, L., Nivollet, P., Ray, D., Storet, Y. and Vindret, J.-F. (1993) European Patent Application 550,361, assigned to Stepan.
27. Godefroy , L. (1993) European Patent Application 580,527, assigned to Stepan.
28. Uphues, G., Ploog, U., Jeschke, R. and Waltenburger, P. (1994) U.S. Patent 5,296,622, assigned to Henkel.
29. Weissen, H.J. and Porta, N. (1996) U.S. Patent 5,543,066, assigned to Akzo Nobel.
30. Yamamura, M., Okabe, K., Sotoya, K. and Murata, M. (1991) U.S. Patent 5,023,003, assigned to Kao.
31. Mermelstein, R., Baker, E., Shaw, J., Jr. and Wahl, E. (1995) W.O. Patent 95/31524, to Procter and Gamble.
32. Sonntag, N.O.V. (1979) Composition and characteristics of individual fats and oils. In Daniel Swern (ed.), *Bailey's Industrial Oil and Fat Products*, vol. 1, 4th edn. Wiley, New york, p. 343.
33. Bonastre, N. and Subirana, R. Pi (1996) European Patent Application 739,976, assigned to Henkel.
34. Sonntag, N.O.V. (1979) Composition and characteristics of individual fats and oils. In Daniel Swern (ed.), *Bailey's Industrial Oil and Fat Products*, vol. 1, 4th edn. Wiley, New York, p. 115.
35. Akzo Nobel Chemicals (1996) *Fatty Acids and Glycerine* product brochure. Akzo Nobel Chemicals, Chicago, IL.
36. Sonntag, N.O.V. (1979) Composition and characteristics of individual fats and oils. In Daniel Swern (ed.), *Bailey's Industrial Oil and Fat Products*, vol. 1, 4th edn. Wiley, New York, p. 113.
37. Sonntag, N.O.V. (1979) Composition and characteristics of individual fats and oils. In Daniel Swern (ed.), *Bailey's Industrial Oil and Fat Products*, vol. 1, 4th edn. Wiley New York, p. 137–53.
38. Iacobucci, P., Franklin, R and Trinh, P.-N. (1999) U.S. Patent 6,004.913 assigned to Akzo Nobel, Inc.
39. Windholz, M. (ed.) (1976) *The Merck Index*, 9th edn, Merck and Co., Inc., Rahway, NJ, p. 1139.
40. Wahl, E.H., Bacon, D.R., Baker, E.S., Bodet, J.-F., Burns, M.E., Bemeyere, H.J.M., Hensley, C.A., Mermelstein, R., Severns, J.C., Shaw, J.H., Jr., Siklosi, M.P., Vogel, A.M. and Watson, J.W. (1994) W.O. Patent 94/20597, assigned to Procter and Gamble.
41. Severns, J.C., Sivik, M.R., Hartman, F.A., Denutte, H.R.G., Costa, J.B. and Chung, A.H. (1996) U.S. Patent 5,531,910, assigned to Procter and Gamble.
42. Uphues, G., Ploog, U., Jeschke, R. and Waltenberger, P. (1994) U.S. Patent 5,296,622, assigned to Henkel.
43. Davies, O.L. and Goldsmith, P.L. (eds) (1972) *Statistical Methods in Research and Production.* McMillan, New York.

44. Walden, M. and Mariahazey, A. (1971) U.S. Patent 3,625,891, assigned to Armour Industrial Company.
45. Ackerman, J., Miller, M. and Whittlinger, D. (1993) U.S. Patent 5,221,794, assigned to Sherex Chemical Co. Inc.
46. Haq, Z., Kahn-Lodhi, A.N. and Sams, J.P. (1995) W.O. Patent Application 95/27771, assigned to Unilever.
47. Straathof, T. and Konig, A. (1988) U.S. Patent 4,767,547, assigned to Procter and Gamble.
48. Baker, E. (1994) Preparation, properties and formulation of DEEDMAC: An environmentally friendly cationic surfactant. *Oral presentation at the 85th Annual AOCS National Meeting*, Atlanta, GA.
49. Ellis, S.R. and Turner, G.A. (1996) U.S. Patent 5,516,437, assigned to Lever Brothers.
50. Grandmaire, J.P. and Hermosilla, A. (1996) U.S. Patent 5,525,245, assigned to Colgate Palmolive.
51. Swartley, D.M., Trinh, T., Wahl, E.H. and Huysse, G.M. (1995) U.S. Patent 5,399,272, assigned to Procter and Gamble.
52. Corona, A. (1996) European Patent Application 704,522, assigned to Procter and Gamble.
53. Lam, A.C., Lin, S.Q., Taylor, T.J. and Winters, J.R. (1996) U.S. Patent 5,480,567, assigned to Unilever.
54. Morita, H. and Oota, S. (1995) Japanese Patent Application, 07018578, assigned to Lion.
55. Iacobucci, P. (1997) Tumble dryer sheets. *Oral Presentation at the 88th AOCS National Meeting*, Seattle, WA.
56. Puchta, R., Sandkühler, P., Schreiber, J. and Völkel, T. (1994) W.O. Patent 94/02676, assigned to Henkel.
57. De Buzzaccarini, F., Farwick, T.J. and Zhen, Y. (1995) W.O. Patent 95/29218, assigned to Procter and Gamble.
58. De Buzzaccarini, F., Farwick, T.J. and Zhen, Y. (1995) W.O. Patent 95/29217, assigned to Procter and Gamble.
59. Franklin, R., Mendello, R., Iacobucci, P.A., Steichen, D., Trinh, P.-N. and Dery, M. (2000) U.S. Patent 6,037,315 assigned to Akzo Nobel, N.V.
60. Wahl, E., Tordil, H.B., Trinh, T., Carr, E.R., Keys, R.O. and Meyer, L.M. (1998) U.S. Patent 5,759,990 assigned to Procter and Gamble.
61. Lenoir, P.M. (2004) U.S. Patent 6,680,290 assigned to Dow Europe S.A.
62. Nakama, Y., Harusawa, F., Otsubo, K., Iwai, T., Tamaki, S. and Ohkoshi, M. (1990) U.S. Patent 4,919,846, assigned to Shiseido Co. Ltd.
63. Pings, K.D. (1996) U.S. Patent 5,482,703, assigned to Procter and Gamble.
64. Pings, K.D. (1997) U.S. Patent 5,482,703, assigned to Procter and Gamble.
65. Robbins, C.R. and Patel, A.M. (1997) Patent 5,415,857, assigned to Colgate Palmolive.
66. Akzo Nobel Chemicals (1996) *Akzo Nobel Personal Care: Product Guide.* Akzo Nobel Chemicals, Chicago, IL.
67. Murphy, A.P. (1981) U.S. Patent 4,259,217, assigned to Procter and Gamble.
68. Franklin, R., Iacobucci, P., Steichen, D., Tang, D. and Trinh, P.-N. (2001) U.S. Patent 6,264,931 assigned to Akzo Nobel N.V.
69. Mehreteab, A. and Loprest, F.J. (1995) U.S. Patent 5,441,541, assigned to Colgate Palmolive.
70. Mehreteab, A. and Loprest, F.J. (1995) U.S. Patent 5,472,455, assigned to Colgate Palmolive.
71. Asono, K., et al. (2000) U.S. Patent 6,136,769 assigned to Procter and Gamble.
72. Rorig, H. and Porta, N. (1992) U.S. Patent 5,078,896, assigned to Akzo N.V.
73. Smith, W.L. (1991) U.S. Patent 5,055,219, assigned to the Clorox Company.
74. Leach, M.J. (1995) W.O. Patent 96/26257, assigned to Unilever.
75. Johnson, A.K. and Franklin, R. (2002) U.S. Patent 6,462,014, assigned to Akzo Nobel, N.V.
76. Knudson, M. and Jones, T. (1992) U.S. Patent 5,160,454, assigned to Southern Clay Products.

77. Tso, Su, Beall, G. and Holthouser, M. (1987) U.S. Patent 4,677,158, assigned to United Catalysts Inc.
78. Ogura, Y. (1993) U.S. Patent 5,202,114, assigned to Shiseido Company, Ltd.
79. Munzy, C., Butler, B., Hanley, H., Tsvetkov, F. and Peiffer, D. (1996) Clay platelet dispersion in a polymer matrix. *Mater. Lett.*, **28**(4–6), 379.
80. Kim, Y.S., Song, D., Jeon, J. and Choi, S. (1996) Adsorption of organic phenols onto hexadecyltrimethylammonium-treated montmorillonite. *Separation Science and Technology*, **31**(20), 2815.
81. Sun, J. and Yu, B. (1996) Studies on the treatment of wastewater from coal gas washing with betonite compound adsorbent. *Feijinshukuang*, 1 37.
82. Dickey, M. and Carron, M. (1996) Raman spectroscopic study of sorption to CTAB-modified montmorillonite. *Langmuir*, **12**(9), 2226.
83. Zhang, Z., Sparks, D. and Scrivner, N. (1993) Sorption and desorption of quaternary amine cations on clays: *Environmental Science and Technology*, **27**, 1625.
84. Hellsten, M. and Harwigsson, I. (1999) U.S. Patent 5,902,784, assigned to Akzo Nobel N.V.
85. Mysaka, J., Lin, Z., Stepanek, P. and Zakin, J.L. (2001) Influence of salts on dynamic properties of drag reducing surfactants. *J. Non-Newton. Fluid Mech.*, **97**, 251–66.
86. Alink, B. A.M.O., and Jovancicevic, V. (2002) WO Patent Application 02/33216 assigned to Baker Hughes Inc.
87. Klomp, U., Kruka, V., Reijnhart, R. and Weisenborn, A.J. (1995) U.S. Patent 5,460,728 assigned to Shell Oil Company.
88. Dahlmann, U. and Feustel, M. (2004) U.S. Patent Application 0159041 assigned to Clariant Corporation.
89. Milburn, C.R. and Sitz, G.M. (2002) U.S. Patent 6,444,852, assigned to Goldschmidt Chemical Corporation.
90. Norman, W.D., Jasinski, R.J. and Nelson, E.B. (1996) U.S. Patent 5,551,516, assigned to Dowell, a division of Schlumberger Technology Corporation.
91. Gadberry, J.F., Hoey, M.D., Franklin, R., del Carmen Vale, G. and Mozayeni, F. (1999) U.S. Patent 5,979,555, assigned to Akzo Nobel N.V.
92. Asamori, K., Tamaki, R., Funada, H., Taniguchi, T., Juarez, F.C., Ortiz, F.C., Muniz, A.G. and Hernandez, H.R. (2000) U.S. Patent 6,013,681, assigned to Kao Corporation.
93. Asamori, K., Nagao, S., Tamaki, R., Taniguchi, T., Tomiako, K. and Koyanagi, K. (2000) U.S. Patent 6,048,905, assigned to Kao Corporation.
94. Lif, A. and Olsson, S. (2002) WO Patent Application 02/48294 assigned to Akzo Nobel N.V.
95. Flick, E.W. (1989) *Advanced Cleaning Product Formulations.* Noyes Publications, Park Ridge, New Jersey.
96. Flick, E.W. (1987) *Fungicides, Biocides and Preservatives for Industrial and Agricultural Applications.* Noyes Publications, Park Ridge, New Jersey.
97. Walker, L. E. (1998) U.S. Patent 5,760,088, assigned to Lonza, Inc.
98. Walker, L. E. (2000) U.S. Patent 6,087,303, assigned to Lonza, Inc.
99. Walker, L. E. (2000) U.S. Patent 6,090,855, assigned to Lonza, Inc.

6.2 Amphoteric Surfactants

Richard Otterson

6.2.1 Introduction

This section concerns surfactants that, at least at some pH, are *zwitterionic*, which means they are both anionic and cationic at the same time so that the hydrophilic portion of the molecule has internally neutralized positive and negative charges. True amphoteric surfactants are those that exhibit a varying charge, from positive, to zwitterionic to just negative, on the hydrophile depending on the pH of the solution in which they are found. The classic amphoteric surfactants, such as amphoacetates, are anionic, cationic or zwitterionic at various points in the pH spectrum. This section will also consider those surfactants that exhibit a zwitterionic form regardless of what other charge states they take with changing pH. This latter class of 'amphoteric' surfactants includes betaines, sultaines and sulfonated amphoterics.

Amphoteric surfactants are generally milder to the skin and eyes than anionic, cationic and some nonionic surfactants. The amphoterics, due to their ability to support both positive and negative charges, usually have large 'head groups', the hydrophilic portion of the molecule that exhibits an affinity for the aqueous phase. This property makes them desirable secondary surfactants because they have the ability to modify micellar structure. Amphoterics are generally used in formulations with anionic or nonionic surfactants to modify the solubility, micelle size, foam stability, detergency and viscosity of various cleansing systems and emulsions.

Being internally neutralized, the amphoterics have minimal impact on the biocidal activity of quaternary ammonium salts. For this reason, like the nonionics, they often find use in antimicrobial preparations that are based on cationic surfactants.

6.2.2 Aminopropionates and iminodipropionates

This group of surfactants is discussed first because they are among the oldest amphoterics in commerce with patents dating back to 1949 and because they best illustrate the amphoteric surfactants. The first products introduced to the market in this category were the 'Deriphat^{TM}' [1] products introduced by General Mills Chemicals.

These materials are the reaction product of a primary amine and either acrylic acid, an ester of acrylic acid such as methyl acrylate, ethyl acrylate or crotonic acid. Either 1 or 2 mol of acrylate is used. If 1 mol is added, an N-alkyl β-alanine is produced (Figure 6.6) and if 2 mol of acrylate per mole of amine is used, the corresponding carboxyethyl β-alanine derivative is produced (Figure 6.7).

N-alkyl β-alanine derivatives, the 'monopropionates', have distinct isoelectric points while most other amphoteric surfactants do not. The isoelectric point is the pH at which the molecule is internally neutralized, existing as a zwitterion. Aqueous solubility and the propensity to foam are lowest at the isoelectric point.

Both alkyl primary amines and 'ether amines' are used to produce this group of surfactants (Figure 6.8).

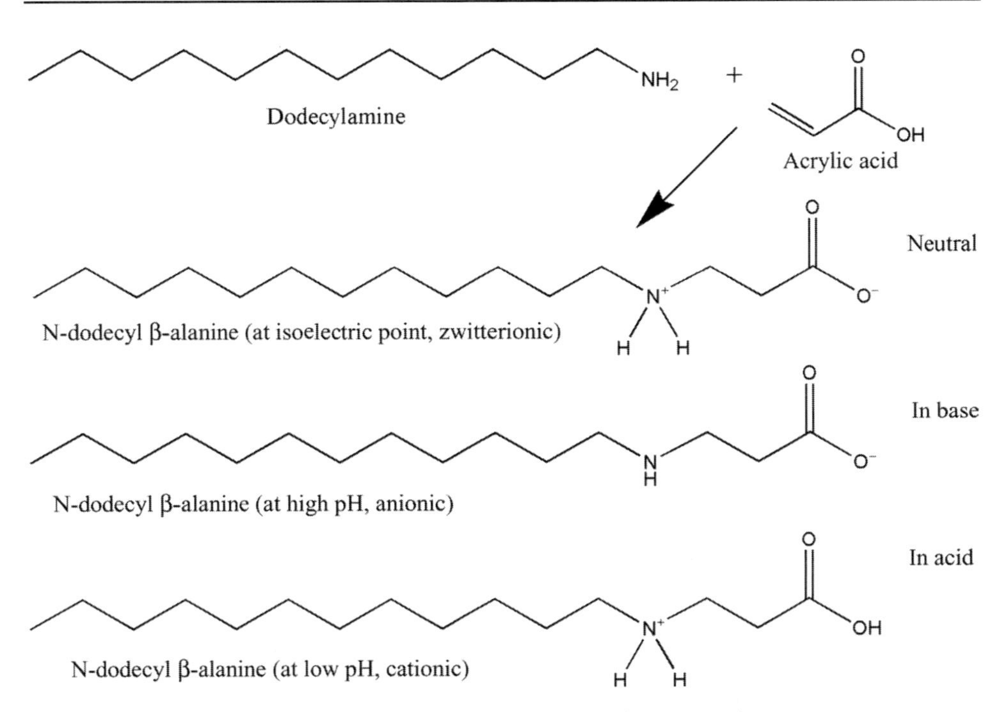

Figure 6.6 N-dodoecyl β-alanine, an example of an amphoteric surfactant.

In both cases, commercial products have alkyl chain lengths from 8 to 18 carbons but the ether amine derived products usually have branched alkyl chains as they are produced from fatty alcohols such as 2-ethylhexanol, isodecyl and tridecyl alcohol. Those produced from alkyl amines usually have linear hydrocarbon chains because they are produced from naturally derived fatty acids except when made from 2-ethylhexylamine.

Monopropionate surfactants, produced from 1 mol of acrylic acid, are fairly rare in commerce. The composition of these products tends to be a mixture of the monopropionate,

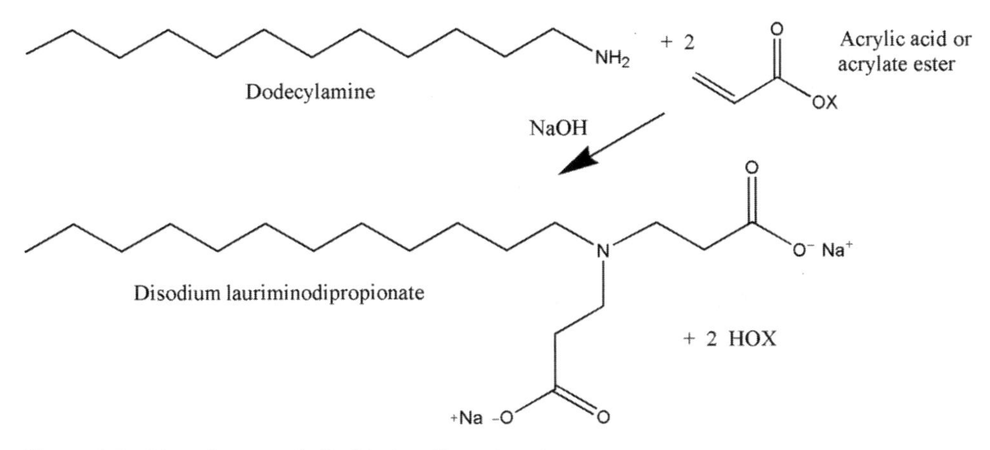

Figure 6.7 Manufacture of alkyl iminodipropionates.

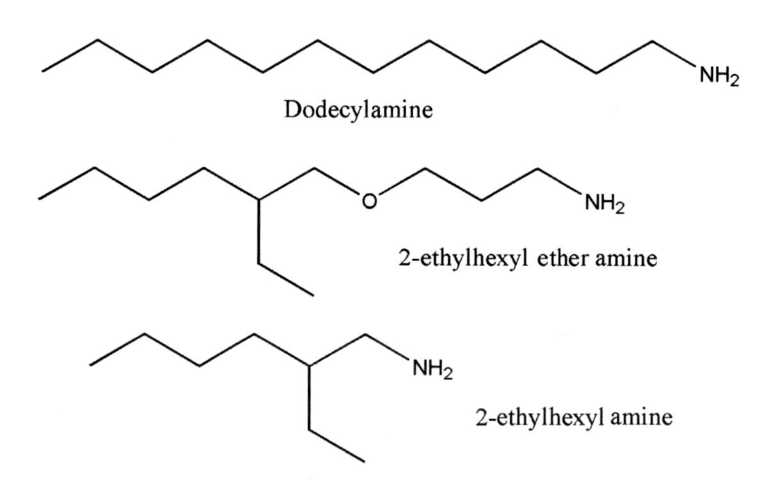

Figure 6.8 Examples of amines used for the synthesis of alkyliminodipropionates.

the dipropionate and unreacted amine. Lauraminopropionic acid is one such commercial material (see Figure 6.8).

Dipropionates, produced with 2 mol of acrylate per mole of amine, are fairly common. Commercially significant products are 2-ethylhexyliminodipropionate, lauriminodipropionate, isodecyloxypropyliminodipropionate and tallowiminodipropionate. All are sold as either monosodium salt or disodium salt. If they are produced from an acrylic acid ester, they are sold as disodium salt and contain 2 moles of either methanol or ethanol as a by-product.

Because the alkyliminodipropionates are hydrolytically stable and more soluble than most surfactants in fairly concentrated solutions of electrolytes, they find use in highly alkaline 'built' detergent formulations and strong acid cleaners. Sodium lauriminodipropionate finds use in personal care applications, despite being among the least mild amphoteric surfactants because of the desirable foam properties it delivers, as well as the conditioning properties it has when formulated at slightly acid pH. It is also used in such applications as fire fighting foam because it generates heavy, wet quality foam.

Being amino acid derivatives, the iminodipropionate surfactants are quite biodegradable and they are compatible with most other surfactants used.

The alkyliminodipropionates described above are somewhat more expensive than most other hydrocarbon-based surfactants. This is due to the cost of the alkyl primary amines used to produce them relative to other available fatty acid amide based amines.

6.2.3 Imidazoline-based amphoteric surfactants

Two major classes of amphoteric surfactants are derived from fatty alkyl hydroxyethyl imidazolines which, in turn, are produced from fatty acids and low molecular weight amines. Because fatty acids are fairly economic, the imidazoline derived amphoacetates tend to be less expensive than the iminodipropionates discussed above. Most imidazoline derived

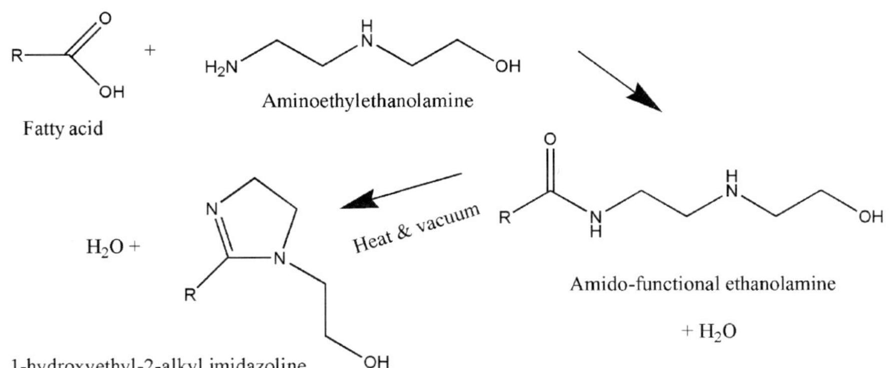

Figure 6.9 Synthesis of hydroxyethyl alkyl imidazolines.

amphoterics are made from fatty acids reacted with aminoethylethanolamine to produce an amido-functional alkanolamine, which then cyclizes to an alkyl hydroxyethyl imidazoline (Figure. 6.9)

These imidazoline compounds have proved very useful as intermediates to amphoteric surfactants. Products made from them, alkylated with sodium chloroacetate or methyl acrylate were patented by Hans Mannheimer who founded Miranol Company in the USA during the 1950s [2]. Miranol Company became the major vendor of imidazoline derived amphoteric surfactants in the world. Other imidazolines are used to produce amphoteric surfactants, such as alkyl aminoethyl imidazoline, but those products are of less economic significance.

6.2.3.1 Amphoacetates

For personal care applications, the major products in this group are 'amphoacetates' or 'amphodiacetates', generally based on alkyl hydroxyethyl imidazolines from either a whole coconut fatty acid distribution or a lauric cut. The 'ampho' portion of their name is a convention established by the International Nomenclature Committee for Cosmetic Products (INCI) to indicate that they are derived from imidazoline structures. The INCI nomenclature applied to these materials, amphoacetate and amphodiacetate, is intended to give an indication of the stoichiometry used to produce them, either 1 or 2 mol of sodium chloroacetate is added to each mole of fatty imidazoline. Modern analytical methods have been used to determine the structure of these products and almost all of them are actually 'monoacetates'. The main difference between amphoacetates and amphodiacetates is the composition of the by-products.

As mentioned above, most commercial products are based on either a lauric (mainly C-12) or a whole coconut distribution (C-8 to C-18, with approximately 50% C-12) since these alkyl distributions give the best detergency. Early on, the imidazoline derived amphoterics were characterized as exceptionally mild to the skin and eyes relative to most surfactants available at the time. This made them excellent candidates for use in baby shampoos, geriatric cleansing products, hand wash for medical facilities and so on.

Johnson & Johnson formulated them into its 'no more tears' baby shampoo and became the first major user of these materials in a consumer product [3].

Baby shampoos remained the largest market for these materials until the 1990s when they found utility in skin cleansers and body washes. When formulated with sodium laureth sulfate in roughly equimolar amounts, gel-like, non-Newtonian consistency cleansers can be produced that tolerate the inclusion of conditioning agents such as vegetable oils without negatively affecting foaming properties. Procter & Gamble and Unilever both introduced body washes utilizing this technology during the 1990s [4]. The use of significant amounts of the imidazoline derived amphoteric surfactants results in less 'defatting' of the skin and the use of these formulations provides an easily perceived conditioning effect to the skin.

The original products introduced by Miranol Company consisted of either 1 or 2 moles of sodium chloroacetate reacted with each mole of alkyl hydroxyethyl imidazoline or two moles. The expectation was that the first mole quaternized the imidazoline ring and the second mole formed an ether with the hydroxyethyl group (see Figure 6.10).

Neither assumption was correct. The original products introduced to the market had either a significant amount of hydrolyzed imidazoline in them, in the case of the 1-mol products, or a significant amount of hydrolyzed sodium chloroacetate, which is glycolic acid, in the case of the 2-mol products (see Figure 6.11).

Though neither the 1 or 2-mol products were optimal in composition, they continue to be produced and used to this day, especially disodium cocoamphodiacetate, due to its long history of use with an excellent safety record.

Over the years, optimized amphoacetates were developed that were intermediate between the 1 and 2-mol products. An effort was made to deliver more amphoteric surfactant and less by-products to the formulator. Most baby shampoos and body washes produced today are formulated with such optimized products, which are produced from an optimum ratio

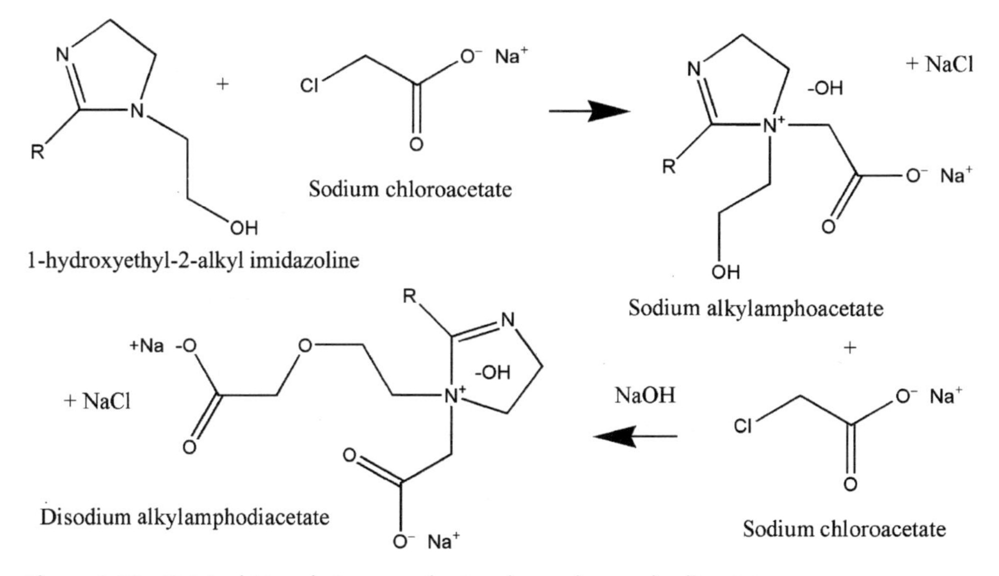

Figure 6.10 Original Mannheimer synthesis scheme for amphodiacetates.

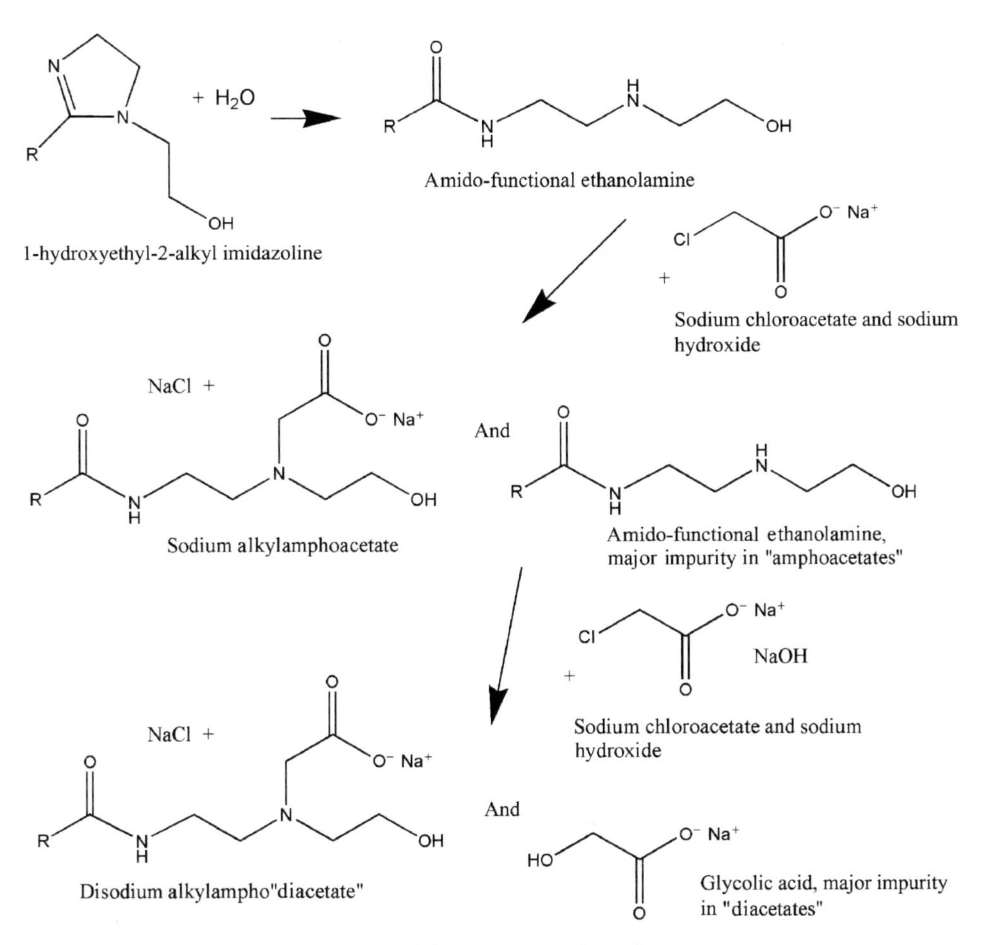

Figure 6.11 Modern understanding of the synthesis of amphoacetates.

of imidazoline to sodium chloroacetate to provide a minimum amount of by-products. The two main imidazoline amphoteric surfactants used in consumer products are sodium lauroamphoacetate or sodium cocoamphoacetate. Modern products typically have less than 2% each of free amine and glycolic acid (see Table 6.8).

Due to the low free amine content of well below 1%, the traditional amphodiacetates, those produced from 2 mol of sodium chloroacetate, such as disodium cocoamphodiacetate, are still the mildest of the various amphoacetate products available. Because they are the mildest, they still find wide use in consumer products such as premoistened wipes, baby wipes and specialized skin cleansers. The low free amine content is accompanied by a high glycolic acid level, typically on the order of 5%. Glycolic acid, an alpha hydroxy acid or AHA, is known to irritate the skin and produce a kerotolytic effect at pH values below 4.5. Chemical 'skin peel' products intended to rejuvenate the skin are often formulated with glycolic acid at a pH of between 3 and 4.5. Because formulations containing amphodiacetates are typically neutral or only slightly acid, the glycolic acid content does not lead to skin irritation.

Table 6.8 Comparison of a traditional cocoamphoacetate and a cocoamphodiacetate with a modern 'optimized' sodium cocoamphoacetate

Property	Amphoacetate	Optimized amphoacetate	Amphodiacetate
Nonvolatiles (% by wt.)	44.0	37.0	49.0
NaCl (%)	7.3	7.2	11.7
Salt (as a % of solids)	16.6	19.5	23.9
Viscosity (cps)	8000	800	3000
Sodium glycolate (%)	0.3	1.6	5.5
Unreacted amidoamine (%)	4.2	0.6	0.1
Actives %	32.2	27.6	31.7
Actives/solids (%)	73	75	65

6.2.3.2 Amphopropionates

The other major class of fatty imidazoline derived amphoteric surfactants is the amphopropionates. Again, the *ampho* portion of the name indicates that they are derived from imidazolines but, rather than being alkylated with sodium chloroacetate, they are 'carboxylated' with an acrylate via the Michael reaction. A primary or secondary amine is added across the double bond of the acrylate to yield the beta-alanine derivative.

Miranol Company introduced a series of these surfactants based on the condensation of alkyl hydroxyethyl imidazolines with methyl acrylate [5]. Similar to the amphoacetates above, the assumption was made that the first mole of methyl acrylate quaternized the imidazoline ring and the second added to the hydroxyethyl group to produce an ether carboxylate (see Figure 6.12).

Most of the amphopropionate surfactants produced are of the amphodipropionate type, 2 mol of methyl acrylate or sodium acrylate added per mole of imidazoline. Depending on the reaction conditions, 1 mol of acrylate can add to the fatty R group at the alpha carbon. Upon hydrolysis of the imidazoline, the second reacts with the liberated secondary amine to produce the beta alanine derivative. If methyl acrylate is used, the methyl ester of the amphoteric surfactant is formed. An equimolar amount of sodium hydroxide is added to effect saponification to the sodium salt of the surfactant. Methanol is formed as a by-product and it is generally left in the final product as part of the solvent system.

The synthesis of amphopropionates is thus similar to the amphoacetates except that, rather than sodium chloride being formed as a by-product, methanol often is. If made from methyl acrylate, there is usually about 5–7% methanol in the surfactant product. If alkylation is conducted with sodium acrylate, methyl acrylate is the more efficient alkylating agent.

Though the chemistry may be similar to the amphoacetates, the current applications for the amphopropionates tend to be quite different. Amphoacetates are generally used in cleansing products for personal care while the amphopropionates find most utility in hard surface cleansing.

The most economically important member of this family of surfactants is disodium cocoamphodipropionate. It is a fairly good detergent in itself but it has other properties that make it very useful for formulated liquid cleaners. It is quite soluble in fairly concentrated

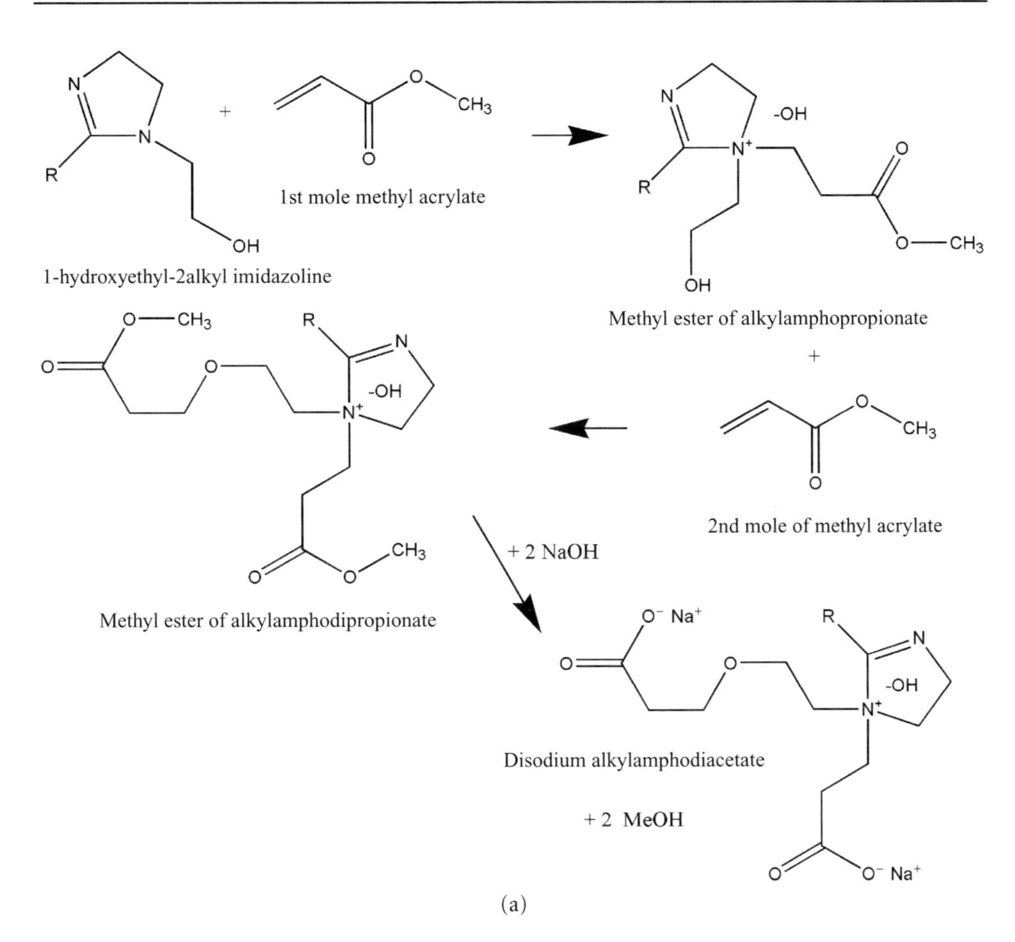

Figure 6.12(a) Original Mannheimer scheme for amphodipropionates.

electrolyte solutions such as the metasilicate and pyrophosphate builders used in these products. In addition, disodium cocoamphodipropionate has the ability to couple nonionic surfactants into strong electrolytes efficiently. Modest corrosion inhibition properties add to the utility of this group of surfactants.

There are also personal care applications but they are somewhat limited as compared with amphoacetates. Salt-free amphoteric surfactants are useful for coupling high amounts of conditioning polymers into 'neutralizing shampoos'. These shampoos are used after the use of alkali-based hair relaxers. Conditioning is accomplished with cationic polymers that are rendered compatible with sodium laureth sulfate by the use of disodium cocoamphodipropionate. Of course, methanol found in this material was undesirable so many of these products are now formulated with products made from sodium acrylate and are thus free of both methanol and sodium chloride. The high amount of sodium chloride present in the amphoacetate products can be problematic when formulating them into anionic-based cleansers, particularly if a high concentration of total surfactant is desired. The sodium

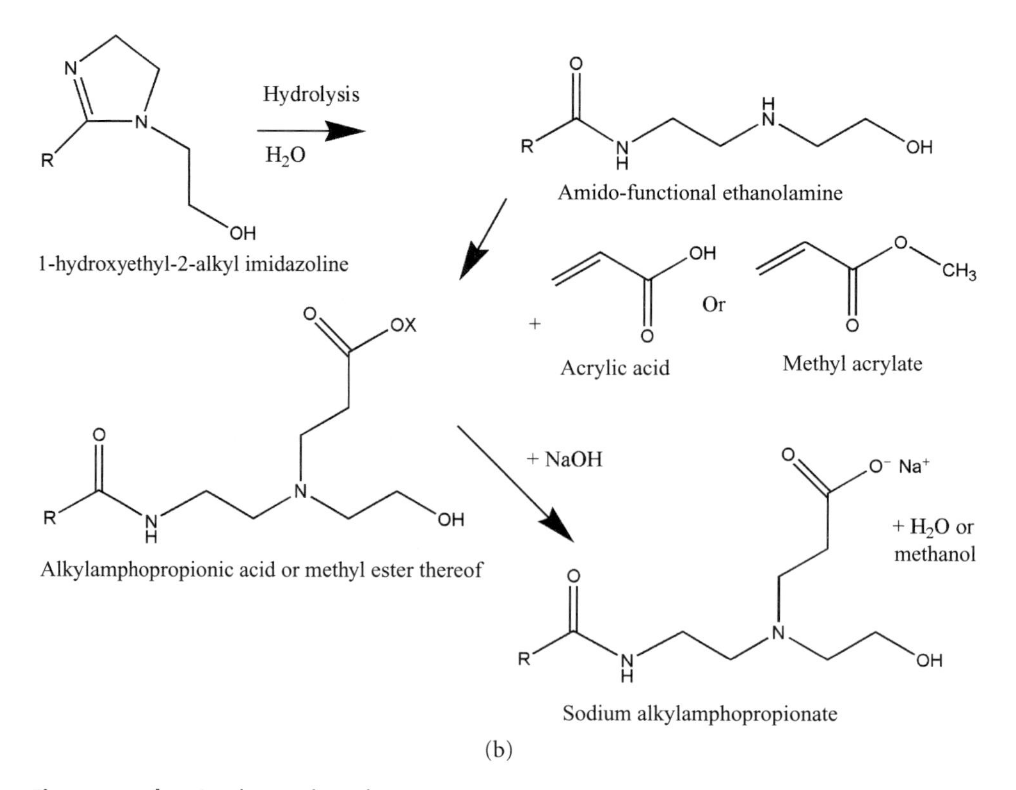

Figure 6.12(b) Synthesis of amphopropionates.

chloride tends to raise the viscosity to unacceptable levels and, in these cases, the salt and methanol-free amphopropionates can serve as effective, especially mild, secondary surfactants.

Hard surface cleaners require a variety of properties such as wetting, detergency, coupling, low or high foaming and so on. In addition to the coco and lauryl derivatives, the caprylic and capric, the C-8 and C-10 derivatives of the amphodipropionates, have proved useful. The caprylic versions afford the formulator the ability to couple nonionic surfactants into a built system without substantially increasing foaming or volatile organic content. The capric derivatives offer modest foaming along with enhanced coupling.

6.2.3.3 Amphohydroxypropylsulfonates

Like the amphoacetates and amphopropionates above, the amphohydroxypropylsulfonates are derived from hydrolyzed alkyl hydroxyethyl imidazolines. In this, amido-functional alkanolamine hydrolysis product is alkylated with sodium 1-chloro-2-hydroxypropane sulfonic acid (see Figure 6.13). These products have good hard water tolerance and have applications including metal cleaning as well as personal cleansing.

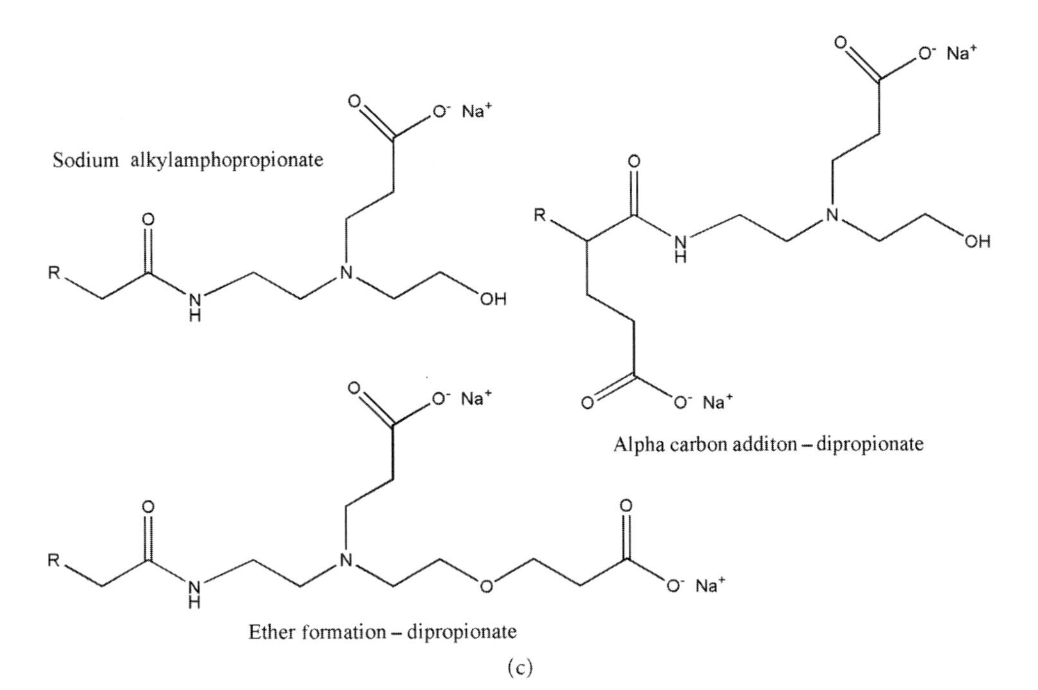

Figure 6.12(c) Species present in amphodipropionates.

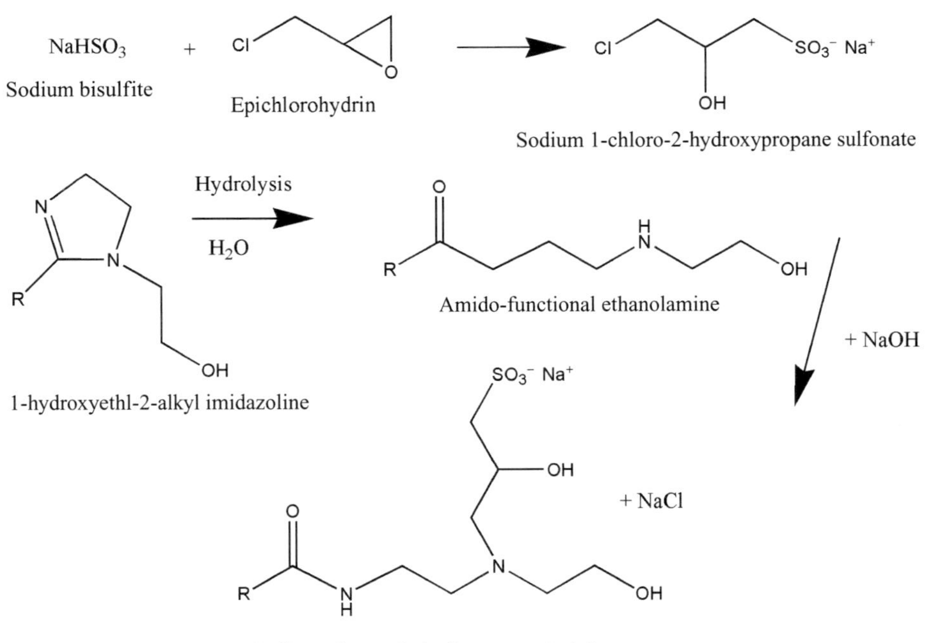

Figure 6.13 Synthesis of 'sulfonated' amphoterics.

6.2.4 Betaine surfactants

Betaine, or trimethyl glycine, is a naturally occurring zwitterionic nutrient that was first discovered in beets but occurs in a wide variety of plants and animals. Surfactant betaines are structurally analogous compounds but they are manufactured synthetically.

The most simple surfactant betaines are alkyl betaines where one of the methyl groups of trimethyl glycine is replaced with a fatty alkyl moiety. Synthetically, they are made by condensing 1 mol of sodium chloroacetate with an alkyldimethylamine to yield the surfactant and 1 mol of sodium chloride by-product (see Figure 6.14).

The large head group of betaine surfactants is a desirable characteristic. They form mixed micelles with anionic surfactants and the large hydrophilic group affects the packing, altering micelle shape. This affects the detergency, foaming and the viscosity of formulations based on them.

The most economically significant alkyl betaines are those with a *stripped or topped* coconut alkyl distribution, which means the C8-10 portion is removed leaving the C-12 to C-18 components and those with a lauryl distribution, predominantly C-12 and C-14, because they offer the greatest detergency, act synergistically with anionic surfactants to stabilize foam, enhance detergency and build viscosity. Octyl betaine is a low foaming wetting agent whilst cetyl and stearyl betaine are mild surfactants used as specialty emulsifiers. The alkyl betaines tend to be used less frequently than the structurally similar alkylamidopropyl betaines discussed below because they are more costly to produce.

The wide availability of relatively inexpensive dimethylaminopropylamine (DMAPA) allows surfactant producers to convert economic triglycerides, fatty acids and methyl esters into amido-functional tertiary amines that may then be quaternized with sodium chloroacetate to produce alkylamidopropyl betaines (see Figure 6.15). The most economically significant of these is cocamidopropyl betaine which can be produced from a variety of feedstocks and lauramidopropyl betaine which is generally produced from lauric acid. These are widely used secondary surfactants in consumer products such as shampoos, bath products, washing up liquids and other cleaners.

Usage of these products as secondary surfactants has greatly increased around the world as cocamide DEA (CDEA) fell into disfavor due to the propensity of free diethanolamine to

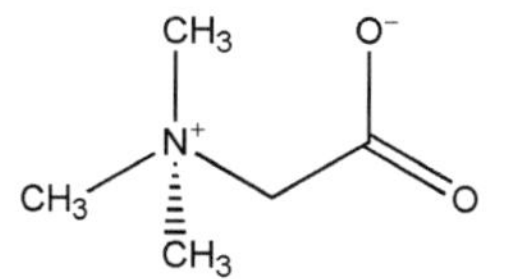

Betaine – the natural product

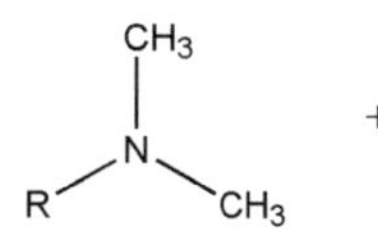

+

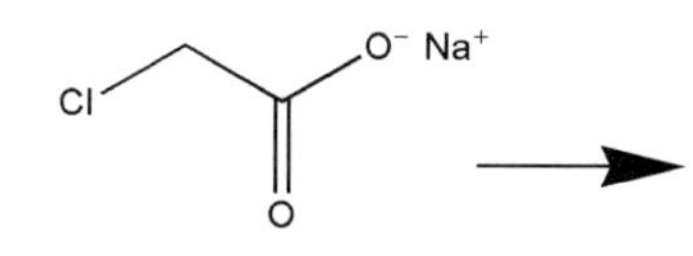

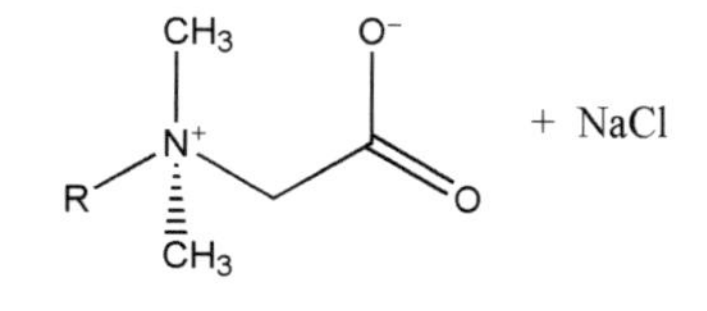

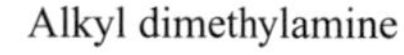

Sodium chloroacetate

alkyl betaine

Figure 6.14 Synthesis of alkyl betaines.

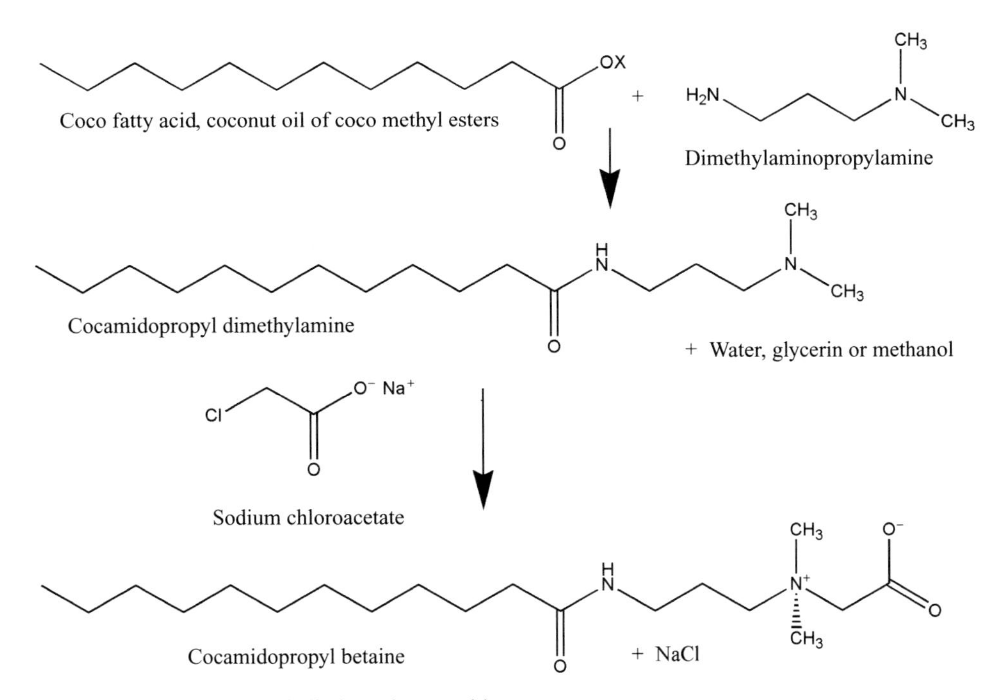

Figure 6.15 Synthesis of alkylamidopropyl betaines.

form carcinogenic nitrosodiethanolamine and studies carried out by the National Toxicology Program (NTP) in the United States which indicated that DEA itself may have carcinogenic activity, at least in laboratory animals. The NTP operates under the auspices of the U.S. Department of Health and Human Services and is intended to identify problematic chemicals in our environment. Like cocamide DEA, the alkylamidopropyl betaines enhance detergency; stabilize foam and aid in viscosity building of formulated products so they are the preferred replacements for CDEA in many applications.

Over the past decade, studies by DeGroot [6] and others have indicated that free amines present in cocamidopropyl betaine appear to be sensitizers. It has not been clear as to whether the problem was due to free DMAPA or cocamidopropyl dimethylamine but recent evidence suggests it may be due to both. The producers of these products reacted to the problem and the typical alkylamidopropyl betaine produced today contains less than 10 ppm of free DMAPA and less than 0.5% of cocamidopropyl dimethylamine (Table 6.9).

Sodium chloroacetate is a reactive and toxic material, so it is hydrolyzed to glycolic acid nearly quantitatively at the end of the production cycle. Chloroacetic acid always contains traces of dichloroacetic acid, a toxic and unreactive material that appears on the California Prop. 65 list. The laws of the U.S. State of California require that the Governor of the State publish, annually, a list of chemicals known to cause cancer and reproductive abnormalities. This list is known by the ballot initiative that brought it into law as the 'Prop. 65 List'. The vendors of surfactant betaines use grades of chloroacetic acid containing minimal amounts of dichloroacetic acid.

Betaine surfactants are relatively mild to the skin. While not as mild as the amphoacetates and amphopropionates discussed earlier, they are significantly milder than the commonly

Table 6.9 Comparison of an older commercial cocamidopropyl betaine product with a modern one

Property	Older cocamidopropyl betaine	Modern cocamidopropyl betaine
Nonvolatiles (% by wt.)	35.0	35.0
Sodium chloride (%)	4.85	5.05
Glycolic acid (%)	0.15	0.35
Amidoamine (%)	1.10	0.25
Chloroacetic acid (ppm)	50	<5
Dichloroacetic acid (ppm)	70	20

used anionic surfactants and most cationic surfactants. Johnson & Johnson baby shampoo, which, for years, incorporated amphoacetate surfactants as well as cocamidopropyl betaine and a mild, high molecular weight polysorbate nonionic, was reformulated for the U.S. market in the past 2 years. In this most recent version, cocamidopropyl betaine is the first ingredient listed after water. Sodium lauroamphoacetate, used for so many decades, has been removed.

The production of cocamidopropyl betaine has traditionally been based on two feedstocks: coconut oil and 'topped' or 'stripped' (C-8 and 10 removed) coconut fatty acid or methyl ester. These products are still widely used but, to achieve better colors and odors, hydrogenated feedstocks are now very frequently used, either fully hydrogenated coconut oil triglyceride or stripped, hydrogenated, distilled coconut fatty acid. These products are most frequently sold as aqueous solutions with 35% nonvolatile matter. If made from triglyceride, the betaine surfactant will contain about 2.5% glycerin by-product and 5% sodium chloride by-product in addition to the active surfactant. Products made from fatty acid or methyl ester are approximately 30% active product and slightly more than 5% sodium chloride.

In addition to the coco and lauric based betaines, there are many other products based on other triglycerides and fatty acids. Caprylamidopropyl betaine is a very low foaming material that solubilizes other low foaming surfactants in specialty cleaners. It is also useful for the formulation of premoistened wipes because it is mild to the skin and does not leave foam trails on the skin. Higher molecular weight alkylamidopropyl betaines are used as specialty emulsifiers and conditioning agents in personal care products. Palmitamidopropyl, stearamidopropyl and behenamidopropyl betaine all find utility in such products.

Cocamidopropyl betaine and cocamidopropyl hydroxysultaine, discussed later, are also used in petroleum production. Their relatively high foaming nature, electrolyte tolerance and hydrolytic stability make them useful for foam acidizing and foam fracturing fluids.

As the use of cocamidopropyl betaine increased as a secondary surfactant in anionic systems, the relatively low concentration of about 35% nonvolatiles at which it is normally sold became an issue. At this concentration, betaines are somewhat susceptible to bacterial growth so a preservative is often needed and the low concentration also increases freight costs so that several patented technologies were developed to address this [7]. Typically, the inclusion of about 2% of one of the patented additives allows the producers to prepare an aqueous solution of 45% nonvolatiles which is hostile to microbial growth without preservation and saves over 25% of the normal freight costs. The additives used to achieve this are coconut fatty acid, citric acid, aminoacids and biodegradable iminodisuccinate based chelate [8].

6.2.4.1 Other amphoterics based on sodium chloroacetate

While the betaines and imidazoline-based amphoterics represent the largest volumes of amphoteric surfactant sold, there are a few others that bear mention. Sodium dihydroxyethyl tallow glycinate is the condensation product of dihydroxyethyl tallow amine and sodium chloroacetate. It is used as a thickening agent and a surfactant. Although this would seem to be an alternative betaine structure, it is really a complex mixture of materials. The tallow ethoxylate is somewhat hindered and does not undergo quaternization easily so three reactions actually occur. There is some quaternization, some etherification of the terminal hydroxyls to form the carboxymethyl derivative and some hydrolysis of chloroacetic acid to glycolic acid. All the species present contribute to thickening of acid and alkaline formulations.

In the 1970s and 1980s, a number of amphoteric surfactants were introduced to the market that were based on alkyl polyamines, primary amines upon which is condensed acrylonitrile, then hydrogenated to produce an alkyl propylenediamine and, with additional cycles, alkyl polyamines. These, in turn, are alkylated with sodium chloroacetate to produce alkyl polyamine polycarboxylates. Some of these products find utility in laundry applications, in personal care products and as industrial foamers.

6.2.4.2 Sultaines (sulfobetaines), hydroxysultaines and sulfonated amphoterics

Analogous to the betaine surfactants are the sultaines and hydroxysultaines. Both are derived from either alkyl dimethylamine or alkylamidopropyl dimethylamine. Rather than being the reaction product of sodium chloroacetate, they are manufactured from a tertiary amine and either propane sultone or sodium propanechlorohydrin sulfonate (CHPS).

Sultaines are made by condensing propane sultone with a tertiary amine. These products are relatively rare in industry. Propane sultone is a known human carcinogen, so it must be handled with great care and thus the cost of these products is fairly high. The properties are similar to those of hydroxysultaines which are made by condensing epichlorohydrin with sodium bisulfite to make a propanechlorohydrin sulfonate which is then reacted with a tertiary amine to make a hydroxysultaine with sodium chloride as a by-product (see Figure 6.16).

Hydroxysultaines find use in personal care products, where they function as secondary surfactants to enhance the properties of anionic-based formulations, in much the same way as betaines. They are also among the best lime soap dispersants known, so they are used effectively in natural soap based products where they make the use of hard water practical. They are also used in petroleum production chemicals were they serve as foaming agents for acid and foam fracturing procedures.

Commercial hydroxysultaines are generally produced from either lauryl dimethylamine, the DMAPA condensate of coconut oil or stripped, hydrogenated, distilled coconut fatty acid. This last grade is the most common one used to formulate personal care products as

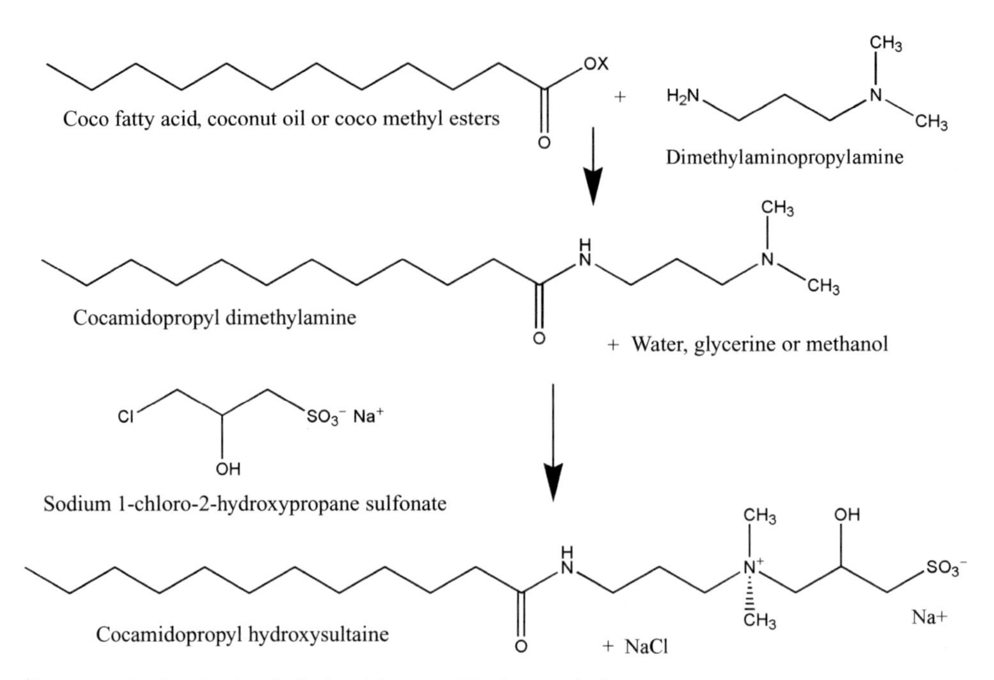

Figure 6.16 Synthesis of alkylamidopropyl hydroxysultaines.

if offers the best color and odor with reasonable cost. Hydroxysultaines produced directly from coconut oil contain about 3% glycerin and they are often used in petroleum production chemicals and in hard surface cleansers though the quality is sufficiently high also to formulate cosmetic products.

Lauryl hydroxysultaine is completely hydrolytically stable, a fairly high foamer and tolerant of electrolyte. It is used as a detergent in both strongly acid and strongly alkaline cleaners where amido-functional surfactants would not have sufficient shelf life due to hydrolysis.

6.2.5 *Other amphoteric surfactants*

The chemical literature and patent art disclose a great variety of amphoteric surface active molecules. The author has attempted to address only those that are of economic importance today. Not all of the current commercial products fall into the categories above, so a few additional ones are mentioned here.

6.2.5.1 *Phosphobetaines and phosphoamphoterics*

Mona Industries received a series of patents in the 1980s for betaines and imidazoline-based surfactants similar to the hydroxysultaines and hydroxypropylsulfonates discussed earlier but alkylated with a propanechlorohydrin phosphate rather than the CHPS [9]. These amphoteric surfactants were demonstrably mild and were thought to have some antimicrobial activity. As a consequence they were used in sensitive applications such as premoistened wipe products.

6.2.6 Summary

Amphoteric surfactants have a variety of roles in industry today. They continue to be among the mildest surfactants available for the formulation of personal care products and are incorporated into products that demand the least irritation potential such as facial cleansers, feminine hygiene washes, no rinse cleansers such as baby wipes, geriatric products and so on.

Their ability to render compatible various surfactants and other types of surfactants into strong solutions of electrolyte make them invaluable in the formulation of strong detergent products where mildness is not an issue.

In addition, because they are generally based on aminoacid structures, they are among the most biodegradable surfactants available to the formulator. Products based on amphoteric surfactants are usually readily biodegradable, thus having a minimal impact on the environment.

The utility of the amphoterics for coupling materials together affords the formulator a powerful tool when designing high active detergent products. That, in addition to their ready biodegradability and general mildness, suggests that this class of surfactants will be of increasing economic importance in the future.

References

1. General Mills Chemicals, Inc. (1972) *Deriphat General Technical Bulletin.* General Mills Chemicals, Minneapolis, MN. *Deriphat* is a trademark of Cognis Corporation.
2. Mannheimer, H.S. U.S. Patent Nos. 2,528,378; 2,773.068; 3,100,799; 3231,580 and 3,408,361 to Miranol.
3. Masci, J.N. and Poirier, N.A. U.S. Patent Nos. 2,999,069 and 3,055,836 to Johnson & Johnson.
4. Giret, M.J., Langlois, A. and Duke, R.P. U.S. Patent No.5,409,640 to P&G. Puvvada, S. U.S. Patent No. 5,965,500 to Levers Brothers Company.
5. Miranol C2M SF Conc. & Miranol H2M SF Conc, *Miranol* is a trademark of Rhodia Corp.
6. De Groot, A.C. (1997) Cocamidopropyl betaine: a "new"important cosmetic allergen. *Dermatosen in Beruf und Umwelt,* **45**(2), 60–3.
7. Armstrong, D.K., Smith, H.R., Ross, J.S. and White, I.R. (1999) Sensitization to cocamidopropylbetaine: an 8-year review. *Contact Dermatitis,* **40**(6), 335–6.
8. Hamann, I., Kohle, H.J. and Wehner, W. U.S. Patent No. 5,962,708 to Witco Surfactants GmbH.
9. Weitemeyer, C., Foitzik, W., Kaseborn, H.D., Gruning, B. and Begoihn, U. U.S. Patent No. 5,354,906 to Th. Goldschmidt AG.
10. Otterson R.J., Berg K.R. and D'Aversa, E.A. U.S. Patent Application No. 20050037942 to McIntyre Group Ltd.
11. O'Lenick, A.J. and Mayhew, R.L. U.S. Patent No. 4,283,542 to Mona Industries.
12. Lindemann, M.K.O., Mayhew, R.L., O'Lenick, A.J. and Verdicchio, R.J. U.S. Patent No. 4,215,064 to Mona Industries and Johnson & Johnson.

6.3 Silicone Surfactants

Randal M. Hill

6.3.1 Introduction

Silicone surfactants consist of a permethylated siloxane group coupled to one or more polar groups [1]. The most common examples are polydimethylsiloxane-polyoxyalkylene graft copolymers but they also include the small-molecule trisiloxane (superwetter) surfactants. They are used in a wide range of applications in which conventional hydrocarbon surfactants are ineffective. They are surface active in both aqueous and nonaqueous systems and they lower surface tension to values as low as 20 mN m^{-1}. They are usually liquids even at high molecular weights. The trisiloxanes promote spreading of aqueous formulations on hydrophobic surfaces such as polyethylene.

All these properties derive from the polydimethylsiloxane (or silicone) component of these surfactants [2]. Silicones are methyl-rich polymers with an unusually flexible backbone [3–5]. The methyl-rich character gives them a low surface energy of about 20 mN m^{-1} and the flexible backbone leads to a very low glass transition temperature, T_g, and liquid form at ambient conditions. The flexible backbone also contributes to the low surface energy by serving as a flexible framework for the methyl groups. The surface energy of a methyl-saturated surface is about 20 mN m^{-1}, which is also the lowest surface tension achievable using a silicone surfactant. In contrast, most hydrocarbon surfactants contain alkyl, or alkylaryl hydrophobic groups that are mostly $-CH_2-$ groups and pack loosely at the air/liquid interface. The surface energy of such a surface is dominated by the methylene groups and for this reason hydrocarbon surfactants typically give surface tension values of about 30 mN m^{-1} or higher.

The surfactant properties of polymeric silicone surfactants are markedly different from those of hydrocarbon polymeric surfactants such as the ethylene oxide/propylene oxide (EO/PO) block copolymers. Comparable silicone surfactants often give lower surface tension and silicone surfactants often self-assemble in aqueous solution to form bilayer phases and vesicles rather than micelles and gel phases. The skin feel and lubricity properties of silicone surfactants do not appear to have any parallel amongst hydrocarbon polymeric surfactants.

In spite of these important differences, silicone surfactants share much in common with conventional surfactants. Equilibrium and dynamic surface tension vary with concentration and molecular architecture in similar ways. Silicone surfactants self-associate in solution to form micelles, vesicles and liquid crystal phases. Self-association follows similar patterns as molecular size and shape are varied and silicone surfactants containing polyoxyalkylene groups exhibit a cloud point. HLB values can be calculated for silicone surfactants, although more useful values can be obtained from calculations that take into account the differences between silicone and hydrocarbon species.

It is a common misunderstanding that silicones and silicone surfactants are incompatible with hydrocarbon oils; this is only partly correct. Small silicone surfactants, such as the trisiloxanes, are very compatible with organic oils. For example, aqueous solutions of the trisiloxane surfactants give very low interfacial tension against alkane oils. The incompatibility between polymeric silicones and some hydrocarbon oils is due more to the polymeric nature of the silicone block rather than to strong phobicity such as that between fluorocarbon and hydrocarbon groups. The compatibility between two species, such as a polymer and a solvent, can be measured by $X * N$, where X is the Flory–Huggins interaction parameter and N is the degree of polymerization (DP) of the polymer. The incompatibility of silicones with hydrocarbon liquids and the surface activity of silicone surfactants in nonaqueous liquids have a strong component due to N and a weaker contribution from X. This is an important consideration in understanding silicone surfactants and developing applications for them – especially in systems that involve hydrocarbon oils and nonaqueous media.

A significant amount of information regarding the uses of silicone surfactants is still found primarily in the patent art, but the major applications have recently been reviewed in journals. Silicone (or siloxane) surfactants are also called silicone polyethers (SPEs), polyalkylene oxide silicone copolymers, silicone poly(oxyalkylene) copolymers and silicone glycols. The International Cosmetic Ingredient Nomenclature and the Cosmetic, Toiletry and Fragrance Association (CTFA) adopted name is dimethicone copolyol.

6.3.2 Structures

The hydrophobic part of a silicone surfactant is a permethylated siloxane or polydimethylsiloxane group. Many different polar or hydrophilic groups have been used, some of which are listed in the table below.

Hydrophilic group	Examples
Nonionic	Polyoxyethylene (pOE or EO) Polyoxyethylene/polyoxypropylene (pOE/pOP or EO/PO) Carbohydrates
Anionic	Sulfate
Cationic	Various quaternary nitrogen groups
Zwitterionic	Betaines

In addition to these polar groups, many types of fluorocarbon and hydrocarbon groups have been attached to the silicone backbone, sometimes in combination with polar groups. Some of these materials are used as compatibilizing agents or for other surface active properties but they lie outside the scope of this chapter.

The most common polymeric silicone surfactants are based on polyoxyalkylene groups. The structures of graft-type (rake-type) and ABA structures are illustrated in Figures 6.17 and 6.18. It should be noted that there are many possible variants of these basic structures. The actual structure of graft-type silicone copolymers is a random copolymer of m and n rather than the 'blocky' structure suggested by the diagram.

A group of well-characterized AB surfactants has been prepared and investigated [6]. Numerous other polymeric structures have been made, including $(AB)_n$, BAB and branched and cross-linked examples. Such materials may provide a unique functionality in a particular application but there has been no systematic investigation of their properties.

Many small-molecule silicone surfactants have been made and their properties (especially wetting) characterized [7–11]. The best known small-molecule silicone surfactants are the trisiloxane surfactants based on 1,1,1,3,5,5,5-heptamethyltrisiloxane, shown in Figure 6.19.

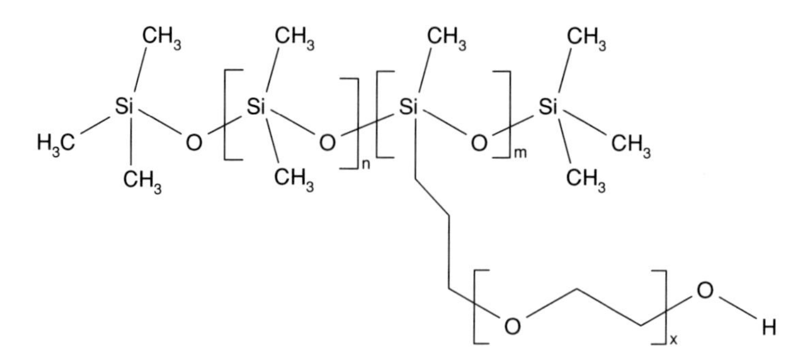

Figure 6.17 Structure of a graft-type silicone-polyoxyethylene copolymer, $MD_n(D'EO_xOH)_mM$.

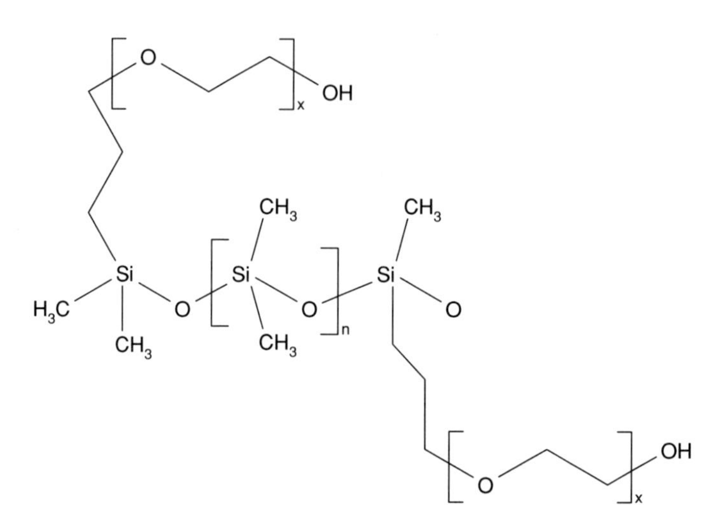

Figure 6.18 Structure of an ABA-type silicone-polyoxyethylene copolymer, $HOEO_xM'D_n$ $M'EO_xOH$.

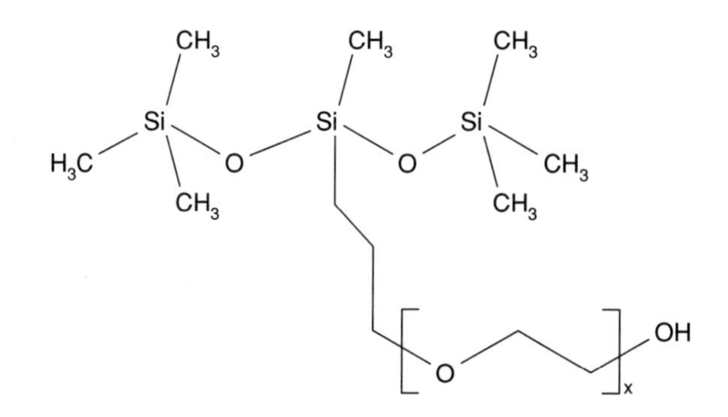

Figure 6.19 Structure of a trisiloxane surfactant, $M(D'EO_xOH)M$.

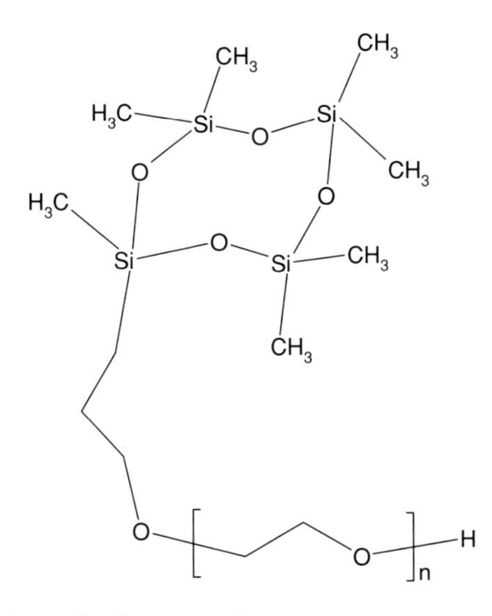

Figure 6.20 Structure of a cyclosiloxane surfactant.

Silicones readily form cyclic structures such as octamethylcyclosiloxane. Small cyclic silicones have also been used to make small-molecule silicone surfactants such as that shown in Figure 6.20 [7].

The wetting properties of the trisiloxane surfactants will be discussed below.

6.3.3 Synthesis

Silicone surfactants are prepared using one of the following three routes:

(a) Transetherification – the reaction of an alkoxy-functional silicone with an alcohol functional polar group
(b) Hydrosilylation – the reaction of an SiH functional silicone with a vinyl functional polar group
(c) Two-step process in which a small reactive group is first attached to the silicone followed by attachment of a polar group through that reactive group.

Relevant details of organosilicon chemistry for preparing silicone surfactants are discussed by Legrow and Petroff [12] and by Gruening and Koerner [13] who also catalogue a large number of possible modifying groups. The first step in all three routes is to prepare a silicone starting material with the appropriate reactive groups – usually SiH or SiOR. This can be done either by co-hydrolysis of chlorosilanes or by the equilibration reaction:

$$\mathrm{MM} + x\mathrm{D} + y\mathrm{D}' \Leftrightarrow \mathrm{MD}_x\mathrm{D}'_y\mathrm{M}$$

Thus, to make a trimethyl end-capped silicone polymer with a total DP of $x + y + 2$, one would start with 1 mol of MM (hexamethyldisiloxane), x mol of D (dimethyl) units, perhaps in the form of octamethylcyclotetrasiloxane and y mol of D′ (tetramethylcyclotetrasiloxane could be used). In the presence of acid or base catalyst, these species rearrange to form the desired polymer.

The MDTQ notation is defined in the following table.

MDTQ notation – the eight siloxane unit building blocks

M	$Me_3SiO_{1/2}$–	A trimethyl end-cap unit
D	–Me_2SiO–	The basic dimethyl unit
T	–$MeSiO_{3/2}$–	A three-way branch point unit
Q	–SiO_2–	A four-way branch point unit
M′	$Me_2(R)SiO_{1/2}$–	A substituted trifunctional end-cap unit
D′	–Me(R)SiO–	A substituted difunctional unit
T′	–$(R)SiO_{3/2}$–	A substituted three-way branch point unit
Me	–CH_3	
R	H, or (after hydrosilylation) some nonmethyl organic group	

The chemistry of the equilibration reaction is explored in detail in several good texts on organosilicon polymer chemistry [3–5]. The reaction can be acid or base catalyzed and represents a convenient route to a nearly infinite variety of silicone structures. For the purposes here, two features of this reaction are important. First, the reaction gives a broad molecular weight distribution in which the D reactive groups are randomly distributed. Second, equilibration of silicones generates cyclic species as well as linear species in an approximately 15/85 ratio [14]. Since the cyclics also contain reactive groups, this means that some polymeric silicone surfactants contain a bimodal molecular weight distribution. This reaction is not a good route to narrow polydispersity low molecular weight silicone groups.

The first silicone surfactants were prepared by reaction of alkoxy-functional silicones with hydroxyl terminated polyethers using the transetherification reaction [7]:

$$[\text{silicone}]\text{SiOR}^{A} + \text{R}^{B}\text{OH} \rightarrow [\text{silicone}]\text{SiOR}^{B} + \text{R}^{A}\text{OH}$$

This reaction yields surfactants in which the polar group is linked to the silicone through an SiOC linkage. These materials have found widespread use in nonaqueous applications such as manufacture of polyurethane foam (PUF) but in an aqueous system the SiOC linkage hydrolyses (rapidly away from pH = 7).

Silicone surfactants that are more hydrolytically stable are prepared by coupling polar groups to the silicone using hydrosilylation [14, 15]:

$$[\text{silicone}]\text{SiH} + \text{CH}_2{=}\text{CHCH}_2(\text{OCH}_2\text{CH}_2)_n\text{OR} \rightarrow [\text{silicone}]\text{Si}(\text{CH}_2)_3(\text{OCH}_2\text{CH}_2)_n\text{OR}$$

This reaction is catalyzed by a platinum catalyst such as Speier's catalyst, chloroplatinic acid. Because the catalyst also isomerizes the terminal double bond, the reaction may be run with an excess of vinyl to make sure all the reactive sites on the silicone are reacted. SiH also reacts with ROH and, for this reason, alkoxy end-capped polyethers may be preferred. Alkoxy end-capped polyethers are also useful when the surfactant will be used in a chemically reactive system such as polyurethane foam manufacture.

The best way to prepare ionic silicone surfactants is a two-step synthesis in which a small reactive group is first attached to the silicone using hydrosilylation and then a polar group is attached to that group [3, 15]. This procedure has been used to prepare cationic [16], anionic [17] and zwitterionic [18] silicone surfactants. Examples of other reactions suitable for the preparation of ionic silicone surfactants are given by Snow et al. [19], Maki et al. [20] and by Gruening and Koerner [13]. Wagner et al. have prepared a series of silicone surfactants with nonionic carbohydrate polar groups [21].

Although not siloxane based, organosilicon surfactants have also been made from permethylated carbosilanes containing an Si-C-Si structure. The simplest version of this is the trimethyl silylated alkyl polyether discussed by Klein [22] and Wagner [11, 23]. These surfactants are more hydrolytically stable wetting agents than the trisiloxanes.

6.3.4 Hydrolytic stability

The polydimethylsiloxane portion of silicone surfactants may hydrolyze in contact with water (this is actually the leftward direction of the equilibration reaction mentioned above). Although not generally considered an issue for polymeric silicone surfactants near neutral pH, for the trisiloxane surfactants it represents a significant barrier to shelf-stable aqueous formulations utilizing these remarkable surfactants. Amongst organosilicon chemists, it is well known that residual acidity or basicity of glassware surfaces catalyzes this reaction – requiring that careful work be done in plasticware or glassware rigorously treated with a hydrophobizing agent such as octyltrichlorosilane. Although the problem is real, there is more qualitative anecdotal description in the literature than quantitative data. For instance, Gradzielski et al. [24] observed that certain trisiloxane surfactants 'hydrolyzed completely' within a few weeks whereas the polymeric silicone surfactants they studied were stable 'for at least a few months'. No quantitative analytical results were presented. Stuermer et al. [25] found that the aqueous phase behavior of their $M(DE_7OH)M$ changed somewhat after a short period of time which they attributed to hydrolysis. Hill and He [8, 9] studied the same class of surfactants and did not observe such short-time changes. Experiments on the trisiloxane surfactants carried out in phosphate buffer solutions in plasticware show no changes in surface tension to indicate degradation [26].

6.3.5 Surface activity

Silicone surfactants in aqueous solutions show the same general behavior as conventional hydrocarbon surfactants – the surface tension decreases with increasing concentration until a densely packed film is formed at the surface. Above this concentration, the surface tension becomes constant. The concentration at the transition is called the critical micelle concentration (CMC) or critical aggregation concentration (CAC). The surface and interfacial activity of silicone surfactants was reviewed by Hoffmann and Ulbricht [27]. Useful discussions of the dependence of the surface activity of polymeric silicone surfactants on molecular weight and structure are given by Vick [28] and for the trisiloxane surfactants by Gentle and Snow [29].

A useful way to compare the surface activity of different surfactants is in terms of their 'efficiency' and 'effectiveness' [30]. *Efficiency* measures the surfactant concentration required to achieve a certain surface tension, while *effectiveness* is measured by the maximum reduction of the surface tension that can be obtained for that particular surfactant. In the following discussion these terms will be used in this specific defined sense.

Silicone surfactants are highly *effective* surfactants in water – that is, they lower aqueous surface tension to *low* values of 20–30 mN m^{-1}. Only the perfluoro-surfactants give lower values – between 15 and 20 mN m^{-1}. The *effectiveness* of silicone surfactants is weakly influenced by the nature of the hydrophilic groups. Nonionic and zwitterionic silicone surfactants are the most *effective* – give minimum values for the surface tension around 20 mN m^{-1}. Ionic silicone surfactants are usually less *effective* but still give surface tension values around 30 mN m^{-1}. The surface tension of solutions of polymeric silicone surfactants increase with the molecular weight of the silicone hydrophobe [28], presumably because of conformational entropy at the interface. Silicone surfactants containing mixed EO/PO groups tend to be less *effective* than all-EO examples [28].

The *efficiency* of the nonionic trisiloxane surfactants is comparable to nonionic hydrocarbon surfactants with a linear dodecyl hydrophobe. The surface properties of a homologous series of trisiloxane surfactants M(DE$_n$OH)M with $n = 4$–20 show that the CAC, the surface tension at the CAC and the area per molecule each vary with molecular structure in a way that is consistent with an 'umbrella' model for the shape of the trisiloxane hydrophobe at the air/water interface [29]. The log(CAC) and the surface tension at the CAC both increased linearly with EO chain length.

The surface activity of silicones in nonaqueous liquids such as the polyols used in polyurethane foam manufacture is due to a combination of the low surface energy of the methyl-rich silicone species and insolubility caused by high molecular weight. Many materials including solvent mixtures and polymer solutions exhibit surface activity as they approach a miscibility boundary. For example, dilute solutions of polydimethylsiloxane in certain organic solvents foam due to marginal solubility. Solubility decreases with increasing molecular weight, which means that a higher molecular weight silicone surfactant will have a greater tendency to segregate to the surface, even with the same proportions of solvo-philic and solvo-phobic groups. In a nonaqueous system this can mean the difference between significant surface activity and solubility. Silicone polyalkyleneoxide copolymers lower the surface tension of a variety of organic liquids including mineral oil and several polyols from values of about 25–30 mN m^{-1} to values near 21 mN m^{-1}.

6.3.6 Wetting

Surfactants assist wetting because they lower surface and interfacial tension. The energy balance that determines spreading is expressed by the spreading coefficient, S, and is illustrated in Figure 6.21.

Surfactants lower γ_{lv}, and usually γ_{ls}, making S more positive and therefore leading to spreading. The term 'wetting' is used here to mean spreading to a thin wetting film with zero contact angle. 'Spreading' is a more general term and describes relaxation of a drop of liquid on a surface to a final contact angle $<90°$. Unlike pure liquids, the surface and interfacial tension of surfactant solutions depend on diffusion of surfactant to the interface – as

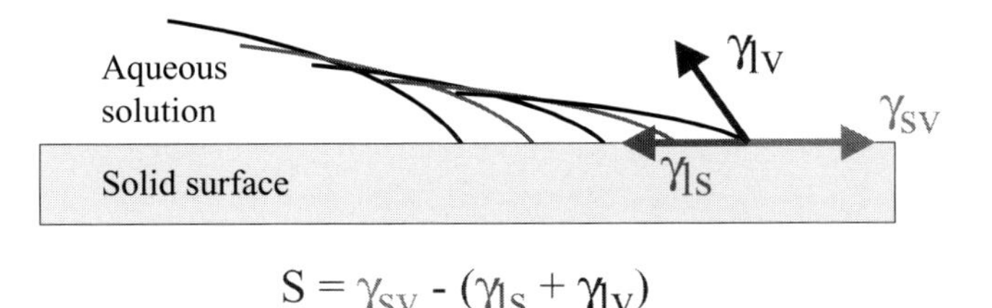

Figure 6.21 Spreading coefficient.

the surface is stretched by spreading, surfactant at the surface must be replenished from the bulk. In addition, since the area is increasing faster at the spreading edge than near the middle of the drop, a surface tension gradient may develop leading to Marangoni effects. The interconnected roles of fluid mechanics, surfactant diffusion and Marangoni tractive forces in controlling surfactant enhanced wetting have not yet been fully accounted for.

The ability of silicone surfactants to promote spreading leads to their use in paints and coatings [31], personal care products [28], textiles [32], the oil industry [33] and as adjuvants for pesticides [34]. The good spreading properties of silicone surfactants have been attributed to low cohesive forces between molecules in the interfacial film. The time to wet by the Draves wetting test depends on the size of the silicone hydrophobe and the length of the polyoxyethylene group – the most rapid wetting is observed for surfactants with the shortest silicone groups and the smaller EO groups [28].

The unique ability of the trisiloxane surfactants to promote complete wetting of aqueous solutions on very hydrophobic surfaces such as polyethylene is called 'superwetting' or 'superspreading'. This property was first documented in the 1960s when it was noted that aqueous solutions of certain small silicone polyethers rapidly spread to a thin film on low energy surfaces. It was determined that the best wetting agents were based on small silicone groups containing 2–5 methyl-siloxane units and that the best wetting was usually observed for surfactants with limited solubility (that is, those that tended to form stable turbid dispersions) [7]. Superwetting has been explained by low dynamic surface tension, the unusual shape of the trisiloxane hydrophobe, the presence of bilayer vesicles in the aqueous phase and Marangoni effects.

Superwetting shares its main characteristics with the wetting of a variety of ionic and nonionic, micelle- and vesicle-forming hydrocarbon and silicone surfactants that spread to a wetting film on non-water-wettable substrates such as polyethylene [35]. Note that many common surfactants, such as sodium dodecyl sulfate, do not spread to a wetting film even on relatively hydrophilic substrates. Of those that do spread, three features are shared in common: (1) there is a maximum in spreading rate as a function of substrate surface energy, (2) surfactant concentration and (3) there is a CWC that is significantly higher than the CAC [36]. Although the surface tension becomes constant at the CAC, the contact angle does not reach zero until a concentration almost 3–5 times higher [37]. On liquid substrates in which the surfactant is insoluble, the rate of surfactant transport to the interfaces controls the spreading [38].

In spite of its obvious practical importance, the spreading behavior of mixtures of silicone surfactants and hydrocarbon surfactants has not been systematically studied. Early patents claimed generally synergistic behavior for a wide range of combinations of silicone surfactants with hydrocarbon surfactants. Recent patents claim enhanced wetting for certain combinations of silicone surfactants with hydrocarbon surfactants [39].

6.3.7 Phase behavior

Besides lowering surface and interfacial tension, surfactants also self-associate in solution to form a variety of self-assembled aggregates ranging from globular, worm-like and disk-shaped micelles to bilayer structures such as vesicles. At higher concentrations these aggregates form liquid crystal phases. What type of aggregate or liquid crystal phase is formed by a particular surfactant, as well as the sequence of liquid crystal phases observed with increasing surfactant concentration, temperature and salt level, is simply related to molecular shape and packing. For example, M(D E_{12}OH)M, which has a relatively large polar group, tends to form highly curved aggregates such as globular micelles.

Understanding surfactant phase behavior is important because it controls physical properties such as rheology and freeze-thaw stability of formulations. It is also closely related to the ability to form and stabilize emulsions and microemulsions. Micelles, vesicles, microemulsions and liquid crystal phases have all been used as delivery vehicles for perfumes or other active ingredients.

The aqueous phase behavior of the trisiloxane surfactants follows a simple pattern – as the polar group decreases in size, there is a progression from globular micelles to worm-like micelles to bilayer structures and then inverse structures [9]. Comparing the phase behavior of the trisiloxane surfactants with that of hydrocarbon surfactants, the trisiloxane hydrophobic group is significantly wider and shorter than (for example) a linear $C_{12}H_{25}$ group. The length of the trisiloxane group is only 9.7 Å compared with 15 Å for $C_{12}H_{25}$ while its volume is larger, 530 Å^3 compared with 350 Å^3 for $C_{12}H_{25}$. This causes the type of aggregates formed by these surfactants to be shifted toward lower curvature. For instance, $C_{12}E_7$ forms globular micelles, whereas M(D E_7OH)M forms bilayer microstructures.

Few studies exist for ionic silicone surfactants. Several trisiloxane anionic, cationic and zwitterionic surfactants have been found to form micelles, vesicles and lamellar liquid crystals. As would be expected, salt shifts the aggregates toward smaller curvature structures [40].

Two other important aspects of surfactant phase behavior are the cloud point and the physical form of the neat surfactant at ambient temperature. Nonionic silicone surfactants are usually liquids at ambient temperature, while ionic silicone surfactants are waxy solids. Nonionic silicone surfactants containing polyoxyalkylene groups become less soluble in water with increasing temperature just as hydrocarbon surfactants do. The temperature at which they become insoluble is called the cloud temperature or cloud point. The general dependence of the cloud point on the weight fraction of polyoxyalkylene is similar to the behavior of hydrocarbon surfactants – the cloud point increases with added EO and decreases with added PO. Some silicone surfactants form stable cloudy dispersions over a wide range of temperatures that are lamellar phase (vesicle) dispersions rather than representing a cloud point. The effects of electrolytes on the cloud point follow the usual patterns. Unlike hydrocarbon surfactants, these relationships cannot be simply related to molecular structure using the HLB system although HLB values are often quoted for silicone surfactants. O'Lenick has developed a 3D HLB system that is claimed to better account for the emulsifying and cloud point properties of silicone surfactants [41].

Several studies of the phase behavior of polymeric silicone surfactants have been published based on commercial grade copolymers with varying levels of characterization [24, 25, 42]. There is no commercially available homologous series of polymeric silicone surfactants. The phase behavior of graft-type and ABA silicone surfactants differs substantially from that of the well-studied EO/PO copolymers (the Pluronics®, for example). EO/PO copolymers tend to form globular micelles and give phase diagrams that contain many small regions of liquid crystal phases, including gel phase and cubic phases.

Polymeric silicone surfactants tend to give simple phase diagrams with one or two larger regions of liquid crystal phase. Bilayer microstructures including vesicles and lamellar phase liquid crystal are common. The phase behavior of these copolymers progresses from lamellar to hexagonal phases with increasing proportion of polyoxyethylene. Salts shift the phases toward inverse curvature and lower temperature. The surprising ability of such complex and polydisperse molecular structures to form ordered structures such as bilayers and liquid crystal phases is attributed to the unusual flexibility of the silicone backbone [43].

The phase behavior of a group of well-defined AB pDMS-pEO diblock copolymer surfactants has recently been reported [6]. The size of the silicone block and the total molecular weight were varied over a wide enough range to find both normal and inverse liquid crystal phases. Because the pDMS block remains liquid to much larger size than do hydrocarbon groups, they were able to explore a wider range of chain lengths than is possible using linear alkyl ethoxylate nonionic surfactants, in effect to access a very wide range of amphiphilicity.

AB block copolymers have been shown to markedly enhance the efficiency of microemulsions [44]. This efficiency boosting effect has recently been found for AB silicone surfactants as well [45]. The paper argues that higher molecular weight AB silicone surfactants represent an ideal material to increase the efficiency of microemulsions because, unlike long chain organic surfactants, silicone copolymers do not have a Krafft point with increasing chain length.

The aggregation behavior of AB silicone surfactants in nonpolar oils including several hydrocarbon oils has been reported by Rodriguez [46]. They found that inverse micelles were formed in all oils, adjacent to the inverse cubic phase formed by the neat copolymers and by concentrated mixtures of copolymer and oil. The CMC depended strongly on the length of the pEO chain but only weakly on the pDMS chain. Inverse hexagonal phase was also observed.

6.3.8 Ternary systems

There is a considerable patent art concerning preparation of transparent mixtures of water with low molecular weight silicone oils using polymeric silicone surfactants. Some representative early references are Keil [47], Gee [48, 49], Gum [50] and Terae [51]. These compositions are called microemulsions in the patents in the sense of being transparent mixtures of water, surfactant and oil – but note that they are transparent because of small particle size or because of index of refraction matching.

Silicone surfactants containing polyoxyalkylene groups are usually soluble in ethers, alcohols, esters, ketones, and aromatic and halogenated solvents. Unlike hydrocarbon nonionic surfactants they are not very soluble in alkanes. Polymeric silicone surfactants are not miscible with polymeric silicone oils [13].

Mixtures of trisiloxane nonionic surfactants with cyclic and short linear silicone oils form extensive regions of microemulsion as well as liquid crystal phases [52, 53]. The phase behavior shifts toward positive curvature (hexagonal and cubic phases) and higher temperatures with increasing EO chain length. Salt shifts microemulsion regions to lower temperature. Higher molecular weight (MW) oils require higher concentrations of surfactant to form one-phase microemulsion. The trisiloxane surfactants are highly efficient – forming one-phase microemulsion of a 1:1 mixture of water and D4 (octamethylcyclotetrasiloxane) at surfactant concentrations of only 6–8 wt.%. Trisiloxane surfactants with larger polyoxyethylene groups form hexagonal and cubic liquid crystal phases with cyclic and short linear silicone oils.

Mixtures of low molecular weight silicon-based surfactants and cosurfactants have been used to prepare a self-dispersing microemulsion of silicone agents applied to building materials to impart water repellency [54, 55]. The structure of the surfactants used was not disclosed but they are described as being themselves reactive so that they bind to the surfaces of the building materials and become part of the water-repellancy treatment.

6.3.9 *Applications*

Silicone surfactants are specialty surfactants that are primarily used in applications that demand their unique properties. Most applications are based on some combination of their (a) low surface tension, (b) surface activity in nonaqueous media, (c) wetting or spreading, (d) low friction or tactile properties, (e) ability to deliver silicone in a water-soluble (or dispersible) form, (f) polymeric nature or (g) low toxicity. The major applications will be discussed briefly in following sections.

6.3.9.1 Polyurethane foam manufacture

The stabilization of foam during manufacture of PUF was the first commercial application of silicone surfactants. Worldwide volume for silicone surfactants in polyurethane foam manufacture was estimated to be about 30 000 metric tons/year in 1994 [56]. Polyurethane foams were first commercialized in the 1950s in Germany – Fritz Hostettler of Union Carbide Corp. was the first to be granted a patent on the use of silicone polyether copolymers in the manufacture of polyurethane foam [57]. PUF is formed by the reaction of polyols and isocyanates. Polyurethane foams range from rigid pneumatic resins to flexible porous elastomers. Rigid PUF is used primarily as an insulating material in construction, piping and packaging whereas flexible PUF is used as a cushioning material in furniture, bedding, carpet underlay, automobiles and packaging. A number of thorough reviews of polyurethane foam processing are available [58, 59]. The silicone surfactant functions to emulsify the mixture of incompatible materials, stabilize the blowing foam, keep urea particles from aggregating and govern film rupture for open-cell foams. The process is complex and exceedingly difficult to study realistically. Mixtures of silicone surfactants are frequently used to achieve properties that cannot be obtained with single materials. The ability of silicone surfactants to stabilize and control the drainage of the thin films separating growing bubbles appears to be a critical property [60, 61].

6.3.9.2 Personal care

The use of silicone surfactants in personal care was reviewed by Floyd [62] but silicone surfactants are only one type of silicone material used in this field. Oil soluble silicones and resins are used to promote spreading and film formation by organic oils and waxes. Silicone copolymers with hydrophilic groups are often viewed as a means to overcome the incompatibility of silicone oil with both water and hydrocarbon oils – allowing the technologist to obtain the 'dry silky' feel of silicones in a water-dispersible form. Silicone polyethers are used mainly in aqueous formulations such as shampoos and shower gels. In shampoos, they improve combing, add gloss and impart a 'dry silky' feel to hair. They have also been shown to reduce the eye irritation of anionic surfactants. Efficacy is determined by solubility – lower HLB versions are more substantive. Incorporation of amine or quaternary nitrogen polar groups improves substantivity in hair conditioning products and silicone polyethers can be added to many types of cosmetics and skin care formulations to impart a 'nongreasy' feel to the skin. In lotions, they impart smoothness and softness to the skin and defoam, thereby minimizing whitening on rub-out. They have also been found to help prevent cracking in soap and syndet bars.

Microscopic closed capsules called vesicles or liposomes can be prepared from phospholipids and many synthetic surfactants including silicone polyether copolymers [43, 63–65]. Vesicles are useful as a delivery vehicle for skin care actives. The ability of phospholipids to form closed structures in water consisting of bilayer sheets, called vesicles or liposomes, was discovered many years ago. Since then, a multibillion dollar market has emerged for skin care products containing vesicles as a delivery vehicle for skin care actives. The advantages of using a vesicle as a delivery vehicle include enhanced delivery of the active into the upper layers of the skin, controlled release and wash-fastness. Silicone surfactants with balanced silicone and hydrophilic groups also form vesicles upon dispersion into water and have been shown to encapsulate water-soluble substances and solubilize lipophilic substances. Silicone vesicles offer improved ease of processing, the ability to combine the aesthetic properties of silicones with effective delivery and chemical stability.

6.3.9.3 Emulsification

Certain comb-type silicone surfactants have been shown to stabilize emulsions in the presence of salts, alcohol and organic solvents that normally cause failure of emulsions stabilized using conventional hydrocarbon surfactants and a study by Wang et al. [66, 67] investigated the cause of this stability. Interaction forces due to silicone surfactants at an interface were measured using AFM. Steric repulsion provided by the SPE molecules persisted up to an 80% or higher ethanol level, much higher than for conventional hydrocarbon surfactants. Nonionic hydrocarbon surfactants lose their surface activity and ability to form micelles in approximately 25% ethanol while the silicone surfactants of the study continued to lower surface tension up to 80% ethanol. The behavior of the silicone surfactants reflects the oleophobic character of the polydimethylsiloxane portion of these molecules.

No systematic studies of the use of silicone surfactants as emulsifiers have yet been published. Silicone polyoxyalkylene copolymers with relatively high molecular weight and a high proportion of silicone are effective water-in-silicone oil emulsifiers and a recent study of these copolymers suggests that they stabilize emulsions by a solid-particle mechanism [68]. This type of silicone surfactant has been used to prepare transparent water-in-oil emulsions (often with an active ingredient in the internal phase) for use as deodorants or antiperspirants as well as cosmetics and other personal care products. Their use as drug delivery vehicles has also been claimed. These copolymers can also be used to prepare multiple emulsions not requiring a two-pot process.

6.3.9.4 Supercritical CO_2

Supercritical CO_2 (scCO2) is an attractive solvent for cleaning and in which to conduct chemical reactions [69]. A number of copolymers have been developed as surfactants for scCO2 including some based on polydimethylsiloxane – for example, poly(dimethylsiloxane)-b-poly(methacrylic acid). Silicone surfactants have been claimed for cleaning applications using scCO2 as well as stabilizers for dispersion polymerization in scCO2 and Fink and Beckman report the phase behavior of a group of silicone-based amphiphiles in scCO2 [70]. The copolymers exhibited upper-critical-solution-temperature (UCST) behavior, and the phase behavior was more sensitive to CO_2-phobic groups than to the size of the silicone group: no ordered liquid crystalline phases were seen. Johnston has published several papers on emulsion formation in CO_2 using silicone surfactants [71–73]. These studies show that mixtures of water and CO_2 with silicone copolymers exhibit a phase inversion temperature accompanied by a minimum in interfacial tension. This behavior can be used to form CO_2 in water emulsions using the phase inversion method.

6.3.9.5 Inks, paints and coatings

Surface active materials perform many functions in inks, paints and coatings. Silicone surfactants in inks and coatings function to defoam, deaerate, improve substrate wetting and enhance slip properties. Although polydimethylsiloxane is used to control foam in many applications, its use in water-based coatings can lead to formation of defects such as fish-eyes and orange-peel. Early use of silicone polyether copolymers also experienced difficulties with residual levels of silicone homopolymer that led to defects. However, the purity of currently available silicone surfactants for use in coatings has eliminated this problem. Two types of coating defects are caused by air entrainment: craters (caused by macrofoam) and pinholes (caused by microfoam). In water-based coatings silicone surfactants are used to address both of these problems but the mechanisms of these two problems are quite different. Macrofoam involves larger bubbles and is best dealt with by promoting rapid rupture (before the coating cures) using a defoaming agent. Relatively insoluble silicone polyethers perform this function best. Microfoam involves very small bubbles that leave pinholes at the coating surface if not eliminated and it is thought that more soluble silicone surfactants cause these small bubbles to rise faster and thereby eliminate the problem. Some silicone surfactants are effective at both functions.

Dynamic interfacial effects and the ability to control surface tension gradients are critical to successful use of surfactants in inks and coatings, particularly for those which are water based. The usefulness of silicone surfactants as wetting agents is due to their ability to lower surface and interfacial tension and thus facilitate spreading. Because silicone surfactants can reduce surface tension to lower values than can hydrocarbon surfactants, they are effective when hydrocarbon surfactants are not. Although the trisiloxane "superwetting" surfactants are the most effective in promoting wetting, polymeric silicone surfactants are also used for this purpose. As the use of water-based coatings and plastic engineering materials increases, the need to spread aqueous coatings on low-energy substrates will require the use of highly effective wetting agents such as the trisiloxane surfactants.

6.3.9.6 Foam control

Foam control process aids are the largest single category of process aids used in the chemical industry and silicone foam control agents are an important segment of this category. Silicone polyalkyleneoxide copolymers are used as foam control agents in diesel fuel defoaming, in the manufacture of plastics such as polyvinyl chloride, in polymer dispersions, in inks, paints and coatings and in some household products [74].

Diesel fuel usually has some moisture in it that affects the function of foam control additives – which must be chosen such that they are able to function in the expected range of moisture contents. The origin of the foaminess is poorly understood but silicone polyalkyleneoxide copolymers are effective defoamers as long as they are neither completely soluble in the fuel, nor absorbed and deactivated by the water. Polyoxypropylene containing copolymers appear to be the most effective.

6.3.9.7 Textiles

Silicone surfactants are used in textile manufacture to facilitate wetting and dispersion of water-insoluble substances and as spinning and sewing lubricants [32, 74]. The silicone surfactants are unique in being thermally stable lubricants with good wetting and low coefficients of friction at high speeds. During fiber production, silicone surfactants enable the lubricant to spread quickly and completely even at very low pickup amounts. The types of silicone surfactants useful for this application are tabulated and discussed by Schmidt [75].

References

1. Hill, R.M. (1999) *Silicone Surfactants.* vol. 86. Dekker, New York. 360 pp.
2. Owen, M.J. (1986) Interfacial activity of polydimethylsiloxane. In K.L. Mittal and P. Bothorel (eds), *Surfactants in Solution.* Plenum, New York, p. 1557.
3. Clarson, S.J. and Semlyen, J.A. (1993) *Siloxane Polymers.* Prentice-Hall, Englewood Cliffs, NJ, p. 673.

4. Brook, M.A. (2000) *Silicon in Organic, Organometallic, and Polymer Chemistry*. Wiley, New York.
5. Jones, R.G., Ando, W. and Chojnowski, J. (2000) *Silicon containing Polymers*. Kluwer, Dordrecht, p. 768.
6. Kunieda, H., Uddin, M.H., Horii, M., Furukawa, H. and Harashima, A. (2001) Effect of hydrophilic- and hydrophobic-chain lengths on the phase behavior of A-B-type silicone surfactants in water. *J. Phys. Chem.* B, **105**(23), 5419–26.
7. Kanner, B., Reid, W.G. and Petersen, I.H. (1967) Synthesis and properties of siloxane-polyether copolymer surfactants. *Ind. Eng. Chem. Prod. Res. Dev.*, **6**(2), 88–92.
8. Hill, R.M., He, M.T., Davis, H.T. and Scriven, L.E. (1994) Comparison of the liquid-crystal phase-behavior of 4 trisiloxane superwetter surfactants. *Langmuir*, **10**(6), 1724–34.
9. He, M., Hill, R.M., Lin, Z., Scriven, L.E. and Davis, H.T. (1993) Phase-behavior and microstructure of polyoxyethylene trisiloxane surfactants in aqueous solution. *J. Phys. Chem.*, **97**(34), 8820–34.
10. Wagner, P., Wu, Y., van Berlepsch, H. and Perepelittchenko, L. (2000) Silicon-modified surfactants and wetting: IV. Spreading behaviour of trisiloxane surfactants on energetically different solid surfaces. *Appl. Organometallic Chem.*, **14**(4), 177–88.
11. Wagner, R., Wu, Y., Berlepsch, H.V., Zastrow, H., Weiland, B. and Perepelittchenko, L. (1999) Silicon-modified surfactants and wetting: V. The spreading behaviour of trimethylsilane surfactants on energetically different solid surfaces. *Appl. Organometallic Chem.*, **13**(11), 845–55.
12. Legrow, G.E. and Petroff, L.J. (1999) Silicone polyether copolymers: synthetic methods and chemical compositions. In R.M. Hill, (edr.), *Surfactant Science Series*. Dekker, New York. pp. 49–64.
13. Gruening, B. and Koerner, G. (1989) Silicone surfactants. *Tenside, Surfactants, Detergents*, **26**, 312–17.
14. Noll, W. (1968) Chemistry and technology of silicones. Translated from the 2nd, rev., and substantially expanded German edn. by B. Hazzard and M. Landau in collaboration with Express Translation Service edn. Academic, New York. p. 702.
15. Plumb, J.B. and Atherton, J.H. (1973). In D.C. Allport and W.H. Janes (eds), Block Copolymers. Applied Science, London, p. 305.
16. Snow, S.A. (1993) Synthesis, characterization, stability, aqueous surface-activity, and aqueous-solution aggregation of the novel, cationic siloxane surfactants (Me3sio)2si(Me)-(Ch2)3+Nme2(Ch2)2or X- (R=H, C(O)Me, C(O)Nh(Ph) X=Cl, Br, I, No3, Meoso3). *Langmuir*, **9**(2), 424–30.
17. Morehouse, E.L. (1972) Siloxane amino hydroxy sulfonates. U.S. 3660452. (Union Carbide Corporation: US. p. 5).
18. Snow, S.A., Fenton, W.N. and Owen, M.J. (1991) Zwitterionic organofunctional siloxanes as aqueous surfactants – synthesis and characterization of betaine functional siloxanes and their comparison to sulfobetaine functional siloxanes. *Langmuir*, 7(5), 868–71.
19. Snow, S.A., Fenton, W.N. and Owen, M.J. (1990) Synthesis and characterization of zwitterionic silicone sulfobetaine surfactants. *Langmuir*, **6**(2), 385–91.
20. Maki, H., Horiguchi, Y., Suga, T. and Komori, S. (1970) Syntheses and properties of organometallic surfactants: VII. Cationic surfactants containing dimethylpolysiloxane. *Yukagaku*, **19**, 1029–33.
21. Wagner, R., Richter, L., Weissmuller, J., Reiners, J., Klein, K.D., Schaefer, D. and Stadtmuller, S. (1997) Silicon-modified carbohydrate surfactants: 4. The impact of substructures on the wetting behaviour of siloxanyl-modified carbohydrate surfactants on low-energy surfaces. *Appl. Organometallic Chem.*, **11**(7), 617–32.
22. Klein, K.-D., Wilkowski, S. and Selby, J. (1996) Silane surfactants – novel adjuvants for agricultural applications. *FRI Bulletin*, **193**, 27–31.
23. Wagner, R. and Strey, R. (1999) Phase behavior of binary water-trimethylsilane surfactant systems: origin of the dilute lamellar phase. *Langmuir*, **15**(4), 902–05.
24. Gradzielski, M., Hoffmann, H., Robisch, P., Ulbricht, W. and Gruening, B. (1990) The aggregation

behavior of silicone surfactants in aqueous solutions. *Tenside, Surfactants, Detergents*, **27**, 366–79.
25. Stuermer, A., Thunig, C., Hoffmann, H. and Gruening, B. (1994) Phase behavior of silicone surfactants with a comblike structure in aqueous solution. *Tenside, Surfactants, Detergents*, **31**, 90–8.
26. Hill, R.M. (1988) Unpublished measurements of hydrolysis of trisiloxane superwetter surfactants in dilute aqueous solutions. (Experiments were carried out in 1998.).
27. Hoffmann, H. and Ulbricht, W. (1999) Surface activity and aggregation behavior of siloxane surfactants. *Surfactant Sci. Ser.*, **86**, 97–136.
28. Vick, S.C. (1984) Structure property relationships for silicone polyalkyleneoxide copolymers and their effects on performance in cosmetics. *Soap Cosmet. Chem. Specialties*, **60**(5), 36-&.
29. Gentle, T.E. and Snow, S.A. (1995) Adsorption of small silicone polyether surfactants at the air/water interface. *Langmuir*, **11**(8), 2905–10.
30. Rosen, M.J. (1989) *Surfactants and Interfacial Phenomena*, 2nd edn. Wiley, New York, p. 431.
31. Fink, H.F. (1991) *Tenside* Surfactants Determinants **28**, 306.
32. Sabia, A.J. (1982) American Dyestuff Reporter, May, p. 45.
33. Callaghan, I.C. (1993) Antifoams for nonaqueous systems in the oil industry. *Surfactant Sci. Ser.*, **45**, 119–50.
34. Stevens, P.J.G. (1993) *Pestic. Sci.*, **38**, 103.
35. Stoebe, T., Lin, Z.X., Hill, R.M., Ward, M.D. and Davis, H.T. (1997) Enhanced spreading of aqueous films containing ethoxylated alcohol surfactants on solid substrates. *Langmuir*, **13**(26), 7270–75.
36. Hill, R.M. (1998) Superspreading. *Curr. Opin. Colloid Interface Sci.*, 3(3), 247–54.
37. Svitova, T., Hill, R.M., Smirnova, Y., Stuermer, A. and Yakubov, G. (1998) Wetting and interfacial transitions in dilute solutions of trisiloxane surfactants. *Langmuir*, **14**(18), 5023–31.
38. Svitova, T.F., Hill, R.M. and Radke, C.J. (2001) Spreading of aqueous trisiloxane surfactant solutions over liquid hydrophobic substrates. *Langmuir*, **17**(2), 335–48.
39. Policello, G. and Stevens, P. (1998) Nonionic siloxane blends with surfactants, as adjuvants in herbicide formulations, Eur. Pat. Appl. 862857. (OSI Specialties, Inc., USA; Crompton Corporation).
40. He, M., Lin, Z., Scriven, L.E., Davis, H.T. and Snow, S.A. (1994) Aggregation behavior and microstructure of cationic trisiloxane surfactants in aqueous solutions. *J. Phys. Chem.*, **98**(24), 6148–57.
41. O'Lenick, A.J., Jr. and Parkinson, J.K. (1997) Applying the three-dimensional HLB system. *Cosmet. Toiletries*, **112**, 59–60, 65.
42. Hoffmann, H. and Stuermer, A. (1993) Solubilization of siloxanes and weakly polar organic additives into rodlike micelles. *Tenside, Surfactants, Detergents*, **30**, 335–41.
43. Hill, R.M., He, M.T., Lin, Z., Davis, H.T. and Scriven, L.E. (1993) Lyotropic liquid-crystal phase-behavior of polymeric siloxane surfactants. *Langmuir*, **9**(11), 2789–98.
44. Jakobs, B., Sottmann, T., Strey, R., Allgaier, J., Willner, L. and Richter, D. (1999) Amphiphilic block copolymers as efficiency boosters for microemulsions. *Langmuir*, **15**(20), 6707–11.
45. Kumar, A., Uddin, H., Kunieda, H., Furukawa, H. and Harashima, A. (2001) Solubilization enhancing effect of A-B-type silicone surfactants in microemulsions. *J. Dispersion Sci. Technol.*, 22(2–3), 245–53.
46. Rodriguez, C., Uddin, M.H., Watanabe, K., Furukawa, H., Harashima, A. and Kunieda, H. (2002) Self-organization, phase behavior, and microstructure of poly(oxyethylene) poly(dimethylsiloxane) surfactants in nonpolar oil. *J. Phys. Chem.* B, **106**(1), 22–9.
47. Keil, J.W. (1981) Antiperspirant emulsion compositions. U.S. 4268499. Dow Corning Corp., U.S.A.
48. Gee, R.P. and Keil, J.W. (1978) Emulsion compositions comprising a siloxane-oxyalkylene copolymer and an organic surfactant, U.S. 4122029. (Dow Corning Corp., USA). Application: US, US. p. 7 pp.

49. Gee, R.P. (1986) Method of preparing silicone emulsions having small particle size. U.S. 4620878. (Dow Corning Corp.: U.S.A).
50. Gum, M.L. (1985) Water-in-volatile silicone emulsifier concentrates for mixing with water to form water-in-volatile silicone emulsions that are useful in personal-care formulations and methods of making same. PCT Int. Appl. 8503641. (Union Carbide Corp., USA: Application: WO). WO. p. 56 pp.
51. Terae, N., Nakazato, M. and Hara, Y. (1992) Manufacture of siloxane microemulsions by polymerization with shearing. Jpn. Kokai Tokkyo Koho 04103632. (Shin-Etsu Chemical Industry Co., Ltd.), Japan: Application: JP, JP. p. 12 pp.
52. Li, X., Washenberger, R.M., Scriven, L.E., Davis, H.T. and Hill, R.M. (1999) Phase behavior and microstructure of water/trisiloxane E-6 and E-10 polyoxyethylene surfactant/silicone oil systems. *Langmuir*, **15**(7), 2278–89.
53. Li, X., Washenberger, R.M., Scriven, L.E., Davis, H.T. and Hill, R.M. (1999) Phase behavior and microstructure of water/trisiloxane E-12 polyoxyethylene surfactant/silicone oil systems. *Langmuir*, **15**(7), 2267–77,
54. Mayer, H. (1994) Silicone microemulsions as aqueous primers and impregnating agents for wall coatings. (Part 1). *Surf. Coatings Int.*, 77, 162–8.
55. Juergensen, P. (1994) Aqueous silicone microemulsions as sealing compositions for impregnation, Ger. Offen. 4230499. (Germany). Application: DE, DE. p. 2 pp.
56. Reed, D. (1995) *Urethanes Technology*. January/February. p. 22.
57. Hostettler, F. (1960) German Patent 1091324.
58. Oertel, G. (ed.) (1985) *Polyurethane Handbook*. Carl Hanser Verlag, Munich.
59. Snow, S.A. and Stevens, R.E. (1999) The science of silicone surfactant application in the formation of polyurethane foam. *Surfactant Sci. Ser.*, **86**, 137–58.
60. Naire, S., Braun, R.J. and Snow, S.A. (2001) An insoluble surfactant model for a vertical draining free film with variable surface viscosity. *Phys. Fluids*, **13**(9), 2492–502.
61. Braun, R.J., Snow, S.A. and Naire, S. (2002) Models for gravitationally-driven free-film drainage. *J. Eng. Math.*, **43**(2–4), 281–314.
62. Floyd, D.T. (1999) Silicone surfactants: applications in the personal care industry. *Surfactant Sci. Ser.*, **86**, 181–207.
63. Hill, R.M. and Snow, S.A. (1995) Silicone vesicles and entrapment. U.S. 5411744. (Dow Corning Corp.: U.S).
64. Ekeland, A.R. and Hill, R.M. (1995) Siloxane MQ resin vesicles for entrapment of water-soluble substances. U.S. 5958448. (Dow Corning Corporation, U.S.A).
65. Hill, R.M. and Snow, S.A. (1994) Silicone vesicles and entrapment. U.S. 5364633. (Dow Corning Corp.: U.S.), p. 8.
66. Wang, A.F., Jiang, L.P., Mao, G.Z. and Liu, Y.H. (2002) Direct force measurement of silicone- and hydrocarbon-based ABA triblock surfactants in alcoholic media by atomic force microscopy. *J. Colloid Interface Sci.*, **256**(2), 331–40.
67. Wang, A.F., Jiang, L.P., Mao, G.Z. and Liu, Y.H. (2001) Direct force measurement of comb silicone surfactants in alcoholic media by atomic force microscopy. *J. Colloid Interface Sci.*, **242**(2), 337–45.
68. Anseth, J.W., Bialek, A., Hill, R.M. and Fuller, G.G. (2003) Interfacial rheology of graft-type polymeric siloxane surfactants. *Langmuir*, **19**(16), 6349–56.
69. Johnston, K.P. (2000) Block copolymers as stabilizers in supercritical fluids. *Curr. Opin. Colloid Interface Sci.*, **5**(5–6), 351–6.
70. Fink, R. and Beckman, E.J. (2000) Phase behavior of siloxane-based amphiphiles in supercritical carbon dioxide. *J. Supercrit. Fluids*, **18**(2), 101–10.
71. Psathas, P.A., Janowiak, M.L., Garcia-Rubio, L.H. and Johnston, K.P. (2002) Formation of carbon dioxide in water miniemulsions using the phase inversion temperature method. *Langmuir*, **18**(8), 3039–46.

72. da Rocha, S.R.P., Psathas, P.A., Klein, E. and Johnston, K.P. (2001) Concentrated CO_2-in-water emulsions with nonionic polymeric surfactants. *J. Colloid Interface Sci.*, **239**(1), 241–53.
73. Psathas, P.A., da Rocha, S.R.P., Lee, C.T., Johnston, K.P., Lim, K.T. and Webber, S. (2000) Water-in-carbon dioxide emulsions with poly(dimethylsiloxane)-based block copolymer ionomers. *Ind. Eng. Chem. Res.*, **39**(8), 2655–64.
74. Hill, R.M. and Fey, K.C. (1999) Silicone polymers for foam control and demulsification. In R.M. Hill, (Edr.), *Silicone Surfactants.* Dekker, New York, pp. 159–80.
75. Schmidt, G. (1990) Silicone surfactants. *Tenside Surfactants Determinants,* **27**, 324.

6.4 Polymerizable Surfactants

Guido Bognolo

6.4.1 Introduction

For a number of years, economic and safety considerations have driven the substitution of solvent-based formulations with aqueous systems in all industrial sectors, and the process has recently received a further impetus from the regulatory activity of the European Union to limit the release of volatile organic compounds in the environment.

In some areas, for example in crop protection, moving away from solvents simply required the development of new forms of delivery, e.g. concentrated emulsions or suspo-emulsions progressively displaced the emulsifiable concentrates. A change in the surfactant system was obviously necessary, but the new formulations could be made using commercially available products. In other instances, for example in the field of emulsion polymers, new colloidal species were developed and are still the subject of extensive research: the reactive surfactants.

Reactive surfactants can covalently bind to the dispersed phase and as such have a distinct advantage over conventional surfactants that are only physically adsorbed and can be displaced from the interface by shear or phase changes with the subsequent loss of emulsion stability. Furthermore, if the substrate is coalesced to produce decorative or protective films, the desorption can result in, e.g. reduced adhesion, increased water sensitivity and modification of the hardness, barrier and optical properties of the film.

Reactive surfactants have also economic and environmental advantages. The binding to the dispersed phase makes these surfactants an integral part of the finished product and enhances the yield in active matter on a weight basis. It furthermore prevents the release of surfactants in the water effluents on production and application and, as such, reduces the environmental impact of intermediate products and commercial formulations.

6.4.2 Reactive surfactants

Depending upon the chemical structure and effects, there are different types of reactive surfactants:

- Functionalized monomers
- Surface active initiators (Inisurfs)
- Surface active transfer agents (Transurfs)
- Polymerizable surfactants (Surfmers)

Inisurfs, Transurfs and Surfmers may be used to reduce/avoid the use of conventional surfactants in emulsion polymerization. However, when Inisurfs and Transurfs are used, the stability of the system cannot be adjusted without affecting either the polymerization rate (Inisurfs) or the molecular weight distribution (Transurfs). Furthermore, the efficiency rate of Inisurfs is low due to the cage effect. It is therefore not obvious yet that these classes will become commercially significant.

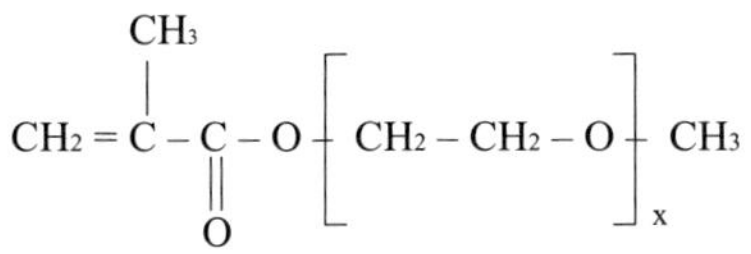

$$CH_2=C(CH_3)-C(=O)-O-[CH_2-CH_2-O]_x-CH_3$$

Figure 6.22 Functionalized monomer.

6.4.2.1 *Functionalized monomers*

Functionalized monomers are sometimes regarded as polymerizable surfactants. Vinyl or allyl monomers are reacted with ethylene oxide (EO), propylene oxide (PO) or butylene oxide (BO) in a sequential or random addition mode. The terminal hydroxyl group can be optionally reacted with methyl or benzyl chloride to produce Williamson ethers (if the hydroxyl group has to be deactivated) or are further sulfated to deliver electrosteric stabilization.

Functionalized monomers are commercialized by e.g. Clariant and BASF. They can be copolymerized with other ethylenically unsaturated monomers for permanent polymer modification (see also Section 6.4.2.4). An important application is the production of derivatized silicone polyols (see Section 6.4.2.1.1).

Table 6.10 gives examples of functionalized monomers and their applications. Examples of chemical structures are given in Figures 6.22–6.27.
The drawback of allylic, acrylic and vinylic polymerizable groups is their tendency to homopolymerize. Allylic derivatives, furthermore, are susceptible to degradative chain transfer.

Silicone surfactants. For the purpose of this work, silicone surfactants will be defined as silicone polyether copolymers with hydrolytically stable silicon–carbon bonds. The manufacturing of these products involves a three-step process:

- Preparation of a silicone hydride intermediate
- Preparation of an allyloxy polyether intermediate
- Hydrosylilation of the silicon hydride with the allyloxy polyether to produce the modified copolymer

Some details of the preparation of allyloxy polyethers are now described. This involves the base-catalyzed ring opening of one or more of the oxiranes ethylene oxide, propylene oxide and butylene oxide in a three-step process as described by Whitmarsh [1].

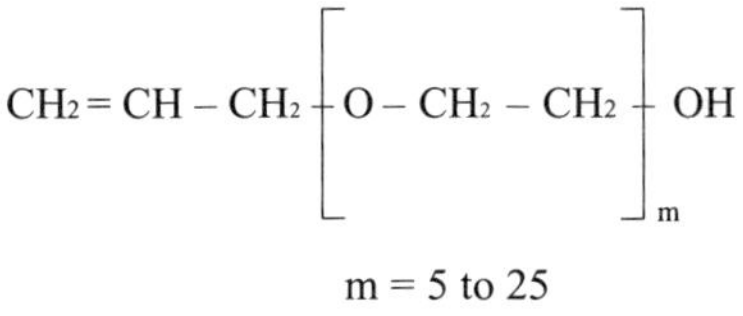

$$CH_2=CH-CH_2-[O-CH_2-CH_2]_m-OH$$

$$m = 5 \text{ to } 25$$

Figure 6.23 Functionalized monomer.

Table 6.10 Examples of functionalized monomers

Chemical composition	Nature	Feature	Comments
Allyl polyalkylene glycol ethers	Nonionic	Copolymerizable emulsifiers for the emulsion polymerization of vinyl acetate, acrylates, styrene/acrylates Addition during emulsion polymerization Improve latex stability and reduce grit levels Reduce water uptake of polymer films	Level of use 1–2% active materials based on monomers
Vinyl polyalkylene glycol ethers	Nonionic	Copolymerizable emulsifiers for the emulsion polymerization of vinyl acetate, acrylates, styrene/acrylates Improve latex stability and reduce grit levels Reduce water uptake of polymer films	Level of use 1–2% active materials based on monomers
Allyl polyalkylene glycol ether sulfate, ammonium salt	Anionic	Copolymerizable emulsifiers for the emulsion polymerization of vinyl acetate, acrylates, styrene/acrylates Anionic monomer with surface activity Can be used without additional emulsifier Improve latex stability and reduce grit levels Reduce water uptake of polymer films	Level of use 1–3% active materials based on monomers
Methacrylic acid esters of alky polyethylene glycol ethers	Nonionic	Hydrophilic monomers for emulsion, inverse emulsion and solution polymerization	Alkyl can be methyl or lauryl
Allyl polyethylene glycol ethers	Nonionic	Used to produce surface active, water-soluble silicone ethers (antifoams)	
Vinyl polyethylene glycol ethers	Nonionic	Used in the copolymerization with acrylates to produce water-soluble polymers	

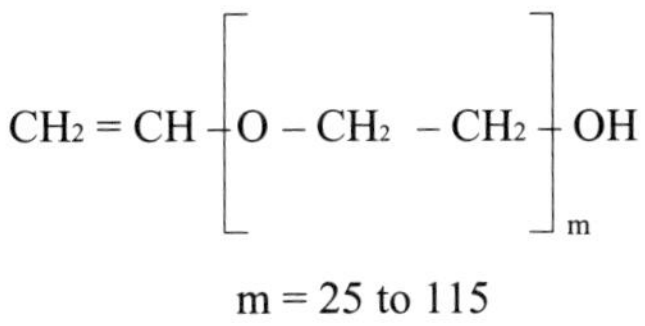

$$CH_2 = CH - \left[O - CH_2 - CH_2 \right]_m - OH$$

m = 25 to 115

Figure 6.24 Functionalized monomer.

$$CH_2 = CH - CH_2 - \left[O - CH_2 - \underset{\displaystyle R}{\underset{|}{CH}} \right]_m \left[O - CH_2 - CH_2 \right]_n OH$$

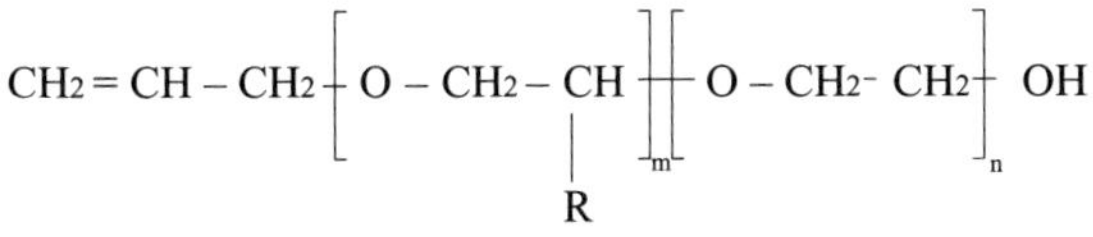

Figure 6.25 Functionalized monomer.

$$CH_2 = CH - O - CH_2 - CH_2 - CH_2 - CH_2 - O - \left[CH_2 - \underset{\displaystyle CH_3}{\underset{|}{CH}} - O \right]_n \left[CH_2 - CH_2 - O \right]_m H$$

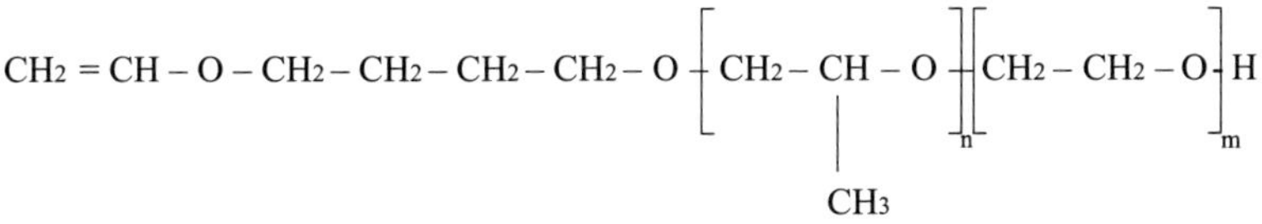

Figure 6.26 Functionalized monomer.

$$CH_2 = CH - CH_2 - \left[O - CH_2 - \underset{\displaystyle R}{\underset{|}{CH}} \right]_m \left[O - CH_2 - CH_2 \right]_n OSO_3^-$$

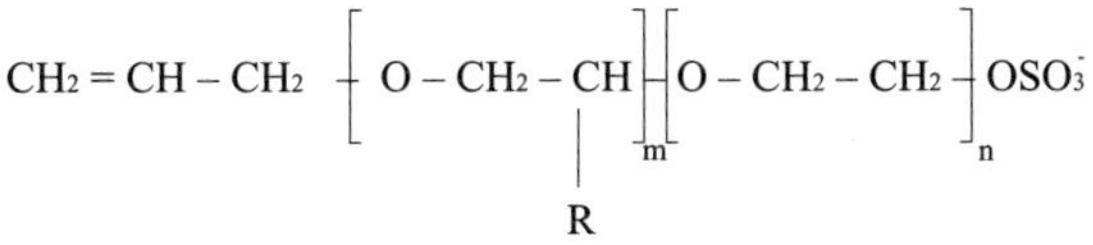

Figure 6.27 Functionalized monomer.

The first step is the formation of an alkoxide anion by the initiating alcohol (allyl alcohol is the initiator most commonly used, although other initiators have been suggested). The appropriate oxide(s) is (are) then added to the alcohol initiator. This causes the opening of the oxirane ring in the oxide and propagates the chain growth of the alkylene oxide on the initiator. The last step is the neutralization of the alkoxide anion to terminate the polymerization.

Derivatives from pure ethylene oxide are water soluble and result in silicon polyols with water solubility or dispersibility. Propylene oxide and even more butylene oxide allow for more compatibility with organic media, e.g. butylene oxide gives compatibility with organic oils. Depending upon the demanded balance of hydrophilicity/hydrophobicity, different proportions and order of addition of the alkoxide can be used. In certain instances blocking of the terminal hydroxyl group may be required, e.g. by reaction with methyl or, less commonly, benzyl chloride.

The preparation of allyl derived alkoxylates is complicated by the toxicity and irritancy of the allyl alcohol initiator which requires particular care for storage and handling. The alkali metal salt formed at the end of the process must be removed (a task often made difficult by the crystalline structure of the precipitate) to avoid causing catalyst poisoning in the next hydrosylilation reactions. Water contamination from the reagents present in the autoclave or produced during the formation of the alcolate initiator must be eliminated to avoid the formation of polymeric diols which cannot be hydrosylilated. They are also practically impossible to separate and will carry through as components in the final product.

Last (but not least) propylene oxide undergoes rearrangement in the presence of base and forms allyl alcohol, thus forming in situ initiators during the alkoxylation process. Unless properly accounted for, this decreases the molecular weight of the polyether produced.
The hydrosylilation of the allyloxy polyether intermediate by a siloxane hydride is catalyzed by e.g. chloroplatinic acid and is exemplified in eqn (6.4.1):

$$(Me_3SiO)_2MeSiH + H_2C{=}CHCH_2O(EO)_xH \rightarrow (Me_3SiO)_2MeSi(CH_2)_3O(EO)_xH \quad (6.4.1)$$

where EO stands for ethylene oxide.
Several reports on hydrosylilation and different forms of catalysis have been produced, and the work of Marciniec and Gulinski [2] provides further references. Schmaucks [3] describes a range of novel siloxane–polyether surfactants produced via the above described method.

6.4.2.2 Surface active initiators (Inisurfs)

Surface active initiators or Inisurfs have the advantage of reducing the number of ingredients in an emulsion polymerization recipe to water, monomer and initiator, at least in the initial stages of the process. However, the surface active properties of the Inisurfs may be reduced on formation of the radicals and additional surfactant must be added to stabilize the latex if high solid levels are wanted.
Inisurf molecules contain three moieties:

- The radical generating moiety, which can be azo or peroxy
- A hydrophobic moiety which is usually a hydrocarbon (alkyl or alkyl phenyl), sometimes extended by the inclusion of propylene oxide
- A hydrophilic moiety which can be anionic, cationic or nonionic

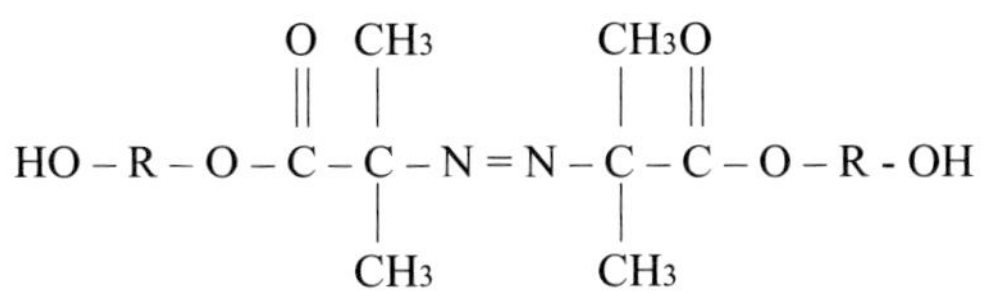

$$HO-R-O-C(=O)-C(CH_3)_2-N=N-C(CH_3)_2-C(=O)-O-R-OH$$

Figure 6.28 Inisurf structure.

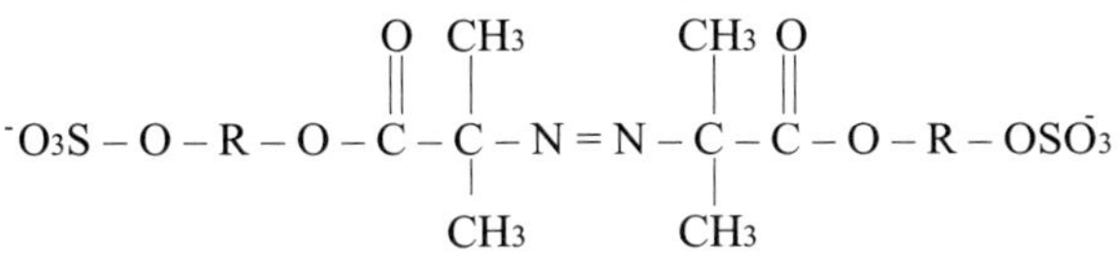

$$^{-}O_3S-O-R-O-C(=O)-C(CH_3)_2-N=N-C(CH_3)_2-C(=O)-O-R-OSO_3^{-}$$

Figure 6.29 Inisurf structure.

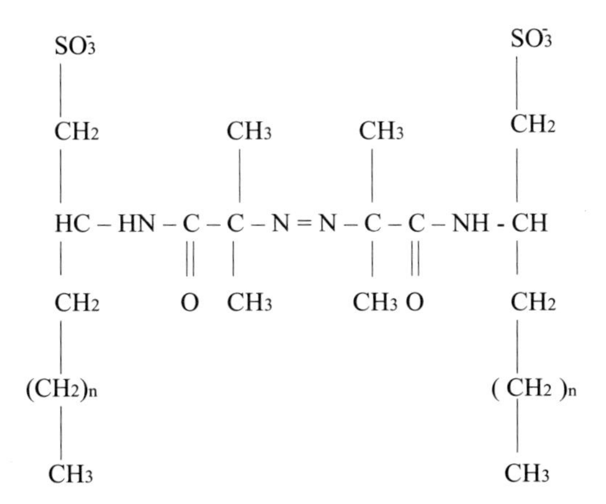

Figure 6.30 Inisurf structure.

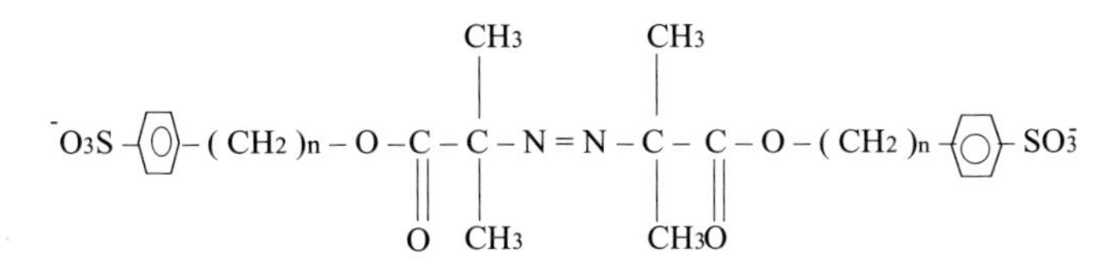

Figure 6.31 Inisurf structure.

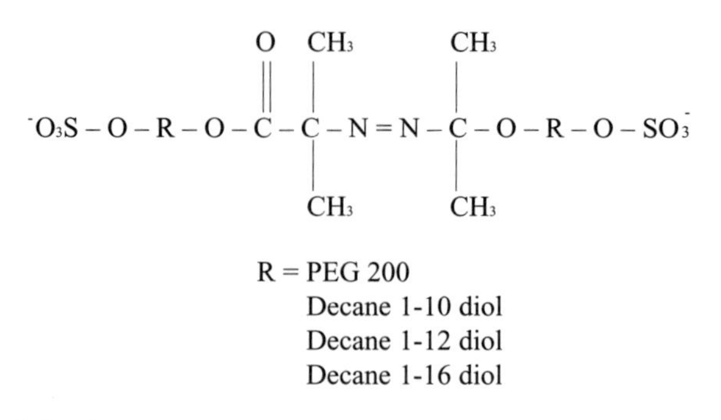

Figure 6.32 Inisurf structure.

The molecules can be symmetrical, i.e. the structural moieties are the same on both sides of the radical-generating group and two surface active radicals are produced on decomposition. If the structure is asymmetrical only one surface active radical is produced on decomposition.

The key feature of Inisurfs is their surfactant behavior. They form micelles and are adsorbed at interfaces, and as such they are characterized by a critical micelle concentration (CMC) and an area/molecule in the adsorbed state. This influences both the decomposition behavior and the radical efficiency, which are much lower than those for conventional, low molecular weight initiators. Tauer and Kosmella [4] have observed that in the emulsion polymerization of styrene, using an Inisurf concentration above the CMC resulted in an increase in the rate constant of the production of free radicals. This was attributed to micellar catalysis effects as described, for example, by Rieger [5]. Conversely, if the Inisurf concentration was below the CMC the rate constant of the production of free radicals decreased with an increase in the Inisurf concentration, which was attributed to enhanced radical recombination. Also note that a similar effect of the dependence of initiator efficiency on concentration was reported by Van Hook and Tobolsky for azobisisobutyronitrile (AIBN) [6].

The Inisurfs originally synthesized were susceptible to hydrolysis and required a multistep synthesis which both added to the manufacturing costs and affected the purity of the products. Products developed more recently have successfully addressed both issues. In

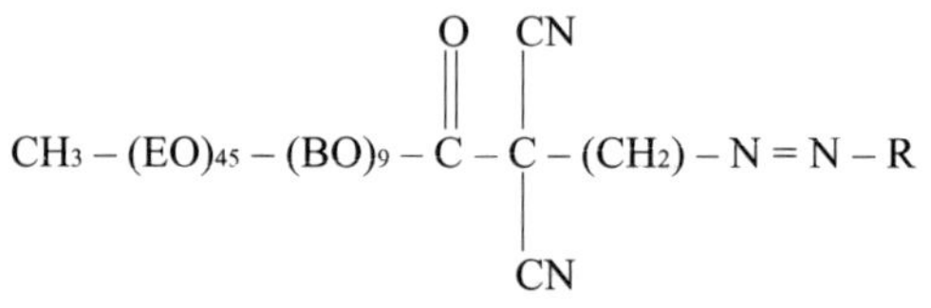

$$CH_3-(EO)_{45}-(BO)_9-\overset{\overset{\displaystyle O}{\|}}{C}-\underset{\underset{\displaystyle CN}{|}}{\overset{\overset{\displaystyle CN}{|}}{C}}-(CH_2)-N=N-R$$

EO = ethylene oxide
BO = butylene oxide
R = t – butyl

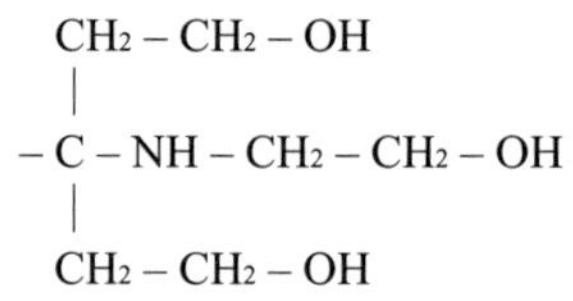

$$-\underset{\underset{\displaystyle CH_2-CH_2-OH}{|}}{\overset{\overset{\displaystyle CH_2-CH_2-OH}{|}}{C}}-NH-CH_2-CH_2-OH$$

Figure 6.33 Inisurf structure.

particular the structure in Figure 6.31 goes first through the synthesis of the bis (phenyl alkyl)-2, 2́-azobisisobutyrate (Pinner reaction) followed by sulfonation of the phenyl ring. The structure in Figure 6.30 can be prepared in a one-step synthesis via a modified Ritter reaction [7].

6.4.2.3 Surface active transfer agents (Transurfs)

Polymerizable surfactants capable of working as transfer agents include thiosulfonates, thioalkoxylates and methyl methacrylate dimer/trimer surfactants. Thioalkoxylates with 17–90 ethylene oxide units were produced from ethoxylated 11 bromo-undecanol by replacing the bromine with a thiol group via the thiazonium salt route [8]. In the presence of water-soluble azo initiator the thio ended Transurfs (used at a concentration above the CMC) gave monodispersed latex particles in emulsion polymerization of styrene. However, the incorporation of the Transurf remained low, irrespective of the process used for the polymerization (batch, semibatch, seeded). The stability of the lattices when the surfactant and the transfer function were incorporated in the same molecule was better than when they were decoupled.

When the same thioalkoxylates were combined with t-butyl hydroxyperoxide initiator the maximum incorporation yield of the Transurf was around 40%. Monomodal or multimodal molecular weight distributions were observed, depending upon the structure of the Transurf, the conversion of the monomer and the process used for feeding the reactor [9].

In 1998, Chiefari et al. [10] attempted to combine the convenience of radical polymerization with the many benefits of living polymerization, e.g. control of the molecular weight and polydispersity and the possibility of synthesizing block copolymers of complex architecture. They used free-radical polymerization reagents of formula (I) to produce a sequence of reversible addition-fragmentation in which the transfer of the S=C (Z) S moiety between active and dormant chains serves to maintain the living character of the polymerization:

Such a mechanism of polymerization was named RAFT (reversible addition-fragmentation chain transfer).

$$\underset{\displaystyle \text{S}}{\overset{\displaystyle \text{Z-C-S-R}}{\|}} \qquad \text{(I)}$$

where Z is phenyl or methyl and R is alkyl phenyl or cyano alkyl, ciano carboxy alkyl or ciano hydroxyalkyl (see Figure 6.31).

The living nature of the RAFT process is confirmed by:

- The narrow polydispersity of the polymers produced
- The linear profile of molecular weight versus conversion
- The predictability of molecular weight from the ratio of monomer consumed to transfer agent
- The ability to produce blocks of higher molecular weight polymers by faster monomer addition

The effectiveness of the reagents of formula (I) in providing a living character is attributed to their very high transfer constants which ensure a rapid rate of exchange between the

$$HS - C_{10} - H_{20} - SO_3^-$$

Figure 6.34 Transurf structure.

dormant and living chains. As a matter of fact, with an appropriate choice of Z and R the transfer constants are too high to be measured with conventional methods. However kinetic modeling experiments suggest that the transfer constant must be higher than 100 in order to obtain polymers with a polydispersity of 1.1 at low conversion. Suitable Z groups are aryl and alkyl. The R groups should be good free-radical leaving groups, and as an expelled radical, R should be effective in reinitiating free-radical polymerization.

A major advantage of RAFT is that it is compatible with a wide range of monomers, including functional monomers containing, for example, acids (e.g. acrylic acid), acid salts (e.g. sodium salt of styrene sulfonic acid), hydroxyl (e.g. hydroxyethyl methacrylate) or tertiary amino (e.g. dimethylaminoethyl methacrylate). It can be used over a broad range of reaction conditions and provides in each case controlled molecular weight polymers with very narrow polydispersion.

Uzulina et al. have found that polymerization of styrene in bulk and emulsion can be better controlled by generating in situ a chain transfer agent produced by using a large excess of azo initiator to the S-thiobenzoyl-thioglycolic acid. The resulting amide is not isolated but added directly to the other components of the polymerization recipe [11].

Monteiro et al. have used a RATF Transurf in the "ab initio" emulsion polymerization of methyl methacrylate at 70°C. The Transurf was synthesized by esterifying a methyl methacrylate dimer with 1, 10 decandiol followed by sulfonation. The authors found that only a small amount of Transurf was incorporated and suggested that, in order to increase the Transurf incorporation, the ratio of monomer to Transurf should be kept as low as possible, as achieved, e.g. in starved-feed conditions [12].

6.4.2.4 *Polymerizable surfactants (Surfmers)*

Polymerizable surfactants may be considered as surface-active monomers and essentially consist of:

- A hydrophilic moiety
- A hydrophobic moiety
- A polymerizable group

In common with conventional surfactants, Inisurfs and Transurfs, Surfmers form micelles in aqueous solutions above the CMC. The organized monomer aggregates of colloidal dimension are microscopically heterogeneous and may affect polymerization kinetics and polymer structure and properties.

$$HS - C_{11} - H_{22} - O -(CH_2 - CH_2 - O)_n - H$$

$$n = 17 \text{ to } 90$$

Figure 6.35 Transurf structure.

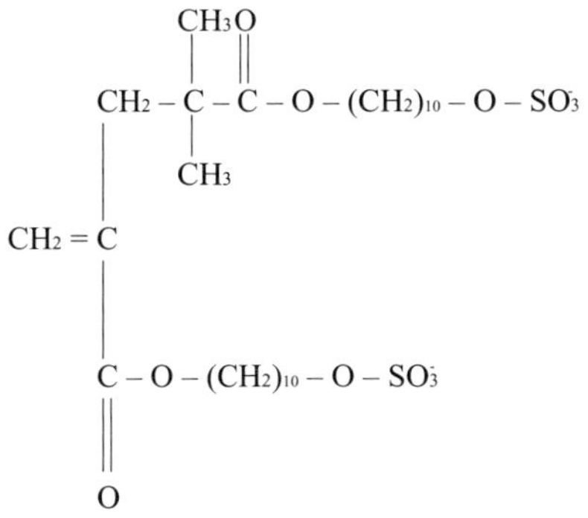

CH_3 O

$CH_2 - C - C - O - (CH_2)_{10} - O - SO_3^-$

CH_3

$CH_2 = C$

$C - O - (CH_2)_{10} - O - SO_3^-$

O

Figure 6.36 Transurf structure

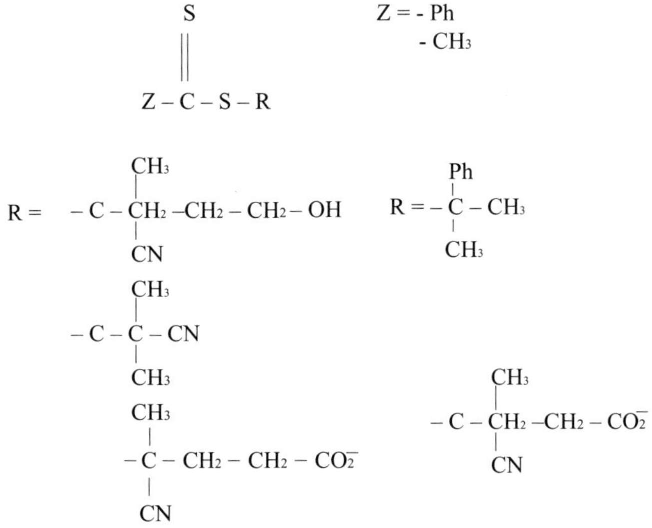

S

$Z - C - S - R$

Z = - Ph
- CH_3

R = $- C(CH_3)(CN) - CH_2 - CH_2 - CH_2 - OH$

R = $- C(Ph)(CH_3) - CH_3$

$- C - C(CH_3)(CN) - CH_3$

$- C(CH_3)(CN) - CH_2 - CH_2 - CO_2^-$

$- C - C(CH_3)(CN)H_2 - CH_2 - CO_2^-$

Figure 6.37 Transurf structure.

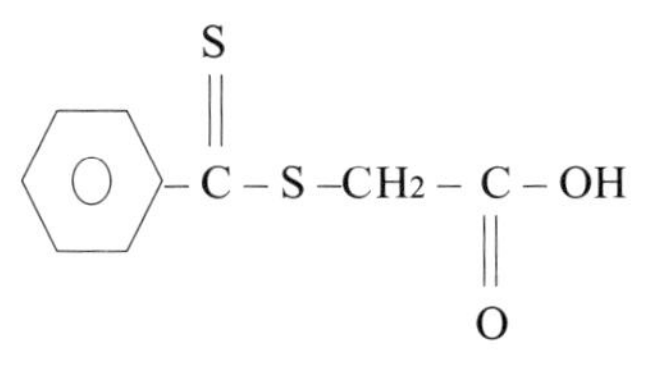

Figure 6.38 Transurf structure.

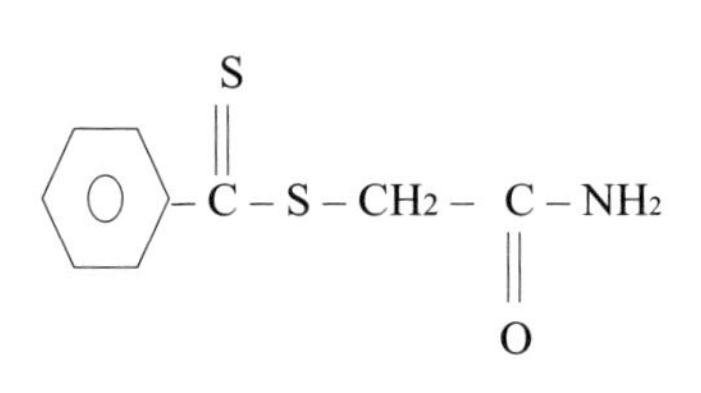

Figure 6.39 Transurf structure.

Advantages of polymerizable surfactants in emulsion polymerization processes include latex stabilization and resistance to electrolyte addition and to freeze-thaw cycles. In film forming polymers the most interesting property is, however, the superior water resistance achievable compared to conventional surfactants. This manifests in an increase in the hydrophobicity of the films because the covalent bonding of the Surfmer to the particles reduces migration to the surface. Water uptake is significantly reduced (weight gains after immersion in water are generally one-third lower than with conventional surfactants). Dimensional stability and mechanical properties (e.g. resistance to elongation) are consequently significantly improved.

By contrast, conventional, nonreactive surfactants, apart from water uptake and the resulting drawbacks, may cause a permanent reduction in surface hardness, poor blocking resistance, inferior sandability and dirt pick-up. These effects were already observed by Vanderhoff in the early 1950s and were confirmed more recently by Hellgren et al. using the atomic force microscopy (AFM) technology [13].

One important requirement in replacing a conventional, nonreactive surfactant with a reactive one is that neither the molecular weight nor the particle size distribution of the latex may significantly change. Also, the Surfmer reactivity is important: if the Surfmer is too reactive compared to the other monomers in the recipe, it will become partially buried inside the growing polymer particles. This will cause poor stability during polymerization and broadening of the particle size distribution.

Most of the reactive surfactants used for emulsion polymerization have the reactive group at the end of the hydrophobic moiety of the molecule, on the assumption that the polymerization process takes place in the latex particle. Work of Ferguson et al. [14] shows indeed a lower stability of lattices produced with Surfmers with an acrylate group attached to the end of the hydrophilic chain than those produced with the equivalent terminated with an ethyl ester group.

Ionic Surfmers. Ionic Surfmers were extensively considered in the early developments of polymerizable surfactants. Examples of products with anionic, cationic and amphoteric moieties are given in Figures 6.40–6.48.

The work of Guyot [7] reviews the use and effects of ionic Surfmers in different polymerization processes.

CH_3
$CH_2=C$ O
$C-O-CH_2-CH-CH_2-O-P-O-C_nH_{2n+1}$
O   OH   O

$n = 8, 12, 14, 16, 18$

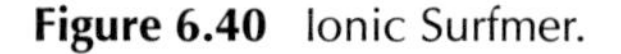
Figure 6.40 Ionic Surfmer.

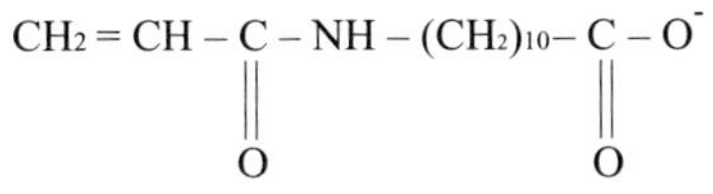

$$CH_2=CH-\underset{\underset{O}{\|}}{C}-NH-(CH_2)_{10}-\underset{\underset{O}{\|}}{C}-O^-$$

Figure 6.41 Ionic Surfmer.

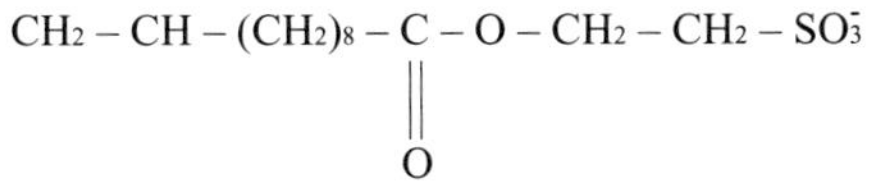

$$CH_2-CH-(CH_2)_8-\underset{\underset{O}{\|}}{C}-O-CH_2-CH_2-SO_3^-$$

Figure 6.42 Ionic Surfmer.

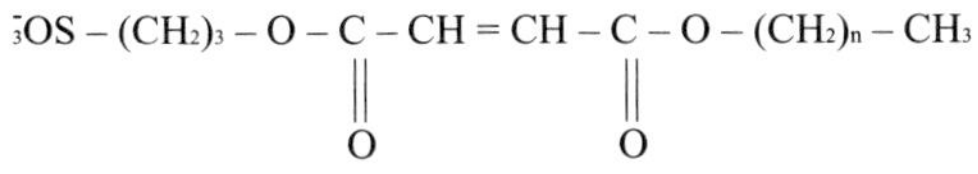

$$^-_3OS-(CH_2)_3-O-\underset{\underset{O}{\|}}{C}-CH=CH-\underset{\underset{O}{\|}}{C}-O-(CH_2)_n-CH_3$$

Figure 6.43 Ionic Surfmer.

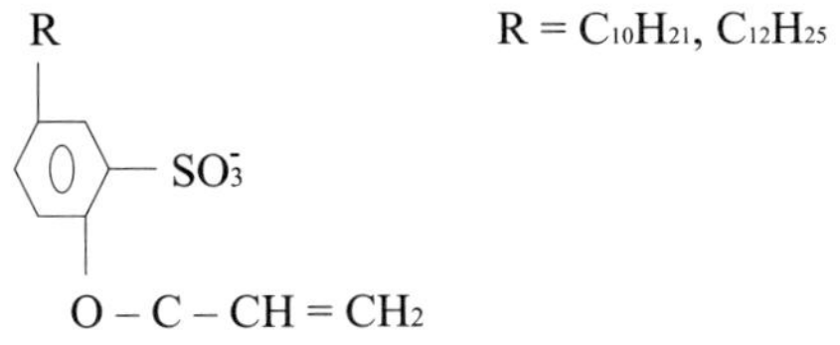

R

SO_3^-

$O-C-CH=CH_2$

$R = C_{10}H_{21}, C_{12}H_{25}$

Figure 6.44 Ionic Surfmer.

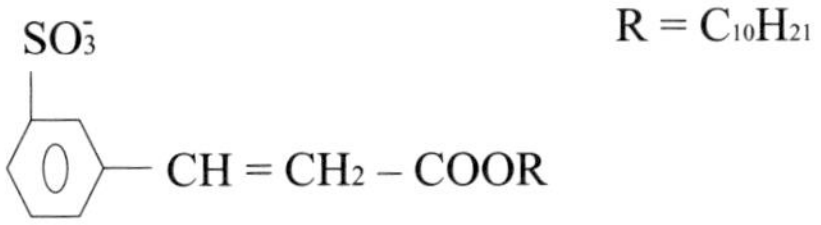

SO_3^-

$CH=CH_2-COOR$

$R = C_{10}H_{21}$

Figure 6.45 Ionic Surfmer.

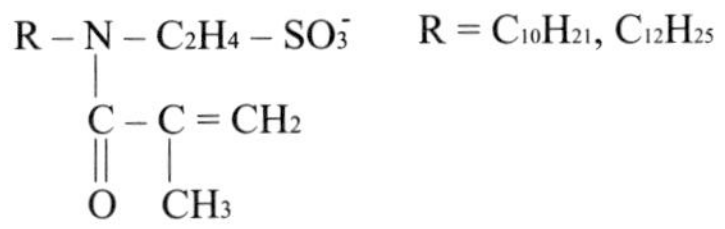

$$R-\underset{\underset{\underset{O}{\|}\;\;\underset{CH_3}{|}}{C-C=CH_2}}{|}{N}-C_2H_4-SO_3^- \quad R = C_{10}H_{21}, C_{12}H_{25}$$

Figure 6.46 Ionic Surfmer.

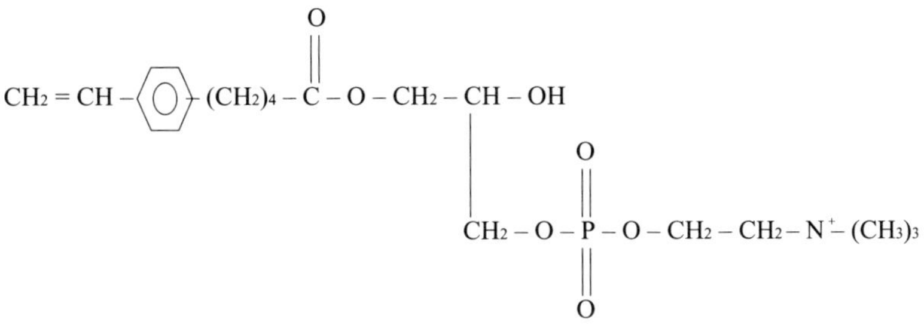

$$CH_2=CH-C_6H_4-(CH_2)_4-\overset{\overset{O}{\|}}{C}-O-CH_2-\underset{|}{CH}-OH$$

$$CH_2-O-\overset{\overset{O}{\|}}{\underset{\underset{O}{\|}}{P}}-O-CH_2-CH_2-N^+-(CH_3)_3$$

Figure 6.47 Ionic Surfmer.

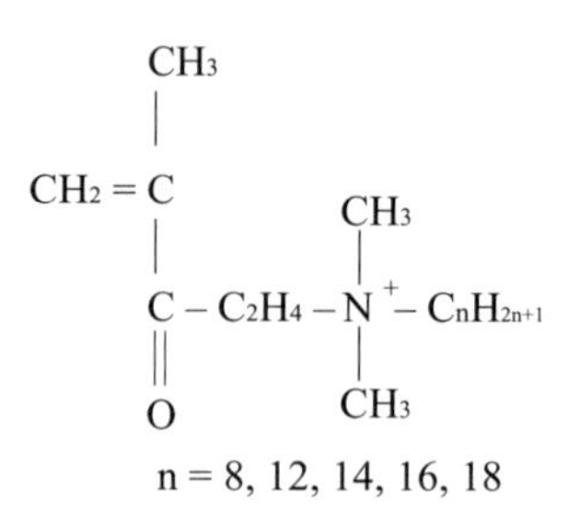

Figure 6.48 Ionic Surfmer.

In the last 10–15 years a number of considerations including performance, cost effectiveness, synthetic complexity and raw material availability have favored the emergence of other classes of Surfmers, and these are discussed below.

Maleate Surfmers. Surfmers with allylic, acrylic and vinylic moieties tend to homopolymerize and produce water-soluble polyelectrolytes if used above their CMC. This has shifted researchers' attention to maleic derivatives that do not homopolymerize at normal temperatures because their ceiling temperature is too low. Tauer and co-workers have pioneered the synthetic work [4, 15] which led originally to compounds like those given in Figure 6.49. An example of maleic-derived Surfmer used in emulsion polymerization lattices is reported in [16] and the advantages provided in commercial paint formulations are discussed later.

Maleic-derived Surfmers have been shown to be quantitatively bound to latex particles. For example the surface tension of the latex serum from the emulsion polymerization of styrene remains above 70 mN m^{-1} after polymerization even if amounts in excess of 100 times the CMC are used (15).

Maleate Surfmers were found to outperform methacrylic and crotonic compounds in the copolymerization of styrene, butyl acrylate and acrylic acid in seeded and nonseeded semicontinuous processes [17]. The maleate Surfmer achieved high conversion without homopolymerization in the aqueous phase which can result in emulsion instability. The methacrylate Surfmer was too reactive as opposed to the crotonate which was not sufficiently reactive. The reported dependence of the maleate Surfmer conversion on the particle diameter is consistent with a reaction at the particle surface.

The simple maleate Surfmer (i.e. the neutralized hemi ester of a fatty alcohol) was used to prepare seeds of polystyrene latex which were grown with a shell of film-forming polymers. The reported incorporation yield was of the order of 75% [18]. The reported latex stability could be further improved by Surfmers in which the ester moiety was substituted for an amide moiety by reaction with a fatty amine. An overall improved stability and a reduced hydrolysis at high temperature were observed [19].

Aramendia et al. [20] have compared the nonreactive sodium lauryl sulfate (SLS) to the polymerizable sodium tetradecyl maleate (M14), synthesized according to the procedure described by Stähler [21] in the seeded polymerization of methyl methacrylate/butyl acrylate/acrylic acid using tert-butyl hydroperoxide and ascorbic acid as initiator. Nonyl phenol 30 EO (NP30) was the nonionic surfactant used in the seed latex. Latex characterization included mean particle size (light scattering), coagulum, Surfmer conversion (HPLC) and glass transition temperature (differential scanning calorimetry). The structure of the top surface of the film was characterized using AFM. The sulfur and sodium content from the SLS at or near the surface of the acrylic film was measured using the Rutherford backscattering spectrometry (RBS).

The conversion of the Surfmer was high (74% for a 55-nm seed, 91% for a 36-nm seed). The latex film made with the SLS showed an irregular film that was removed following immersion in water and that is supposed to consist of SLS. By contrast, the film prepared from the reactive surfactant had a surface in which the particle identity was not obscured by a surface layer. After annealing, the surface was very flat and the particles were fully coalesced. These observations were supported by the RBS analysis. Differences between the SLS and the M14 films were even more evident when the films were annealed. There was very little change in the surfactant surface excess in the M14 film, but surfactant excess concentration was found to increase with temperature in the SLS films and reached thicknesses of up to 100 nm according to the RBS measurements. The pronounced increase in surfactant concentration after annealing the SLS film at 125°C was possibly attributed to a higher surfactant exudation above the glass transition temperature or to the loss of trace amounts of water. The results indicate that surface migration of physically adsorbed surfactants can be significant, especially after annealing at elevated temperatures. By contrast, polymerizable surfactants can be permanently attached to the polymer and, in this case, a minimal surfactant surface effect is found.

Recent changes in the regulations concerning the emission of organic solvents in the European Union have led to a growing interest in waterborne coating systems at ambient temperature. One-pack systems using the state-of-the-art technology combine carbonyl-hydrazide cross-linking with acrylic latexes of controlled morphology and properties, especially the film formation resulting from the coalescence and polymer interpenetration due to the thermal movement of the macromolecule segments. Since conventional, nonreactive surfactants interfere in the process, attention focused on a polymerizable surfactant that was easy to prepare with cheap and readily available chemicals [17]. A hemi maleate ester as described by Sindt et al. [18] was used as the sole surfactant in the seeded, semicontinuous emulsion polymerization of acrylate monomers, including functional carbonyl monomers suitable for cross-linking with bis-hydrazide. In optimum polymerization conditions the grafting was of the order of 60–70%. As only very small amounts of the Surfmer were recovered in the ultra-filtrate, it was assumed that the remainder of the Surfmer is strongly associated with the polymer particles. The final particle size of the latex is smaller than that achieved with conventional surfactants and the particle size distribution is monomodal, indicating that there is no flocculation or reseeding during monomer addition. The films produced from the Surfmer-polymerized acrylates have better cohesion, mechanical properties, gloss retention and water barrier properties. In particular, the water barrier properties are important for the intended use of the polymer in wood protection and it was observed that after 15 days of immersion in water the weight gain of the Surfmer film was 30–40% lower than that achievable with a nonreactive surfactant. A coating product using the hemi maleate Surfmer has been introduced to the market with the trade name of Setalux 6774 EPL.

Examples of maleate Surfmers are given in Figures 6.49–6.53.

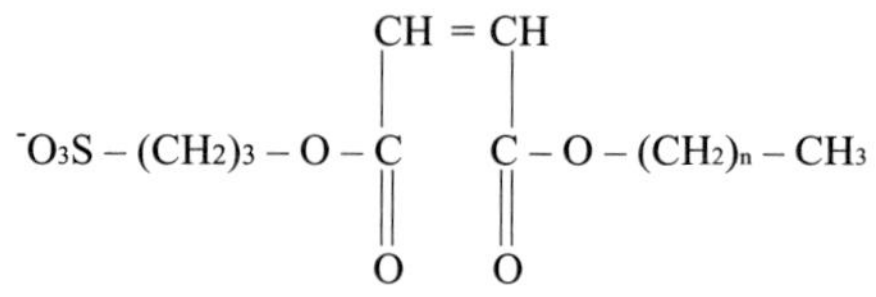

$$
\begin{array}{c}
CH=CH \\
^{-}O_3S-(CH_2)_3-O-\overset{|}{\underset{\|\atop O}{C}}\quad \overset{|}{\underset{\|\atop O}{C}}-O-(CH_2)_n-CH_3
\end{array}
$$

Figure 6.49 Maleate Surfmer.

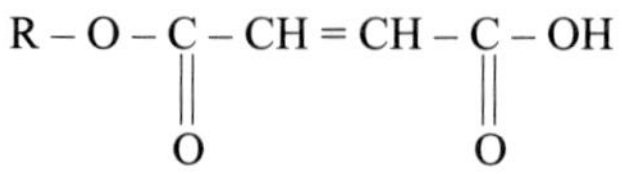

$$
R-O-\underset{\|\atop O}{C}-CH=CH-\underset{\|\atop O}{C}-OH
$$

$R = C_4H_9, C_8H_{17}, C_{12}H_{25}, C_{16}H_{33}$

Figure 6.50 Maleate Surfmer.

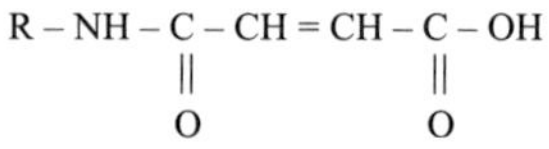

$$
R-NH-\underset{\|\atop O}{C}-CH=CH-\underset{\|\atop O}{C}-OH
$$

$R = C_4H_9, C_8H_{17}, C_{12}H_{25}, C_{16}H_{33}$

Figure 6.51 Maleate Surfmer.

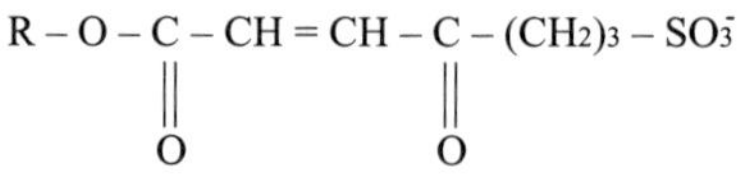

$$
R-O-\underset{\|\atop O}{C}-CH=CH-\underset{\|\atop O}{C}-(CH_2)_3-SO_3^-
$$

Figure 6.52 Maleate Surfmers.

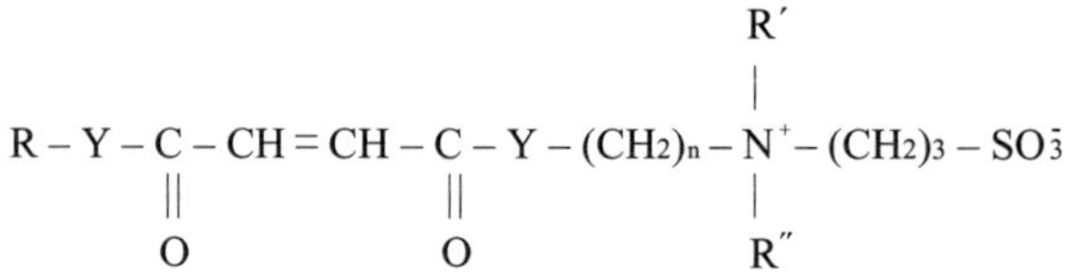

$$
R-Y-\underset{\|\atop O}{C}-CH=CH-\underset{\|\atop O}{C}-Y-(CH_2)_n-\overset{R'\atop |}{\underset{|\atop R''}{N^+}}-(CH_2)_3-SO_3^-
$$

$Y = O, NH_2$
$R = C_{12}H_{25}, C_{16}H_{33}$, Ph, Bu, Ar
$R' = R'' = CH_3, C_2H_5$

Figure 6.53 Maleate Surfmers.

Performance enhancement of maleate Surfmers. Several options have been proposed to enhance the performance of maleate Surfmers. In particular the modulation of the reactivity has been considered, to achieve a controlled and moderate reactivity during most of the polymerization and a high conversion at the end of the process. These requirements limit the useful range of values of the reactivity ratios of the Surfmer/monomer systems [22].

One way to achieve this result relies on the change in the relative monomer reactivity following composition drifts. Thus, in a combination of high and low reactivity monomers, the former will preferentially react first, leaving a considerable proportion of the latter for copolymerization when the supply of the high reactive monomer is depleted. This has been confirmed in the terpolymerization of methyl methacrylate/butyl acrylate/vinyl acetate in the presence of the maleate Surfmer reported in Figure 6.49.

It has been argued however that, despite the experimental observation under specific conditions, this approach may be too optimistic, or at least problematic to implement in industrial processes. If a Surfmer has a low reactivity it is logical to expect a low incorporation in the latex and therefore a reduced contribution to the latex stabilization. Also it can be expected that reactivity across the polymerization will cause burying into the latex particles. Although these are not insurmountable issues, they are however important enough to highlight that substantial applied research work has to be put in place to develop industrial processes to fully exploit the advantages of Surfmers against conventional surfactants established in emulsion polymerization.

Another suggested approach is to provide for a larger surface area for the Surfmers in the late stage of the polymerization process, for example by introducing a new seed of particles with a small amount of monomer. It is reported that this approach increased the conversion of a Surfmer from about 50% to nearly 100% [22].

Cationic and amphoteric Surfmers were synthesized from the hemi ester or hemi amide with a C_{12}–C_{20} alkyl chain and diethyl (chloroethyl) amine followed by quaternization with conventional agents or by reaction with propanesultone [23].

The cationic Surfmers produced much smaller particle sizes in the emulsion polymerization of styrene and styrene/butyl acrylate than the amphoterics (20–50 nm versus 100–300 nm). Some of the latter, however, conferred to the copolymer lattices stability to electrolytes and freeze-thaw [24]. Similar, but nonreactive surfactants produced from succinic anhydride gave similar stability but had much inferior water resistance [25].

The acylation of alcohol-containing monomers, e.g. hydroxyethyl acrylates or vinyl benzyl alcohol with maleic, succinic or sulfosuccinic anhydride leads to bifunctional polymerizable surfactants. A range of such products has been synthesized and tested in batch polymerization and core-shell polymerization of styrene and butyl acrylate [26]. In both cases good stability, high conversion and little burying of the Surfmers were observed. Water rebound was also limited. These advantageous features were however offset by an unacceptable resistance to electrolytes and to freeze-thaw.

Nonionic Surfmers. The alkoxylation of polymerizable substrates has been for many years a source of building blocks for innovative surfactant species as well as for the synthesis of high performance Surfmers. Some of the early experimental prototypes have evolved into commercial products and the growing understanding of structure–performance relationships allows the development of new molecules. The reasons for this success can be attributed to:

- The simplicity, flexibility and versatility of alkoxylation, which allow an extraordinarily large number of synthetic options
- The comparative ease of scaling-up the lab processes to semitechnical and full scale manufacturing
- Low manufacturing costs, as a consequence of the point above
- Wide and ready availability of raw materials
- The advantageous regulatory position, which, due to the 'polymer exemption', simplifies the protocols for the introduction of new products to the market
- The lattice stabilization through steric or electrosteric mechanisms, resulting in electrolyte and freeze-thaw stability

However one constraint of alkoxylated Surfmers is their cloud point versus the polymerization temperature. If the former is lower than the latter, salting-out of the Surfmer occurs, with loss of surface activity and reactivity. The cloud point of nonionic alkoxylates can be adjusted to a certain extent by the choice of the alkoxylation initiator, the relative percentage of hydrophilic and hydrophobic alkoxylation moieties and their order of addition. Also, introducing some ionic character in the molecule (e.g. by weak polar groups that do not substantially affect the nonionic behavior of the molecule) may prove useful. Nevertheless there have been and there can be instances where nonionic Surfmers cannot be used.

In the second half of the 1980s, polyethylene glycol (PEG) methacrylic ester monomers [27] and a triallyl penthaerithritol alkoxylated ester based on BO and EO were studied alone and in combination with conventional nonionic nonreactive surfactants. The low reactivity of the allyl groups limited the scope of application for the BO/EO Surfmer, but it was shown that the methacrylic esters were settling preferentially at the surfaces of the lattices, replacing/displacing the other nonionic surfactants. Noncharged lattices were stable to electrolytes and freeze-thaw [28].

The commercial availability of the PEG methacrylic esters encouraged further investigation to explore potential benefits under conditions different from those that were originally targeted. Ottewill and Satgurunathan [29] studied the influence of the addition of the monomers at different stages of surfactant-free latex initiated with potassium persulfate at different stages of polymerization. It was observed that the particle size and their distribution in the base system (essentially monodispersed particles) changed depending upon the stage at which the PEG-modified monomer was added. An early addition results in a bimodal distribution of both the original and larger particle size. A late addition shifts the particle size distribution toward the original distribution. Irrespective of the addition stage, the electrolyte and freeze-thaw stability suggests that a significant proportion of the monomer is grafted.

'Reverse' ethylene oxide/propylene oxide block copolymers (in which a hydrophilic core of PEO is terminated at both ends with hydrophobic PO moieties) are used in industrial applications. This is because of the different and unique performance properties compared to the 'conventional' block copolymers, where a hydrophobic PO core is block copolymerized with EO. Dufour and Guyot [30] have built on this observation and synthesized Surfmers in which a PEG core (about 37 EO units) was tipped with about 10 PO units to further react with a chlorine-carrying polymerizable group or with maleic anhydride to produce reactive Surfmers.

The maleic Surfmers were tested in core-shell emulsion polymerization of styrene/butyl acrylate in comparison with a standard nonreactive surfactant (nonyl phenol reacted with 30 mol of EO – NP30). While the methacrylic-derived Surfmer was completely incorporated during the polymerization (although about one-third of it was buried inside the particles) the NP30, the maleic Surfmer and the allylic and vinyl Surfmers were not incorporated and could be extracted with acetone (for the last two probably because of the formation of acetone-extractable oligomers due to a chain transfer behavior) [31].

Styrenic Surfmers of nonionic and anionic structures can be prepared using a vinyl benzene alcolate as initiator for the sequential reaction of BO and EO. Upon controlled addition of the alkoxylating moieties, the reaction was chilled either with methanolic HCl (thus producing a nonionic alkoxylate) or by reaction with propane sultone to give an anionic Surfmer [32]. Both the anionic and nonionic Surfmers gave very stable lattices in the seeded copolymerization of methyl methacrylate and butyl methacrylate. Lattices produced using nonreactive surfactants of similar structure were shear unstable, although the stability to freeze-thaw suggests a strong adsorption on the particle surface [33].

Recently Uniqema has introduced commercially a Surfmer under the trade name of Maxemul 5011. Maxemul is produced by esterification of an unsaturated fatty anhydride with a methoxy PEG such that the reactive group is close to the hydrophilic moiety [34]. Stable latexes with a solid content of 52% were produced in the seeded emulsion polymerization of film-forming methyl methacrylate/butyl acrylate/acrylic acid (3% Surfmer on monomers, constant monomer feeding rate over 4 h, potassium persulfate/sodium metabisulfate redox initiator). The latexes were stable to electrolytes but not to freeze-thaw.

It was estimated that, if all the Surfmers contributed to stabilization, the surface coverage would be close to 20% at the end of the process. When Surfmer burial is considered, the minimum surface coverage is in the region of 14.7–15.0 % [35]. The authors have also studied the influence of the addition procedure on the evolution of the Surfmer conversion and concluded that, despite the low reactivity due to the presence of the alkenyl double bond, the incorporation could be increased to 72% from the original 58% obtained with a constant feeding rate. A mathematical model able to describe Surfmer polymerization was used in the optimization process [36].

Other than through alkylene oxide chemistry, monomeric Surfmers have been produced from polyvinyl alcohol [37] and saccharides [38].

6.4.3 Emulsion polymerization

Emulsion polymerization is one of the major processes for the production of industrial polymers. It represents a sizable application for surface active agents, although manufacturers tend to minimize their use because of economic and environmental considerations (surfactants are usually more expensive compared to monomers and are mostly left in the liquor) and because of the negative effects on the final properties of the polymers and of their coalesced films.

The concept of using reactive surfactants in emulsion polymerization processes is relatively recent and aims at eliminating the drawbacks associated with the use of conventional, monomeric, nonreactive surfactants. Despite the demonstrated advantages and the availability of commercial products, reactive surfactants have not yet reached a widespread penetration in Europe, whereas they are much more popular in Japan. In Europe commercial products have been proposed, for example by Uniqema (Maxemul), Cognis (Sidobre-Sinnova) (C16 maleate hemi ester and ethoxylated derivatives) and Clariant (allyl and vinyl ethers of oxirane block copolymers). Condensation products of hydroxyethyl methacrylate with maleic and succinic anhydride are listed in the Aldrich catalog and are presumably available for laboratory work.

The foundations of emulsion polymerization were described originally by Harkins [39]. The first theoretical treatment was proposed by Smith and Ewart [40]. The theory was later modified to some extent by O'Toole [41] and more fundamentally by Garden [42], who proposed an unsteady-state mechanism for the concentration of free radicals in the emulsion particles. Tauer [43], Gilbert [44] and Lovell and El-Aasser [45] have produced recent reviews.

6.4.3.1 Industrial processes and applications

Polymer emulsions can be produced by the direct and the inverse emulsion process. The direct emulsion polymerization can be performed in a batch, semibatch and continuous process.

In the batch process low-water-solubility monomers are emulsified in water by water-soluble surfactants, purged and heated at the initiation temperature (for energy saving this is usually lower than the reaction temperature to benefit from the reaction exotherm) and the initiator added. Temperature is then maintained for the reaction period, which can range from 1 to 24 h. Reactions are driven to the maximum conversion compatible with the system and the residual monomer and other volatile compounds are removed either by stripping or by chemical treatment.

The main drawback of batch polymerization, i.e. the risk of runaway reaction because of the high volume of monomer present at the beginning of the polymerization, can be overcome by the semibatch process. This has the additional advantage of a higher reactor capacity because of the volume shrinkage during polymerization and of the reduction in side reactions of the monomers that may lead to the formation of off odor by-products. Semibatch processes allow a better scope for maximizing the polymerization rate, but are unsuitable if polymers with high linearity and good tensile strength are required.

The process usually starts with the polymerization of a small proportion of the reagents at a very low monomer to water ratio (the seed stage), followed by the feeding of the remaining monomer (which may take several hours) and of other materials (if needed) once the conversion in the reactor has reached 70% or more. The in-reactor conversion will then depend upon the rate of polymerization compared to the rate of feed. If the reaction is continued under the so-called monomer-starved conditions, the in-reactor conversion is kept at a high 80–90%, which reduces the polymerization rate. To compensate, temperature is raised: however, then the initiator depletes faster and more has to be added during the reaction.

In the continuous processes, all ingredients and all the reagents are fed at one end of a train of 6–12 reactors and are recovered at the other end as full emulsion polymers. Extra ingredients may be added at any point. The start-up of the system is complex and therefore only suitable for large-volume/single recipe polymers. Alternatively, the polymerization can be carried out in a tubular loop reactor. The ingredients are premixed, fed to the reactor and recirculated through the loop for a preset time period. The emulsion is then discharged and the reaction allowed to reach full conversion outside the loop reactor.

The main uses of emulsion lattices are:

- Film formers in paints, inks, coatings, paper coatings, textile sizing, nonwoven textiles, glass fiber binders
- Synthetic elastomers, e.g. styrene-butadiene rubber, polychloroprene rubber, nitrile-butadiene rubber
- Thermoplastic polymers, notably acrylonitrile-butadiene-styrene and polyvinyl chloride

Applications include:

- Packaging and wood adhesives
- Pressure sensitive adhesives
- Caulks/sealants
- Bitumen
- Paper coatings and bonding
- Textile-screen printing
- Latex paints
- Rubber articles
- Woven and nonwoven fabrics
- Carpet backings
- Thickeners
- Structured particles, e.g. impact modifiers

In inverse polymerization, water-soluble monomers are emulsified with low HLB (hydrophilic-lipophilic balance) surfactants in an organic medium and the reaction is initiated with water-soluble or oil-soluble initiators. A review of the subject can be found in a recent publication of Greenshields [46].

The polymers manufactured are high molecular weight acrylamide derivatives used in water purification, enhanced oil recovery and solid flocculation.

6.4.3.2 Surfmers in emulsion polymerization

The emulsion polymerization process has been and is being extensively investigated and the information gained has significantly contributed to the understanding of the role of Surfmers and of their structure–performance relationships. Progress was, however, hindered by the limited availability of commercial Surfmers and by the difficulty in determining with a reasonable degree of accuracy the amount of reacted Surfmers and the location of the reacted molecules in the polymer particles.

The copolymerization of Surfmers with monomers differs from the copolymerization of conventional monomers because of, among other reasons, the comparatively large size of the Surfmers molecules. The kinetics and reactivity of non-surface-active macromonomers have been reviewed [47, 48] and it is believed that the factors affecting the reactivity of macromonomers will play a similar role in the reactivity of Surfmers. These factors are:

- The chemical nature of the polymerizable group in the macromonomer
- The degree of compatibility of the macromonomer with the propagating comonomer chain
- The molecular weight of the comonomer
- The polymerization medium
- The conversion of the polymerization

In addition there is evidence that the position of the unsaturation in the Surfmer's molecule is also expected to affect its behavior [14].

Asua and Schoonbrood [49] have produced an extensive review of the literature dealing with copolymerization of Surfmers, of the Surfmers polymerization loci and the influence of Surfmers on particle nucleation and growth. From this, they concluded that the main feature of a Surfmer is its intrinsic reactivity and provided suggestions for the choice of the reactive group in a Surfmer. They also made proposals to maximize Surfmer performance and effectiveness, namely:

- Change in the main monomer activity, e.g. by adding a monomer that is highly reactive to the Surfmer at the end of the polymerization process or by an intrinsic change in the comonomer activity because of concentration effects
- Change in the Surfmer reactivity because of a change in the operating conditions
- Addition profile of the Surfmers
- Suppression of the particle size growth

The review provides recommendations to prevent early Surfmer polymerization and the consequent burying, so as to achieve a high degree of Surfmer incorporation at the end of the polymerization process. There are also hints on the possible use of Surfmers in dispersion and micro emulsion polymerization.

Acknowledgments

The author is indebted to the Uniqema Information Team Gouda and to Jacob Van den Berg for their assistance in the bibliographic references and to Jo Grade, Uniqema Customer Service, for his comments and advice.

References

1. Whitmarsh, R. (1996) Nonionic surfactants. In. V. Nace (ed.), *Polyoxyalkylene Block Copolymers.* Dekker, New York.
2. Marciniec, B. and Gulinski, J. (1993) Recent advances in catalytic hydrosilylation. *J. Organometal. Chem.*, **446**, 15.
3. Schmaucks, G. (1999) Silicon surfactants. In R.M. Hill (ed.), *Novel Siloxane Surfactants Structures.* Dekker, New York.
4. Tauer, K. and Kosmella, S. (1993) Synthesis, characterisation and application of surface active initiators. *Polym. Int.*, **30**, 253.
5. Rieger, M.M. (1986) Skin irritation: Physical and chemical considerations. *Cosmet. Toiletries*, **101**, 85–6, 88, 90–2.
6. Van Hook, J.P. and Tobolski, A.V. (1958) The thermal decomposition of 2,2′ azo-bis-isobutyronitrile. *J. Am. Chem. Soc.*, **80**, 779–82.
7. Guyot, A. (2003) Polymerizable surfactants. In Krister Holmberg (ed.), *Novel Surfactants Preparations, Applications and Biodegradability*, 2nd ed. Dekker, New York, p. 499.
8. Vidal, F., Guillot, J. and Guyot, A. (1995) Surfactants with transfer agent properties (transurfs) in styrene emulsion polymerisation. *Colloid Polym. Sci.*, **273**, 999–1007.
9. Guyot, A. and Vidal, F. (1995) Inifer surfactants in emulsion polymerisation. *Polym. Bull.*, **34**, 569–76.
10. Chiefari, J., Chong, Y.K., Ercole, F., Krstina, J., Le., T.P.T., Mayadunne, R.T.A., Meijs, G.F., Moad, C.L., Moad, G., Rizzardo, E. and Thang, S.H. (1998) Living free-radical polymerisation by reversible addition-fragmentation chain trasfer: The RAFT process. *Macromolecules*, **31**, 5559–62.

11. Uzulina, I., Kanagasapatty, S. and Claverie, J. (2000) Reversible addition fragmentation transfer (RAFT) polymerisation in emulsion. *Macromol. Symp.*, **150**, 33–8.
12. Monteiro, M.J., Brussels, R. and Wilkinson, T.S. (2001) Emulsion polymerization of methyl methacrylate in the presence of novel addition-fragmentation chain-transfer reactive surfactant (transurf). *J. Polym. Sci.*, A **39**, 2813–20.
13. Hellgren, A.C., Weissenborn, P. and Holmberg, K. (1999) Surfactants in water-borne paints. *Progr. Org. Coat.*, **35**, 79–87.
14. Ferguson, P., Sherrington, D.C. and Gough, A. (1993) Preparation, characterization and use in emulsion polymerization of acrylated alkyl ethoxylate surface-active monomers. *Polymer*, **34**, 3281–92.
15. Tauer, K., Wedel, A. and Mosozova, M. (1992) Synthesis of nitro-group-containing copolymers by radical-initiated copolymerisation. *Macromol. Chem.*, **193**, 1387–98.
16. Communication from Dr. Dirk Mestach, Akzo Nobel Resins bv, Synthesebaan 1, P.O. Box 79, 4600 AB, Bergen op Zoom, the Netherlands, 'New High Performanc Materials for Waterborne Acrylic Surface Coatings'.
17. Schoonbrood, H.A.S., Unzue, M.J., Beck, O. and Asua, J.M. (1997) Reactive surfactants in heterophase polymerization. 7. Emulsion copolymerization mechanism involving three anionic polymerizable surfactants (surfmers) with styrene-butyl acrylate acrylic acid. *Macromolecules*, **30**, 6024–33.
18. Sindt, O., Gauthier, C., Hamaide, T. and Guyot, A.J. (2000) Reactive surfactants in heterophase polymerization. XVI. Emulsion copolymerization of styrene-butyl acrylate-acrylic acid in the presence of simple maleate reactive surfactants. *J. Appl. Polym. Sci.*, **77**, 2768–76.
19. Abele, S., Graillat, C., Zigmanis, A. and Guyot, A. (1999) Hemiesters and hemiamides of maleic and succimic acid: synthesis and application of surfactants in emulsion plymerization with styrene and butyl acrylate. *Polym. Adv. Technol.*, **10**, 301–10.
20. Aramendia, E., Mallégol, J., Jeynes, C., Barandiaran, M.J., Keddie, J.L. and Asua, J.M. (2003) Distribution of surfactants near acrylic latex film surfaces: A comparison of conventional and reactive surfactants (surfmers). *Langmuir*, **19**, 3212–21.
21. Stähler, K. (1994) Einfluss von Monomer Emulgatoren anf die AIBN-initiierte Emulsion Polymerisation von Styren. Ph.D. Thesis, Postdam University, Germany.
22. Schoonbrood, J.M. and Asua, J.M. (1997) Reactive surfactants in heterophase polymerisation. 9. Optimum surfmer behavior in emulsion polymerization. *Macromolecules*, **30**, 6034–41.
23. Zicmanis, A., Hamaide, T., Graillat, C., Monnet, C., Abele, S. and Guyot, A. (1997) Synthesis of new alkyl maleates ammonium derivatives and their use in emulsion polymerisation. *Colloid Polym. Sci.*, **275**, 1–8.
24. Abele, S., Zicmanis, A., Graillat, C., Monnet, C. and Guyot, A. (1999) Cationic and zwitterionic polymerizable surfactants: Quaternary ammonium dialkyl maleates. 1. Synthesis and characterization. *Langmuir*, **15**, 1033–44.
25. Abele, S., Gauthier, C., Graillat, C. and Guyot, A. (2000) Films from styrene-butyl acrylate lattices using maleic or succinic surfactants: mechanical properties, water rebound and grafting of the surfactants. *Polymer*, **41**, 1147–55.
26. Uzulina, I., Zicmanis, A., Graillat, C., Claverie, J. and Guyot, A. (2001) Synthesis of polymer colloids using polymerizable surfactants. *Macromol. Chem. Phys.*, **202**, 3126–35.
27. Ottewill, R.H. and Satgurunathan, R. (1987) Nonionic lattices in aqueous media. 1. Preparation and characterisation of polystyrene lattices. *Colloid Polym. Sci.*, **265**, 845–53.
28. Ottewill, R.H., Satgurunathan, R., Walte, A. and Wetsby, M.J. (1987) Nonionic polystyrene lattices in aqueous media. *Br. Polym. J.*, **19**, 435–40.
29. Ottewill, R.H. and Satgurunathan, R. (1995) Nonionic lattices in aqueous media. 4. Preparation and characterisation of electrosterically stabilized particles. *Colloid Polym. Sci.*, **273**, 379–86.
30. Dufour, M.G. and Guyot, A. (2003) Nonionic reactive surfactants. I. Synthesis and characterization. *Colloid Polym. Sci.*, **281**, 97–104.

31. Dufour, M.G. and Guyot, A, (2003) Nonionic reactive surfactants. Part 2. Core-shell latexes from emulsion polymerization. *Colloid Polym. Sci.*, **285**, 105–12.
32. Soula, O. and Guyot, A. (1999) Styrenic surfmer in emulsion copolymerization of acrylic monomers. I. Synthesis and characterization of polymerizable surfactants. *Langmuir*, **15**, 7956–62.
33. Soula, O., Guyot, A., Williams, N., Grade, J. and Blease, T. (1999) Styrenic surfmer in emulsion copolymerization of acrylic monomers. II. Copolymerization and film properties. *J. Polym. Sci.* A, **37**(22), 4205–17.
34. Aramendia, E., Barandiaran, M.J., Grade, J., Blease, T. and Asua J.M. (2002) Polymerisation of high-solids-content acrylic latexes using a nonionic polymerisable surfactant. *J. Polym. Sci.* A, **40**, 1552–59.
35. Aramendia, E., Barandiaran, M.J. and Asua J.M. (2003) On the optimal surfmer addition profile in emulsion polymerisation. *C.R. Chimie*, **6**, 1313–17.
36. De la Cal, J.C. and Asua, J.M. (2001) Modeling emulsion polymerization stabilized by polymerizable surfactants. *J. Polym. Sci.* A, **39**, 585.
37. Charleux, B. and Pichot, C . (1993) Styrene-terminated poly(vinyl alcohol) macromonomers. 1. Synthesis by aldol group transfer polymerization. *Polymer*, **34**, 195.
38. Revilla, J., Delair, T., Pichot, C. and Gallot, B. (1995) Preparation and properties of comb-like polymers obtained by radical homo- and copolymerization of a liposaccharidic monomer with styrene. *Polymer*, **37**, 687–98.
39. Harkins, W.D. (1947) A general theory of the mechanism of emulsion polymerization. *J. Am. Chem. Soc.*, **69**, 1428–44.
40. Smith, W.V. and Ewart, R.H. (1948) Kinetics of emulsion polymerisation. *J. Chem. Phys.*, **16**, 592.
41. O' Toole, J.T. (1965) Kinetics of emulsion polymerisation. *J. Appl. Polym. Sci.*, **9**, 1291-65.
42. Gordon, J.L. (1968) Emulsion polymerisation. 1. Recalculation and extension of Smith-Ewart theory. *J. Polym. Sci.* A1, **6**(3PA1), 623.
    Gordon, J.L. (1968) Emulsion polymerisation. 2. Review of experimental data in context of revised Smith-Ewart theory. *J. Polym. Sci.* A1, **6** (3PA1) 643.
    Gordon, J.L. (1968) Emulsion polymerisation. 3. Theoretical prediction of effects of slow termination rate within latex particles. *J. Polym. Sci.* A1, **6**(3PA1), 665.
    Gordon, J.L. (1968) Emulsion polymerisation. 4. Experimental verification of theory based on slow termination rate within latex particles. *J. Polym. Sci.* A1, **6**(3PA1) 687.
    Gordon, J.L. Emulsion polymerisation. 5. Lowest theoretical limits of ratio KT/KP. *J. Polym. Sci.* A1, **6**(10PA), 2853.
    Gordon, J.L. Emulsion polymerisation. 6. Concentration of monomers in latex particles. *J. Polym. Sci.* A1, **6**(10PA), 2859.
43. Tauer, K. (2003) The role of emulsifiers in the kinetics and mechanism of emulsion polymerisation. In D.R. Karsa (ed.), *Surfactants in Polymers, Coatings, Inks and Adhesives.* Blackwell Publishing, CRC Press, Boca Raton, FL.
44. Gilbert, R.G. (1995) *Emulsion polymerization: A Mechanistic Approach.* Academic, New York.
45. Lovell, P.A. and El-Aasser, M.S. (ed.) (1997) *Emulsion polymerization and Emulsion Polymers.* Wiley, New York.
46. Greenshields, J.N. (2000) Surfactants in inverse (water-in-oil) emulsion polymers of acrylamide. In D.R. Karsa (ed.), *Surface Active Behavior of Performance Surfactants, Annual Surfactants Review* vol. 3. Sheffield Academic, CRC Press, Boca Raton, FL.
47. Meijs, G.F. and Rizzardo, E. (1990) Reactivity of macromonomers in free-radial polymerisation. *JMS-Rev. Macromol. Chem. Phys.*, **C30**(3&4), 305-77.
48. Capek, I. and Akashi, M. (1993) On the kinetics of free-radial polymerisation of macromonomers. *JMS-Rev. Macromol. Chem. Phys.* **C33**(4), 369–436.
49. Asua, J.M and Schoonbrood, H.A.S. (1998) Reactive surfactants in heterophase polymerisation. *Acta Polymerica*, **49**, 671–86.

6.5 Fluorinated Surfactants

Richard R. Thomas

6.5.1 Introduction

Commercially, the production and use of surfactants is dominated by modified hydrocarbon-based chemicals. In a number of instances, however, a hydrocarbon-type surfactant will not provide the desired product attributes or performance and, in such cases, two options are presented. One involves reformulation of the product to accommodate a hydrocarbon-type surfactant and the other is the use of a fluorosurfactant. Fluorosurfactants behave typically as would a hydrocarbon type except that properties such as surface tension reduction are larger in magnitude. Furthermore, the presence of fluorine in the hydrophobic portion of the molecule causes them to differ from their hydrocarbon counterparts in more subtle ways that have commercial importance. An example of a difference would be the reduced dielectric constant or index of refraction of a fluorosurfactant compared to its hydrocarbon analog. While this may be of no consequence when formulating cleaners, it most certainly exists in a number of electronics applications.

The large majority of surfactants can be classified as hydrocarbon types, which means that the hydrophobe is a hydrocarbon. There is another class of surfactants, differentiated by the name fluorosurfactant, that uses a fluorocarbon instead of a hydrocarbon as the hydrophobe. Typically, the fluorocarbon is based on $-(CF_2)_nF$ where the number-averaged value of $n \approx 8$ and effectiveness and efficiency of fluorosurfactants are sensitive functions of n. Commercially, a value for $n \approx 8$ is chosen to give maximum effectiveness and efficiency. When compared to hydrocarbon surfactants, similar fluorocarbon surfactants have a higher efficiency and effectiveness. The interested reader is referred to an excellent review on the structure and properties of fluorosurfactants [1].

6.5.2 Uses

Clearly, due to the price differential between a hydrocarbon and analogous fluorocarbon surfactant ($\sim$10–100$\times$), the latter is used often as a 'last resort' when nothing else will perform adequately. Not only does the fluorosurfactant provide a lower surface tension than a hydrocarbon analog on a molecule-to-molecule basis, but many other important differences are used advantageously. Often, the user is searching for a material that not only will dominate the surface but also impart unique properties to the material. Several such uses are worthy of recognition. Hydraulic fluid used in aircraft contains a fluorosurfactant and, although there is disagreement about the actual mechanism, it is claimed that the presence of the fluorosurfactant is necessary for the proper functioning of valves in the aircraft hydraulic system. Very thin layers ($\sim$5–10 Å) of fluorosurfactants are used as antireflection layers in the photolithographic process in microelectronics fabrication. Typical photoresists have a relatively high index of refraction and, as the light used to process the photoresist reflects off the substrate, standing wave patterns are exposed in the photoresist due to multiple, coherent internal reflections. These standing wave patterns affect the critical tolerance of the desired pattern. The lowered index of refraction of a thin layer of fluorosurfactant on the photoresist negates internal reflection of the light off the substrate, therefore, allowing for greater control of critical tolerances. Gelatin is used in large quantities in the photographic film industry as a stabilizer for the colloid responsible for the latent image. The completed film is then wound rapidly on metal spools causing a great deal of triboelectric charging. Discharging exposes the film and is undesirable and the addition of a fluorosurfactant to the gelatin layer mitigates cathodic charging on the native material. Advanced fire fighting foams (AFFF) are another example that exploits the inherent properties of fluorosurfactants versus hydrocarbon surfactants. AFFF materials are aqueous-based products used to combat fire in critical applications such as aircraft. The material must foam, contain the fire from spreading, not become fuel itself, help to extinguish the fire by preventing oxidant from entering the combustion zone and not damage sensitive components. Another application that exploits the differences between hydrocarbon and fluorocarbon surfactants is electroplating. Here, the problem lies in the very aggressive conditions of low pH and air-sparging of the bath. Fluorosurfactants, typically, have a much larger pH usage range than hydrocarbon analogs and can tolerate exposure to pH values in the 1–2 range that would be representative of an electroplating bath. In addition, the fluorosurfactant-rich foam present on an air-sparged electroplating bath is a more effective barrier to evaporation and aerosolization of a corrosive mist in a manufacturing environment.

6.5.3 Applied theory

There is nothing magical about fluorocarbons or, specifically, the $-(CF_2)_nF$ group [2]. The $-(CF_2)_nF$ is similar to $-(CH_2)_nH$ in many ways. These include dipole moments and polarizabilities that are related to intermolecular forces and, hence, surface tension. Where they do differ is in size, specifically diameter, and a relative comparison for a typical hydrocarbon and similar fluorocarbon surfactant is shown in Figure 6.54.

The terminal $-CF_3$ group $\approx$ 20% larger than the $-CH_3$ group and the same rationale applies for the $-CF_2-$ group versus the $-CH_2-$ group. This means that intermolecular forces/unit volume for fluorocarbons are less than that for hydrocarbons of similar structure.

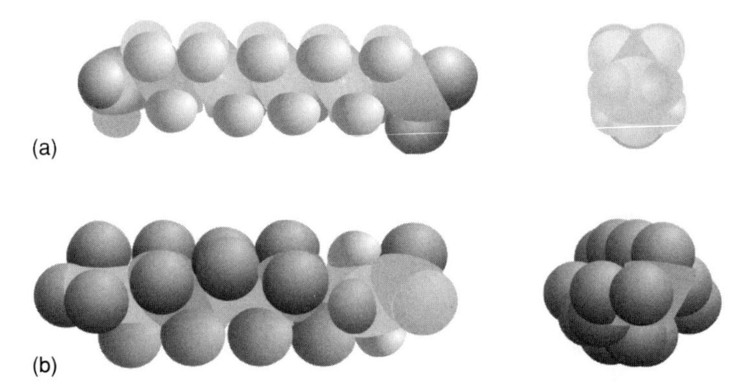

Figure 6.54 Space-filling models of typical hydrocarbon ($H(CH_2)_{10}CO_2^-$) (a) and fluorocarbon ($F(CF_2)_8CH_2CH_2CO_2^-$) (b) surfactants. Hydrophobes are on left side of molecule in profile and at terminus in end view.

A term to describe the aforementioned quotient is cohesive energy density (CED; heat of vaporization/unit volume). To a first approximation, the lower the CED, the lower will be the surface tension and this is the source of the increased efficiency in surface tension reduction of fluorosurfactants versus hydrocarbon surfactants. Therefore, fluorosurfactants are often the choice for applications demanding ultimately low surface tension. Furthermore, fluorosurfactants are far less compatible with water than are hydrocarbon surfactants. This is the origin of the increased effectiveness compared to hydrocarbon surfactants.

How surface tension translates to commercial applications will now be examined. Surfactants are often added to reduce surface tension of a liquid enabling it to wet a surface and the equation governing this phenomenon is attributed to Young [3]:

$$\gamma_{lv} \cos\theta = \gamma_{sv} - \gamma_{sl} \tag{6.5.1}$$

where γ is interfacial tension of liquid/vapor (lv), solid/vapor (sv) and solid/liquid (sl) interfaces and θ is the contact angle of a liquid droplet on a surface. This equation states that if a liquid has a higher interfacial tension than a solid, the liquid will not wet the solid. A familiar case would be a water droplet resting on a surface of Teflon where the water 'beads' on the surface and, for the curious, this is the basis for repellency. In most coating systems, for example, one desires to have the surface tension of the coating to be lower than that of the substrate surface. This ensures that the coating wets the surface and the proper choice of surfactant can aid the coating in wetting the substrate. If the substrate has a very low surface tension (e.g., polyolefin), then an even lower surface tension surfactant, such as a fluorosurfactant, must be used. An example of this effect is shown in Figure 6.55.

As an example, consider a clean floor tile with a surface tension of 32 mN m^{-1} and the same tile that has been soiled (27 mN m^{-1}). This is an example of a very realistic possibility. One can observe clearly that the contact angle increases dramatically once the surface tension of the liquid (for example, floor polish with surfactant) increases above that of the substrate. Shown also in Figure 6.55 are bars that span typical surface tension expected

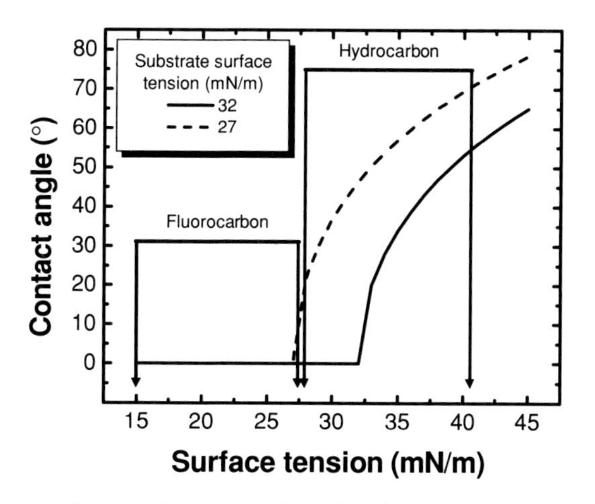

Figure 6.55 Contact angles as a function of liquid surface tension for substrates with surface tension of 32 (—) and 27 (- - -) mN m^{-1}.

at use conditions for solutions containing fluorocarbon and hydrocarbon surfactants. The anticipated contact angles would be seen from the intersection of the vertical lines with the two curves. It is obvious that the fluorosurfactant would provide a better guarantee of the liquid (floor polish in this case) wetting (contact angle $\rightarrow$ 0) the substrate (floor tile whether clean or soiled) than would a hydrocarbon surfactant.

Interfacial rheology is another issue [4]. Much like the bulk, interfaces (specifically, the liquid/air interface) have their own rheological issues. For example, foaming is a consequence of interfacial rheology and low surface tension is a necessary, but not sufficient cause for foaming. In order to support a foam, an interface must have some elasticity. Consider a balloon and imagine immersing the balloon in liquid nitrogen and then adding a great deal of pressure to expand further. The balloon breaks. At liquid nitrogen temperatures, the polymer constituting the balloon is no longer elastic, but rather a rigid solid so it breaks rather than expands. Flow and leveling is an even more complicated issue and, according to first principles, leveling stress, λ, is given by [5]:

$$\lambda = -\frac{\gamma}{r} \tag{6.5.2}$$

where γ is surface tension and r is radius of interfacial curvature. Equation 6.5.2 states that leveling stress is directly proportional to surface tension and implies that lowering the surface tension of a coating will make it level *even less* with added surfactant than without. This seems contrary to what is observed in reality and obviously, the situation is not that simple. Many factors, as yet understood poorly in practical coating systems, are at play and a complete theoretical description has not been forwarded. In summary, flow and leveling in coatings is a dynamic event that defies description by a single static variable such as surface tension. Surface tension alone is not a very good indicator of the performance of materials in most situations and applications and must be measured as some facsimile of the application to get meaningful information.

For many uses, hydrocarbon surfactants often behave satisfactorily. There simply is no need to use expensive fluorosurfactants. However, one of the basic requirements of a coating is that it wets the substrate and secondly, is that it flows and levels to give required optical properties such as gloss or distinctness of image. This is where fluorosurfactants come into play. Some coating systems inherently wet, flow and level well without fluorosurfactants whilst others do not. Wetting is troublesome particularly when trying to coat low tension surfaces such as polyolefins (polyethylene or polypropylene, for example) or surfaces contaminated with low tension materials such as silicone oils or greases. The relevant equation governing wetting is given in eqn 6.5.1 and is well known. The surface tension of the coating has to be less than that of the substrate in order to wet and fluorosurfactants afford lower surface tension than comparable hydrocarbon surfactants. Thus, if wetting of a low tension surface is desired, particularly from a relatively high tension aqueous coating, then the use of a fluorosurfactant is warranted. In contrast, flow and leveling is much less understood without a detailed understanding of all the material parameters. To know this a priori would take far too long to be practical and the formulator is faced with a simple question: does the coating flow and level satisfactorily or not? If it does, then there is no need for a fluorosurfactant. If it does not, then a fluorosurfactant may well provide the required flow and leveling attributes desired.

The economics of fluorosurfactant use will now be examined briefly by categorizing the price of chemicals on a $\log_{10}$ scale (US currency; 2004):

(i) Fillers, hydrocarbon monomers, *hydrocarbon surfactants*, etc. $\$0–10^0\ \text{lb}^{-1}$
(ii) Additives such as HALS (hindered amine light stabilizer), bulk rheology modifiers, etc. $\$10^0–10^1\ \text{lb}^{-1}$
(iii) Pharmaceuticals, *fluorosurfactants*, etc. $\$10^2–10^3\ \text{lb}^{-1}$

Note the discrepancy in price between hydrocarbon and fluorocarbon surfactants. Fluorosurfactants are in league with pharmaceuticals in terms of price. Even at $10^2–10^3$ ppm usage levels, fluorosurfactants can add substantial cost. Price tolerance must be evaluated carefully. For example, inexpensive alkyd paints sell for ~US$10/gal. The addition of several hundred ppm of fluorosurfactant would increase cost and price substantially to maintain the same margin for the formulator. On the other hand, consider 2K urethane coatings that sell for ~US $10–100/gal. The term '2K' refers to a two-component system: one as the base resin and the other as the curing agent. A typical example would be a base resin of an acrylic polyol with a polyisocyanate as a curing agent. The addition of a similar amount of fluorosurfactant would result in a price or cost increase that is a mere fraction of that in alkyd paints. This is much more attractive in addition to the fact that higher performance is required typically of 'high end' coatings. In summary, cost conscious formulators would use a fluorosurfactant only if it is absolutely necessary.

6.5.4 Environmental considerations

In 2001, 3M Company announced the withdrawal of a very successful line of fluorochemicals including fluorosurfactants because of potential issues around bioaccumulation and environmental impact. For most common commercially available fluorosurfactants, minimum surface tension is observed when n in $F(CF_2)_n$– is approximately 8 and the molecule is relatively small. The value of $n \approx 8$ makes these types of molecules potentially bioaccumulative in the environment[6–12]. The small molecule nature of these fluorosurfactants also makes them potentially bioaccessible. The general class of fluorosurfactants is often referred to as PFOS (perfluorooctyl sulfonate) and PFOA (perfluorooctanoic acid) depending on the specifics of the technology. The United States Environmental Protection Agency (EPA) is now considering regulatory action on these types of materials. Furthermore, numerous other international regulatory agencies are considering the fate of these types of materials in commerce.

6.5.5 Latest developments

As a result of governmental regulatory scrutiny, few new fluorosurfactants have been introduced commercially after 2001. One new type is based on a nonafluorobutanesulphonyl fluoride intermediate that is converted into fluorosurfactants through a sulphonamide process. These new materials have a perfluoroalkyl group with $n \leq 4$ and are not of as much concern from a regulatory perspective as are fluorochemicals with $n > 4$. They have been

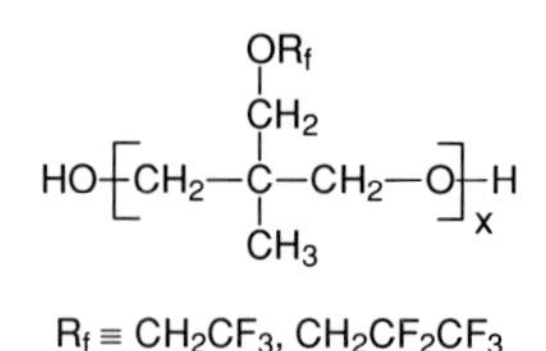

Figure 6.56 Generic PolyFox polymer structure.

developed by 3M and are polymers with the previously used Fluorad trademark replaced by Novec^{TM}.

A novel concept in fluorosurfactant technology was introduced by OMNOVA Solutions, Inc. with a range of fluorosurfactants such as wetting, flow and leveling agents for coatings and for other applications mentioned previously. This range of fluorosurfactants, marketed under the PolyFox trademark, was designed specifically to address the US EPA's concerns and to provide a satisfactory alternative to the consumer. The PolyFox range of surfactants is based on a platform of poly(oxetane) polymers [13] and the generic poly(fluorooxetane) structure is shown in Figure 6.56.

A schematic diagram for the synthesis of the poly(fluorooxetane) platform chemistry is shown in Figure 6.57.

The fluorinated oxetane monomer is prepared by Williamson ether synthesis using the appropriate fluorinated alcohol and bromomethyl methyloxetane under phase transfer catalytic conditions. The poly(fluorooxetane) precursor polymer is prepared by cationic ring-opening polymerization catalyzed by $BF_3 \bullet$ THF (tetrahydrofuran) with a hydroxyl-containing initiator (neopentyl glycol in the present case). Typically, $2 \leq x + y \leq 20$. Surface activity is introduced through chemical modification of end groups, such as sulfation on the polymer making them, in essence, hydrophilic while the perfluoroalkyl group, R_f, is considered the hydrophobe.

Generally, polymers are less bioaccessible to the environment and organisms based simply on size and are, therefore, of less concern to the U.S. EPA. The degree of polymerization, $x + y$, and functionality can be controlled readily allowing for the tailoring of the poly(fluorooxetane) platform to a particular application. More importantly, shorter

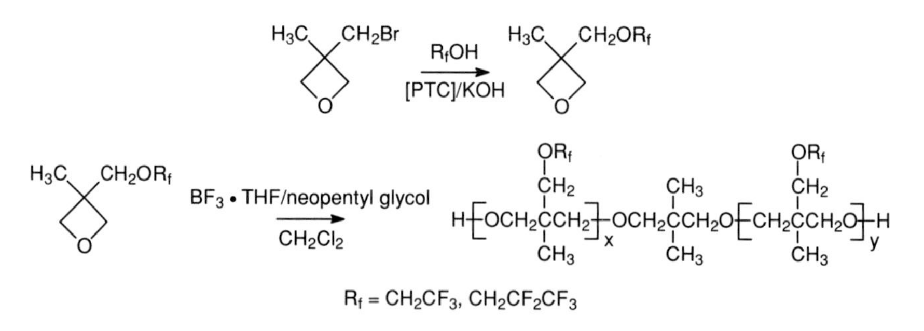

Figure 6.57 Synthesis of poly(fluorooxetane) surfactant precursor.

perfluoroalkyl chain derivatives of the poly(fluorooxetane) platform are used and these shorter perfluoroalkyl chain derivatives are of less concern to the US EPA. Due to low interfacial elasticities, poly(fluorooxetane) fluorosurfactants have little tendency to foam. If slight foaming is present, it does not persist long enough to introduce a defect in most coating systems but the same cannot be said for many commercially available, small molecule fluorosurfactants. Fluorosurfactants are often added to coatings as an insurance policy against certain defects such as cratering that can be caused by surface tension gradients in the coating. Surface tension gradients can be established by components that are not entirely compatible with the coating package. Often, fluorosurfactants are not entirely compatible with the coating package and, they themselves, can become a source of defects such as craters. This is observed commonly when an excess above what is needed for performance is added accidentally to a coating. PolyFox fluorosurfactants can and are designed to be compatible with specific coating systems reducing the potential of becoming defect nucleation sites.

Poly(fluorooxetane) fluorosurfactants have been shown to be effective wetting, flow and leveling additives in a number of aqueous and solvent-borne coatings. The poly(fluorooxetane) platform allows for the construction of molecules with unique architectures that are able to deliver very low surface tension without having to resort to longer perfluoroalkyl groups of concern to the EPA [14, 15]. Shown in Figure 6.58 are surface tension isotherms for the poly(fluorooxetane) fluorosurfactants in pH 8-buffered aqueous solutions along with a 'typical' fluorosurfactant 3M Fluorad™ FC-129 (removed from commerce). Structures are shown in Figure 6.59.

The experimental surface tension is far lower than predicted theoretically for fluorosurfactants based on short perfluoroalkyl groups.

Normally, fluorosurfactants used as wetting, flow and leveling agents are fugitive. This implies that they no longer serve a useful function after performing their primary function and can move about the bulk or surface of any object into which they are formulated. Ultimately, the final residence of a fluorochemical in this situation will be decided by the kinetics and thermodynamics of the particular formulation. This can have undesirable consequences as the unexpected presence of a surfactant at an inappropriate interface can

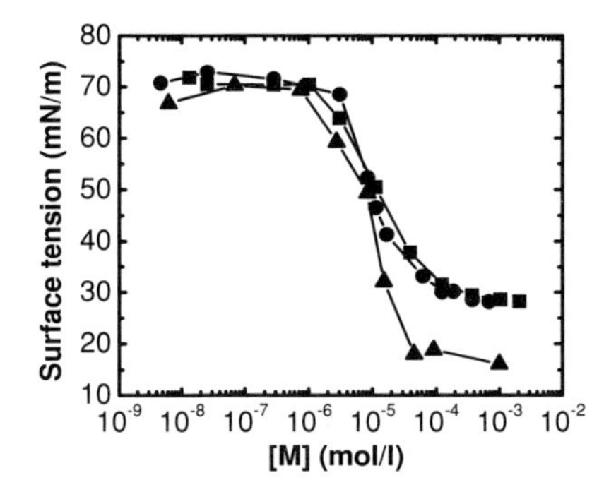

Figure 6.58 Surface tension isotherms for poly(fluorooxetane)s 1(■), 2(•), and 3M Fluorad™ FC-129 (▲) in pH 8-buffered aqueous solution. Structures are shown in Figure 6.59.

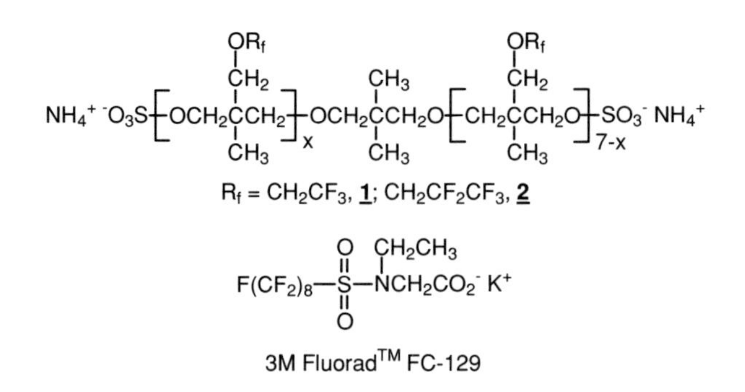

Figure 6.59 Structures for poly(fluorooxetane) surfactants 1, 2, and Fluorad™ FC-129.

cause the surface to be much more hydrophobic or hydrophilic than expected for the base formulation. Cratering, dewetting or delamination of composite coatings are one particular negative example that can be caused by the undesirable migration of a surfactant. Whilst some users desire only the wetting, flow and leveling attributes afforded by fluorosurfactants, others may decide to take advantage of the surface enrichment of a formulation by the permanent presence of a fluorosurfactant. Surface enrichment of a fluorosurfactant can lower the surface tension of the coating and provide additional benefits such as increasing the inherent stain resistance of a coating. Kinetically, the path to permanence is through the use of a reactive fluorosurfactant. This is accomplished by addition of a curable functional group such as a reactive acrylate that can homopolymerize (forming an interpenetrating network) or react with other functional groups present in the coating resin. Commercially, several reactive fluorochemicals are available; however, $n \approx 8$, as for typical fluorosurfactants and the environmental issues still exist. A reactive fluorosurfactant based on poly(fluorooxetane) chemistry with $n < 4$ has been introduced and the structure is shown in Figure 6.60.

It has been demonstrated that the incorporation of this type of reactive fluorosurfactant into a radiation curable acrylic system provides for a high gloss coating due to wetting, flow and leveling attributes, but results also in increased stain resistance of the coating.

In summary, these poly(fluorooxetane) fluorosurfactants have been designed specifically to take advantage of the ability to control the architecture of a molecule to exploit specific properties without leaving an undesirable environmental legacy.

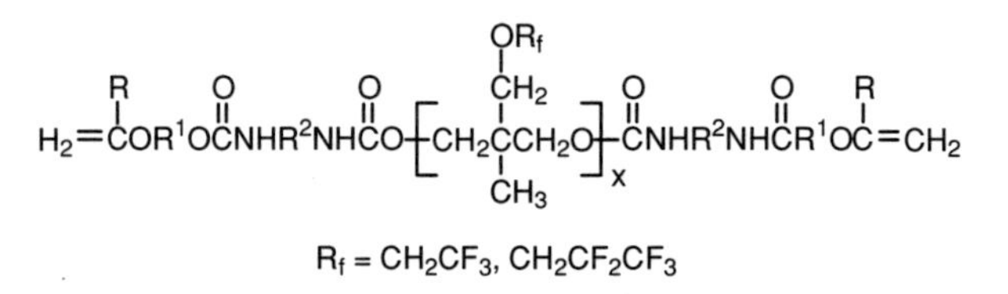

Figure 6.60 Generic acrylate-modified poly(fluorooxetane) reactive surfactant.

References

1. Kissa, E. (1994) *Fluorinated Surfactants-Synthesis, Properties and Applications.* Dekker, New York.
2. Thomas, R.R. (1999) Material properties of fluoropolymers and perfluoroalkyl-based polymers. In G. Hougham, P.E. Cassidy, K. Johns and T. Davidson (eds), *Fluoropolymers 2 Properties.* Kluwer/Plenum, New York, pp. 47–67.
3. Adamson, A.W. (1990) *Physical Chemistry of Surfaces*, 5th edn. Wiley, New York, p. 385.
4. Edwards, D.A., Brenner, H. and Wasan, D.T. (1991) *Interfacial Transport Processes and Rheology.* Butterworth-Heinemann, Boston.
5. Kornum, L.O. and Raaschou Nielsen, H.K. (1980) Surface defects in drying point films. *Prog. Org. Coat.*, **8**, 275.
6. Moody, C.A., Kwan, W.C., Martin, J.W., Muir, D.C.G. and Mabury, S.A. (2001) Determination of perfluorinated surfactants in surface water samples by two independent analytical techniques: liquid chromatography/tandem mass spectrometry and ^{19}F NMR. *Anal. Chem.*, **73**, 2200.
7. Ellis, D.A., Mabury, S.A., Martin, J. and Muir, D.C.G. (2001) Thermolysis of fluoropolymers as a potential source of halogenated acids in the environment. *Nature*, **412**, 321.
8. Moody, C. A., Martin, J. W., Kwan, W. C., Muir, D. C. G. and Mabury, S. A. (2002) Monitoring perfluorinated surfactants in biota and surface water samples following an accidental release of fire-fighting foam into Etobicoke Creek. *Environ. Sci. Technol.*, **36**, 545.
9. Martin, J. W., Mabury, S. A., Solomon, K. R. and Muir, D. C. G. (2003) Bioconcentration and tissue distribution of perfluorinated acids in rainbow trout (Oncorphynchus Mykiss). *Environ. Toxicol. Chem.*, **22**, 196.
10. Martin, J. W., Mabury, S. A., Solomon, K. R. and Muir, D. C. G. (2003). Dietary accumulation of perfluorinated acids in juvenile rainbow trout (Oncorphynchus Mykiss). *Environ. Toxicol. Chem.*, **22**, 189.
11. Dinglasan, M. J. A., Yun, Y., Edwards, E. A. and Mabury, S. A. (2004) Fluorotelomer alcohol biodegradation yields poly- and perfluorinated acids. *Environ. Sci. Technol.*, **38**, 2857.
12. Ellis, D. A., Martin, J. W., De Silva, A. O., Mabury, S. A., Hurley, M. D., Andersen, M. P. S. and Wallington, T. J. (2004) Degradation of fluorotelomer alcohols: a likely atmospheric source of perfluorinated carboxylic acids. *Environ. Sci. Technol.*, **38**, 3316.
13. Kausch, C. M., Leising, J. E., Medsker, R. E., Russell, V. M., Thomas, R. R. and Malik, A. A. (2002) Synthesis, characterization, and unusual surface activity of a series of novel architecture, water-dispersible poly(fluorooxetane). *Langmuir*, **18**, 5933.
14. Kausch, C. M., Kim, Y., Russell, V. M., Medsker, R. E. and Thomas, R. R. (2003) Surface tension and adsorption properties of a series of bolaamphiphilic poly(fluorooxetane)s. *Langmuir*, **19**, 7182.
15. Kausch, C. M., Kim, Y., Russell, V. M., Medsker, R. E. and Thomas, R. R. (2003) Interfacial rheological properties of a series of bolaamphiphilic poly(fluorooxetane)s. *Langmuir*, **19**, 7354.

Chapter 7
Relevant European Legislation

7.1 Biodegradability

Paul J Slater

The objective of this section is to provide sufficient background to support an understanding of the current and developing legislative framework with respect to surfactants and their biodegradability. In view of the high volumes of surfactants manufactured and used in detergents together with their high visibility, the legislation review will have a strong focus on the activity within this area. The legislation has been driven from the European Union so, for continuity, the European Directives and Regulations will be referenced rather than the Statutory Instruments used for implementation in the UK.

7.1.1 Biodegradation of surfactants

This process describes the breakdown of any organic substance by living organisms. It is a biological process and is influenced by the chemical composition, the prevailing conditions and the selection of the end point to be measured.

The major organic components of surfactants will support biodegradation: the only question is to what extent and how quickly this is achieved. In view of the large volumes of surfactants which find their way via direct or diffuse routes into the aquatic environments, it is important to understand biodegradability in the context of the potential risks to the environment.

The use of sulphonation chemistry together with the availability of cost effective hydrocarbon feedstock led to the production and wide-scale use of synthetic surfactants in detergents during the 1950s. In particular, tetrapropylene benzene sulphonate rapidly replaced soap-based products in many applications because it does not react with calcium and magnesium in hard water to form 'scum'. The benefits of this new surfactant together with an increasing consumer demand resulted in a substantial diffuse release of this substance into the environment.

The result was the appearance of significant amounts of foam at water treatment plants and in rivers where there was sufficient agitation. This image of foam in rivers with some being blown in the wind has had a marked impact both on the responsibilities taken by the detergent industry as well as on subsequent legislation.

The foam generated from the use of tetrapropylene benzene sulphonate was caused by a combination of two factors. The first of these was failure of the micro-organisms in the water treatment plants to break down the surfactant activity: in other words the lack of biodegradability. The second was the increased solubility of this surfactant compared to soap which made it available to generate foam. Persistent surfactant activity raises a less visible, but more basic potential environmental issue that is linked to bioavailability. The surfactant properties needed for effective cleaning can potentially disrupt critical mechanisms in aquatic organisms causing both acute and chronic toxicity problems. Ecotoxicity in most cases is associated with the parent surfactant compound. However, in a very limited number of cases, the breakdown to intermediates that are also toxic needs to be considered. The ability of surfactants to biodegrade rapidly to inert components is an important factor in considering their potential impacts related to ecotoxicity.

7.1.2 Sewage treatment plants

The level of sewage treatment varies significantly in different continents from virtually none up to 90% in North America (WHO Global Assessment of Water Supply and Sanitation 2000) [1]. Within the European Union, the reported figure of 66% is rapidly increasing as a consequence of the Urban Waste-Water Treatment Directive 91/271/EEC [2] amended by 98/15/EC [3]. Increasingly it can be assumed that diffuse disposal of surfactants in North America and Europe will pass through a sewage treatment facility.

During sewage treatment, surfactants, which in some cases will have already partly biodegraded 'en route', will be exposed to micro-organisms under aerobic conditions as they flow through the facility. The composition of the micro-organisms will adapt to the average composition of the sewage received so that, under normal conditions, the process is optimised for maximum breakdown of the organic material in the liquid phase.

Surfactants which can absorb on to the sludge fraction may be exposed to predominantly anaerobic conditions where the breakdown of active groups, such as sulphonates, can be restricted. Disposal of sludge to agricultural land can thus lead to certain surfactants reaching the terrestrial environment without significant levels of biodegradation. These potential concerns have led to increased legislative interest in anaerobic biodegradability within the detergent regulations (648/2004/EC) [4] and in local consent limits on sludge disposal to land. Karsa and Porter [5] have provided a detailed review of surfactant biodegradation. The key factor in relation to environmental impact and legislative control is the speed at which this process occurs for individual surfactants. In general, there are a number of factors that can influence the rate of biodegradation. The structure of the surfactant, which in most cases is represented by the general homologue distribution, not surprisingly, can have significant influence. The chemical bond structure, the degree of branching, alkyl chain length and level or type of alkoxylation can be seen to influence biodegradation even within similar surfactant groups.

The composition of micro-organisms needs to be optimal for the biodegradability of specific surfactant structures. However, the populations can adapt on longer term exposure to certain surfactant loads. There is evidence to suggest that modern bacterial populations can now break down surfactants which were chosen a few decades ago as negative standards in biodegradability testing. Certain surfactants can also demonstrate bacteriostatic or bactericidal properties that can impact the biodegradability process if they are present at high levels. As with all biological processes, temperature and time will impact the rate of biodegradation. This can alter due to the performance of sewage treatment plants during different seasons. In the event of storms, the residence time under controlled biodegradation conditions can be significantly shortened. Increased rates and levels of biodegradability will therefore reduce the risk of any environmental impact. This is even more critical where there may be limited treatment after disposal.

7.1.3 Measurement of biodegradability

The development of reliable measures of potential biodegradability is a key requirement in order to identify the potential environmental impacts of the increasing use of more sophisticated organic compounds.

In essence, the test procedure requires introducing a sample of the organic material into a healthy bacterial suspension under controlled conditions and measuring the time or rate at which a suitable end point is achieved.

As with all biological systems, controlling variables to give reproducible results requires a lot of standardisation. A detailed background of the development of biodegradability testing in relation to surfactants is given in Karsa and Porter [5], Painter [6] and Swisher [7].

In relation to surfactant biodegradability testing there are a number of terms which are used specifically in the legislative context. Two endpoints are commonly used in measuring the biodegradability of surfactants. One relates to measuring loss of functionality which, in the case of surfactants, is the 'surface active properties': this is also referred to as 'primary biodegradation'. The second is 'ultimate biodegradability' often interchanged with the term 'mineralisation' which is the complete breakdown of the organic component of surfactants to carbon dioxide, water and mineral salts.

The test methods used in current and proposed legislation are based primarily on the simulation of aerobic freshwater environments. There are essentially two types of biodegradability test methods, the 'die away type' and the 'sewage works simulation type'. The former operates with lower micro-organism levels which offers a more stringent measure of biodegradability and is more favoured in legislative criteria. The simulation type operates with higher levels of micro-organisms which provides more optimum conditions and is more often used in confirmatory measurements and in risk assessment situations.

The measurement of primary biodegradability is predominantly based on the use of 'die away' methodology over 19 days. The endpoint is currently based on the use of non-specific indicators such as methylene blue to indicate anionic active groups (MBAS) or bismuth to indicate non-ionic active groups (BiAS). While these indicators provide a generic endpoint, they do not respond to all the anionic or non-ionic surfactant types. Complete loss of measured functionality will tend to give high test result values of 80–90% providing the surfactant can easily undergo biodegradation.

Ultimate biodegradability measurement is based on oxygen use, carbon dioxide generation or carbon removal from the aqueous phase during the test. A complication in these tests is that the bacterial population will assimilate carbon as it grows, using the organic material as a food source. Ultimate 'die-away', tests used to define substances that are 'ready biodegradability', cannot therefore achieve the theoretical values for complete breakdown during the test. In practice, 60% for the measurement of oxygen/carbon dioxide and 70% for carbon removal after 28 days are seen as the lowest endpoints that confer 'ready biodegradability', a definition that has legislative implications. Test substance degradation in the 'ready' tests is only possible if the micro-organisms present are capable of using it as their primary substrate. The rate of biodegradation in the test has an initial lag phase where the micro-organism population is increasing substantially: this is followed by a linear increase to the point at which the rate reaches a plateau due to limiting factors. A positive '10 day window' is often required in the test protocol to confer 'ready biodegradability' to a substance. This requires the biodegradability to reach 10% and its pass level in the 'ready test' within a '10 day window'. This additional requirement can fail surfactants that may break down to fragments having different rates of biodegradation but, overall, would meet the 'ready pass' requirement after 28 days.

In addition to 'primary' and 'ready' tests, within legislative compliance there are two other types of ultimate biodegradability tests. 'Inherent' and 'simulation' tests provide valuable information on the level of biodegradability expected if the test material does not meet the 'ready' criteria. This provides essential information when carrying out risk assessments particularly on surfactants which can have high levels of ecotoxicity.

A list of the different test methodologies referred to in European Legislation is provided in Tables 7.1 and 7.2. In the case of surfactants, the choice of test method is determined by the physical properties of the homologues present. The potential toxicity of the surfactant in relation to the micro-organisms is a key consideration particularly where the surfactants are sufficiently toxic as to be considered as biocides. In addition, solubility and absorption need to be considered in the selection of the test method and in analysing the results. The use of dissolved organic carbon as an indicator of biodegradability has lost favour in newer legislation as there is a risk that loss of test material may be due to absorption on to either the equipment or sediment within the test vessel.

7.1.4 Legislation

The development of legislation specifically for surfactants was focused primarily on the detergent industry. In direct response to the concern over surfactants and foam on rivers, European legislation was developed in the early 1970s. The initial legislation 73/404/EEC [8] was intended to be a framework directive to address biodegradability in the four major surfactant groups, anionic, non-ionic, cationic and amphoteric. The enforcement of the legislation was intended to rest with the Member States which had the responsibility to determine compliance with the directives from analysis of formulated products. The extraction process, in practice, provided additional technical complications to establishing non-compliance with the subsequent legislation.

One of the intended 'daughter directives' was published on the same date: directive 73/405/EEC [9] provided the required test methodology for anionic surfactant primary biodegradability based on reduction of MBAS and set a test level of 80% or greater for the use of these surfactants in detergents. The second 'daughter' directive, 82/242/EEC [10], was published in the early 1980s and covered non-ionic surfactants based on reduction

Table 7.1 Regulatory biodegradability testing procedures used in surfactant legislation including the Detergents Regulation

Type	Test name	Test number	Parameter measured	Pass
Primary	Screening test (648/2000/EC)		Indicator or specific instrumental analysis of surfactant active substances	80% or greater
Primary	Reference method (confirmatory test 648/2000/EC)	EN ISO 11733	Indicator or specific instrumental analysis of surfactant active substances	80% or greater
Ready	Reference method	EN ISO 14593: 1999	CO_2	60% or greater
Ready	Modified Sturm test (67/548/EEC)	EC V.C.4-C ISO 9439 OECD 301B	CO_2 only	60% or greater
Ready	Closed bottle test (67/548/EEC)	EC V.C.4-E ISO N. 160 OECD 301D	O_2 only	60% or greater
Ready	Manometric respirometry (67/548/EEC)	EC V.C.4-D ISO 9408 OECD 301 F	O_2	60% or greater
Ready	MITI test (67/548/EEC)	EC V.C. 4-F OECD 301C	O_2	60% or greater
Ready	DOC die-away (67/548/EEC) Justification for use required	EC V.C.4-A ISO 7827 OECD 301A	DOC – dissolved organic carbon	70% or greater
Ready	Modified OECD screening (67/548/EEC) Justification for use required	EC V.C.4-B ISO 7827 OECD 301 E	DOC – dissolved organic carbon	70% or greater

of BiAS. This directive also recognised that certain non-ionic surfactants, used in very specific low foam industrial and institutional applications, would not meet the new primary biodegradability requirements. A derogation provided time for technological advances and reformulation work to meet both performance and biodegradability criteria. This important derogation was maintained until 31 December 1989 (86/94/EEC) [11] after which this exemption was never renewed. Directive 82/243/EEC [12] amended 73/405/EEC [9] providing updated methodology for anionic surfactants. No further EU legislation on surfactant biodegradability in detergents was ever published to support the original framework

Table 7.2 Expanded or additional regulatory biodegradability testing procedures used in dangerous substance legislation risk assessment

End point	Test name	Test number	Parameter measured	Pass
Ready	Modified Sturm test (67/548/EEC)	EC V.C.4-C ISO 9439 OECD 301B	CO_2/dissolved organic carbon	60/70% or greater
Ready	Closed bottle test (67/548/EEC)	EC V.C.4-E ISO N.160 OECD 301 D	O_2/dissolved organic carbon	60/70% or greater
Inherent	Semicontinuous activated sludge (SCAS) Off. J.E.C. 1988	ISO 9887 OECD 302A	Dissolved organic carbon	20% or greater for inherent
Inherent	Zahn-Wellens Off. J.E.C. 1988	ISO 9888 OECD 302B	Dissolved organic carbon	70% or greater for ultimate
Simulation	Activated sludge	ISO N.140 OECD 303A	Dissolved organic carbon	70% or greater expected

directive in the control of cationic or amphoteric surfactants. An outline of the obligations in force prior to the implementation of the Detergents Regulation (648/2004/EC) [4] is shown in Figure 7.1.

Following on from the existing legislation, detergent manufacturers worked with the EU to ensure that information on the content of surfactant types as well as correct dosage rates were available at the point of sale resulting in a voluntary agreement on detergent labelling, 89/542/EEC [13].

Prior to and following the development of legislation, leading surfactant suppliers and detergent manufacturers took major steps to implement developments based on improved biodegradability. The past two decades have seen responsible companies continuing to address potential environmental issues with improved methodologies in both testing and in developing risk assessment capabilities. There is however a significant diversity of surfactant types, proprietary manufacturing processes and specific applications that present potential economic barriers to data collection.

The perceived weakness in the current legislation for surfactants in detergents is that it does not cover all types of surfactants and it only considers primary rather than ultimate biodegradability. It is estimated that 50% of surfactants in use fall outside the scope of the current legislation. However, based on information collected from surfactant and detergent manufacturers prior to the detergents regulation publication, less than 100,000 tonnes or 3% of all the surfactants currently used in this sector within the EC are unlikely to meet the 'ready biodegradable' threshold.

Scientific reports of oestrogenic activity from short chain alkyl phenols [14], together with environmental hazard classifications and increasing voluntary pressures, have resulted in

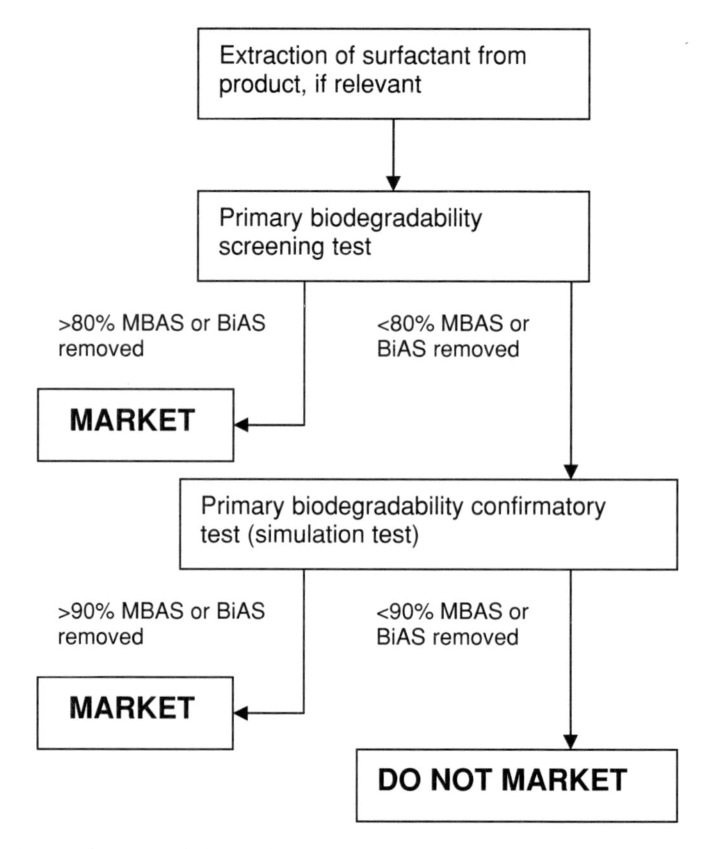

Figure 7.1 Testing framework for surfactants in washing, rinsing and cleaning products prior to the Detergents Regulation.

increased pressure for higher biodegradability standards in the recently published Detergents Regulation (648/2004/EC) [4].

Over the last decade, there has been a significant increase in the voluntary controls on surfactant biodegradability. The Paris Commission (PARCOM) was set up to protect the North-East Atlantic marine environment from land based contamination. The Commission, now known as OSPAR, is a representative body of governments from countries which border the North Sea. The PARCOM 92/8 (1992) [15] recommendation based on the persistent nature of nonylphenol ethoxylates triggered a series of voluntary agreements across Europe. A similar situation occurred following PARCOM 93/4 (1993) [16] recommendation for the phasing out of certain persistent cationic fabric conditioners.

Product ecolabels based on third party validation were starting to develop in the late 1970s with the German federal ministry setting up the 'Blue Angel' scheme. Currently, around 20 schemes are active on a global basis, covering a range of mainly consumer goods including chemical preparations. A review of major schemes and product ranges can be found in the Eco-labelling Guide (GEN 1999) [17]. A common factor of all the schemes in the criteria for chemical cleaning products is a ban on the use of alkyl phenol ethoxylate surfactants. More demanding requirements in relation to aerobic and, in some cases, anaerobic biodegradability of surfactants are needed for licensing under the EC Nordic Swan and Falcon schemes. The European Detergents Ingredient Database [18] supports these ecolabels with validated information on the biodegradability of common high volume surfactants. In markets where ecolabelled products have a high market share, this places pressure on formulators and manufacturers to use well documented 'ready biodegradable' surfactants. Increased activity in environmental procurement programmes and the recent directive on Integrated Product Policy (COM(2001)68) [19] are likely to increase interest in ecolabel schemes over the next few years.

The surfactant and detergent manufacturers have responded to try and coordinate action which meets the needs behind these voluntary activities. Voluntary programmes such as the International Association for Soaps, Detergents and Maintenance Products (AISE) 'Code of Good Environmental Practice' (98/480/EC) [20] achieved targets for reduction in poorly biodegradable ingredients. This is being carried forward in the Charter for Sustainable Development currently under development by AISE.

7.1.5 Detergents Regulation

The Detergents Regulation (648/2004/EC) [4] was published in March 2004 and is intended to come into effect in Member States on 8 October 2005. This legislation will replace all the existing detergent specific legislation and elevate the formal surfactant biodegradability criteria for placing detergents on the market. The scope of the new legislation is clarified with a definition for surfactants as well as a comprehensive list of what are considered to be detergent applications. This scope is far more prescriptive and inclusive than what existed in the previous legislation. The Regulation is not retrospective and therefore will apply to all products within its scope, placed on the market by the manufacturer or importers once it comes into effect. The Regulation is additional to other horizontal legislation on chemical substances or preparations which will be addressed later in this chapter.

The intention of the new legislation is to try to protect the aquatic environment by ensuring that only 'ready biodegradable' surfactants are used in 'non-exceptional' detergent applications. There is a derogation option which is open to low dispersive, industrial or institutional applications where there is a socio-economic benefit which exceeds the environmental risk. There is no derogation option for surfactants which fail to meet the criteria for 'primary biodegradability'. A schematic representation of the new legislation is shown in Figure 7.2 which can be compared with the previous legislation in Figure 7.1. This new legislation will be self-regulating with the manufacturers or importers being required to provide, on request to authorities, proof that their products conform to the requirements.

The test methods used have been selected from existing methods already within the legislative framework, updated and improved based on technical progress. The tests used in this legislation are shown in Table 7.1; the additional tests in Table 7.2 are those currently recognised in existing risk assessment procedures. In the case of 'primary biodegradability' test methods, the analysis options have been expanded so that they now cover all surfactants that fall within the scope. International testing standards (EN ISO 17025) [21] or good laboratory practice (GLP) is now specified for biodegradability testing although existing data will be accepted if they provide a comparable level of scientific quality.

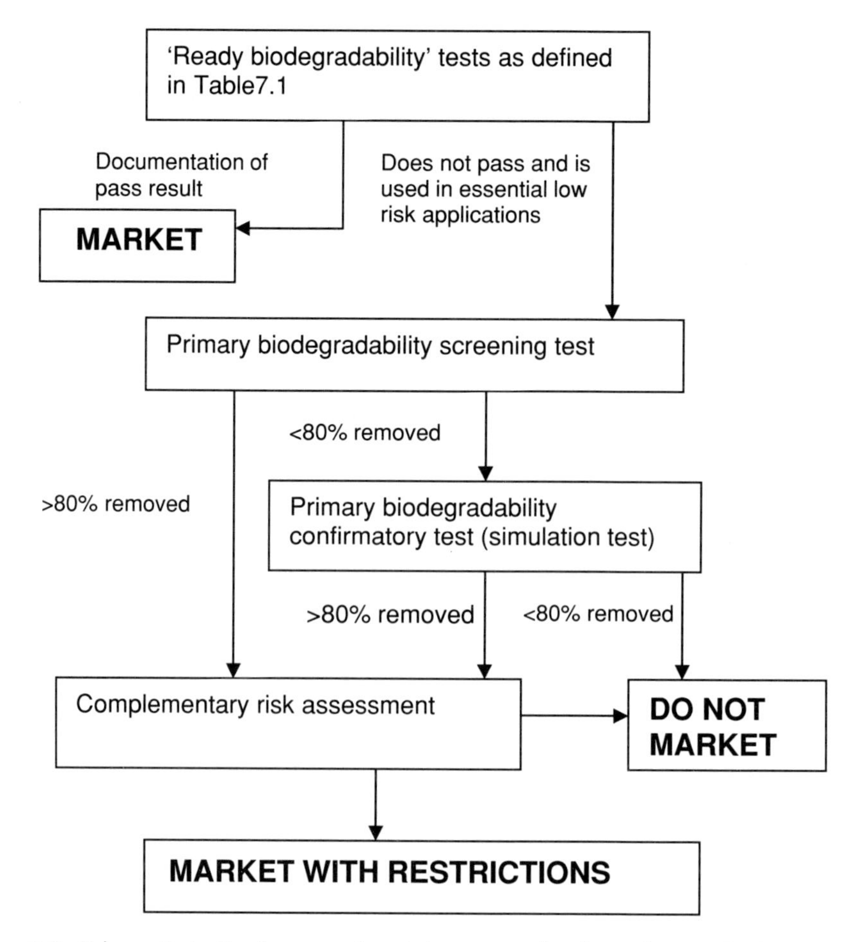

Figure 7.2 Schematic testing framework and documentation for surfactants in detergent products following the 'Detergent Regulation'.

Any derogation will require a 'complimentary risk assessment' which must include information on potential recalcitrant metabolites, rate of biodegradability and potential toxicity effects for target organisms. A tiered approach is recommended but the guidelines for Member States which are yet to be developed will hopefully differentiate this process from a 'Full Risk Assessment' (93/67/EEC) [22] within the context of the Dangerous Substances Directive (67/548/EEC) [23].

The Full Risk Assessment is a key part of the notification of new chemical substances (92/32/EEC) [24] and the risk assessment procedure deals with the risks to humans and the environment. The notification of new chemicals (EINECS) is based on required data sets depending on marketed volumes. A number of Technical Guidance Documents (TGD2, 2003) [25] have been produced to support the risk assessment process. These documents provide guidelines for predicting the environmental concentration in relation to the predicted no-effect concentration. The level and detail of biodegradation test data are critical factors in establishing these values. In light of the potential aquatic toxicity of the parent surfactant compounds and their predicted rate of biodegradation, it is important to balance environmental and economic needs within the risk assessment with real environmental improvements. This area has been the subject of considerable study including the investigation of exposure modelling for major surfactants in Dutch rivers [26]. The value of the risk assessment control step in the new legislation will depend largely on how the 'Complimentary Risk Assessment' will be implemented.

It is intended that even a successful derogation will carry restrictions on its use and that these will be subject to periodic review. The surfactants subject to derogation will be published in updates to this Regulation together with those that are banned or restricted. The banned and restricted surfactants will link this legislation to other horizontal legislation such as the Marketing and Use Restrictions (76/769/EEC) [27]. This is used to place restrictions or prevent the use of chemicals within the EU based on the risk assessment carried out within the dangerous substance legislation (67/548/EEC) [23]. This instrument has been used to formalise the ban on nonylphenol ethoxylates (2003/53/EC) [28] in a wide range of high volume applications such as industrial and domestic cleaning products, cosmetic products and pulp/paper manufacturing.

There is a timeline set of 6–18 months for the review of a submitted derogation and during this period the surfactant can still be marketed. It is expected that a phase out period of up to 2 years could be allowed in the event of an unsuccessful application or change in the derogation conditions. The manufacturers or importers will be required to maintain a technical file which identifies the source and validity of the test data for review by competent authorities.

In addition to the increased biodegradability criteria, this Regulation also requires full declaration of ingredients for all detergent preparations which should be made available to medical practitioners on request.

The existing voluntary labelling of detergents now forms part of this Regulation and this has been expanded to list any preservative used in the product and the presence of allergenic perfume ingredients (as defined by the Scientific Committee on Cosmetics and Non Food Products) in excess of 0.01%.

There is also intent within the Regulation that, in 2009, potential further restrictions could apply to surfactants based on their anaerobic biodegradability.

The potential impact of this legislation is most likely to be in specialised industrial areas such as bottle washing, food processing and metal cleaning. In these application areas there is a considerable number of low volume proprietary surfactants used in bespoke cleaning operations where the value of the cleaning application carries significant hygiene and/or economic benefits that are likely to support a potential derogation application. In reality, the cost of supporting even a simplified environmental risk assessment with the potential restrictions and regular review period is likely to rationalise considerably the number of these surfactants now on the market. This is currently being seen in response to the biocides directive (98/8/EC) [29] where the number of final registrations has dropped significantly.

A greater concern is the potential cost of product reformulation and process modifications required to accommodate alternatives that have different performance characteristics. This is likely to have significant impacts on small to medium sized businesses in this sector that will bear the brunt of the changes required in this Regulation.

The rationale behind the development of legislation based on surfactant biodegradability and potential risk is perfectly sound. Any concerns relate to the cost benefit analysis of the legislation in relation to the intended environmental improvement. The bigger challenge for both authorities and manufacturers is to address surfactant biodegradability within the context of sustainable activity. It is anticipated that subsequent progress in this area is more likely to come through voluntary activity and changes in market values over the next few years rather than through new legislation.

References

1. WHO Global Water Supply and Sanitation Assessment 2000 Report http://www.who.int/docstore/water_sanitation_health/Globassessment/GlobalTOC.htm (3 September 2004).
2. 91/271/EEC (1991) Urban waste-water treatment directive. Official Journal of the European Communities, L135 (30th May), 41–52.
3. 98/15/EC (1998) Amendment to the urban waste-water treatment directive. Official Journal of the European Communities, L067 (7th March), 29–30.
4. 648/2004/EC (2004) Regulation on detergents. Official Journal of the European Communities, L104 (8th April), 1–35.
5. Karsa, D.R. and Porter, M.R. (eds) (1995) *Biodegradability of Surfactants.* Blackie Academic & Professional, Glasgow.
   Painter, H.A. (1995) Testing strategy and legal requirements. In D.R. Karsa and M.R. Porter (eds), *Biodegradability of Surfactants.* Blackie Academic & Professional, Glasgow.
6. Painter, H.A. (1992) Anionic surfactants. In N.T. de Oude (ed.), *Detergents.* Springer-Verlag, Berlin Heidelberg, pp. 1–88.
7. Swisher, R.D. (1987) *Surfactant Biodegradation*, 2nd edn. Dekker, New York.
8. 73/404/EEC (1973) Biodegradability of surfactants in detergents. Official Journal of the European Communities, L347 (17 December 1973), 51–52.
9. 73/405/EEC (1973) Testing of anionic surfactants. Official Journal of the European Communities, L347 (17 December 1973), 53–64.
10. 82/242/EEC (1982) Biodegradability of non-ionic surfactants. Official Journal of the European Communities, L109 (22 April 1982), 1–17.
11. 86/94/EEC (1986) Second amendment to Directive 73/404/EEC. Official Journal of the European Communities, L80 (25 March 1986), 51.
12. 82/243/EEC (1982) Biodegradability of non-ionic surfactants. Official Journal of the European Communities, L109 (22 April 1982), 18–30.
13. 89/542/EEC (1989) Labelling of detergents and cleaning products. Official Journal of the European Communities, L291 (10 October 1989), 55–6.
14. Routledge, E.J and Sumpter, J.P. (1996) Oestrogenic activity of surfactants and some of their degradation products assessed using recombinant yeast screen. *Environ. Toxicol. Chem.*, **15**(3), 241–8.
15. PARCOM 92/8 (1992) Nonyl phenol ethoxylates. OSPAR Commission for the Protection of the Marine Environment of the North-East Atlantic. http://www.ospar.org/documents/dbase/decrecs/recommendations/pr92-08e.doc (19 Aug 2004).
16. PARCOM 93/4 (1993) Phasing out of cationic detergents DTDMAC, DSDMAC and DHTDMAC in fabric softeners. OSPAR Commission for the Protection of the Marine Environment of the North-East Atlantic. http://www.ospar.org/documents/dbase/decrecs/recommendations/pr93-04e.doc (19 Aug 2004).
17. GEN (1999) *The Ecolabelling Guide.* October 1999. Published by The Global Ecolabelling Network (GEN) Distributed by Terrachoice, Ottowa, Canada. www.terrachoice.com (19 August 2004).

18. DID List (2004) Detergents ingredient database. http://europa.eu.int/comm/environment/ecolabel/product/pg_did_list_en.htm (25 August 2004).
19. COM(2001)68 final, Green paper on integrated product policy. http://europa.eu.int/comm/environment/ipp/2001developments.htm (25 August 2004).
20. 98/480/EC (1998) Good environmental practice for household laundry detergents. Official Journal of the European Communities, L215 (1 August 1998), 73–5.
21. EN ISO 17025 (1999) General requirements for the competence of testing and calibration laboratories. International Organisation for Standardisation.
22. 93/67/EEC (1993) Commission Directive laying down the principles for assessment of risks to man and the environment of substances notified in accordance with Council Directive 67/548/EEC. Official Journal of the European Communities, L227 (8 September 1993), 9–18.
23. 67/548/EEC (1967) Classification, packaging and labelling of dangerous substances. Official Journal of the European Communities, L196 (16 August 1967), 1.
24. 92/32/EEC (1992) The seventh amendment. Official Journal of the European Communities, L154 (5 June 1992), 1–29.
25. TGD2 (2003): European Chemicals Bureau. Web Site http://ecb.jrc.it/tgdoc (2 September 2004).
26. Feijtel, T.C.J. Struijs, J. and Matthijs, E. (1999) Exposure modelling of detergent surfactants—prediction of 90th percentile concentrations in the Netherlands. *Environ. Toxicol. Chem.* **18**(11), 2645–52.
27. 76/769/EEC (1976) Restrictions on the marketing and use of certain dangerous substances and preparations. Official Journal of the European Communities, L262, (27 September 1976), 201–3.
28. 2003/53/EC (2003) Restrictions on the marketing and use of certain dangerous substances and preparations, nonyl phenol, nonyl phenol ethoxylates and cement. Official Journal of the European Communities, L178 (17 July 2003), 24–7.
29. 98/8/EC (1998) Placing biocidal products on the market. Official Journal of the European Communities, L123 (24 April 1998), 1–63.

7.2 Classification and Labelling of Surfactants

Richard J Farn

Over a period of 20 years, CESIO [1] has conducted reviews of the toxicological data available on marketed surfactants in order to provide guidance to its member companies on classification and labelling in accordance with European legislation and, initially, the requirements of Annex VI of the Fifth Adaptation to Technical Progress [2] of the Dangerous Substances Directive. The first review was carried out in 1984 [3] and the second several years later, being completed in 1990 [4]. Recommendations were made based on data available at that time on acute oral toxicity, skin and eye irritation and skin sensitisation studies.

In 1993, the European Union revised its criteria for classification and labelling of substances and preparations based on their potential to cause ocular lesions [5] and in 1995 new experimental data on the irritation potential of surfactant raw materials became available. This led CESIO, once again, to review its guidance on classification and labelling of anionic and non-ionic surfactants. It resulted in the increase in several classification and labelling recommendations. These revisions, together with the classifications for quaternary ammonium compounds and fatty amines and derivatives which remained unchanged from the 1990 Report, are contained in the latest report published in January 2000 [6].

The 2000 report gives details of why changes in classification have been made and lists the new proposed classifications together with the old ones for comparison purposes. It notes that the recommended classifications apply only to individual surfactants. When used in preparations, some antagonism may occur and such situations must be dealt with on an individual basis.

The 12th Adaptation to Technical Progress [7] of the Dangerous Substances Directive contained, for the first time, the requirements and criteria for classification and labelling of chemical substances 'dangerous for the environment' and the 7th Amendment [8] to the Dangerous Substances Directive gave the symbol 'N' for this which depicts a dead fish and a dead tree.

With regard to preparations, these were regulated by the Dangerous Preparations Directive [9] but the classification dangerous for the environment was not provided until the revision [10] was published in 1999 and which entered into force on 30 July 2002.

In order to advise its members on this new environmental classification requirement, CESIO joined forces with AISE [11] and subsequently distributed initial recommendations to its members in 1994. Further consultations took place in 2001 and 2002 to review the latest data and these resulted in the CESIO Report dated April 2003 [12] in which recommended classifications for the various categories of surfactants are listed together with supporting data.

Both CESIO reports [6, 12] are available for reading and can be downloaded from the Internet at www.cefic.be/cesio or obtained in hard copy from the CESIO Secretariat [13].

Acknowledgement

Permission from CESIO to refer to its publications is gratefully acknowledged.

References

1. Comite Europeen des Agents de Surface et leurs Intermediaires Organiques
2. (1983) Commission Directive 83/467/EEC. Official Journal of the European Communities, L257 (16 September), 1–33.
3. (1985) Tenside Detergents 4, 22.
4. (1990) Classification and Labelling of Surfactants, CESIO Report, 12 October 1990.
5. (1993) Commission Directive 93/21/EEC. Official Journal of the European Communities, L110 (4 May), 20–21.
6. (2000) Classification and Labelling of Surfactants for Human Health Hazards According to the Dangerous Substances Directive. CESIO Report, January 2000.
7. (1991) Commission Directive 91/325/EEC. Official Journal of the European Communities, L180 (8 July), 1–78.
8. (1992) Commission Directive 92/32/EEC. Official Journal of the European Communities, L154 (5 June), 1–29.
9. (1988) Commission Directive 88/379/EEC. Official Journal of the European Communities, L187 (16 July), 14–30.
10. (1999) Commission Directive 1999/45/EC. Official Journal of the European Communities, L200 (30 July), 1–68.
11. Association Internationale de la Savonnerie, de la Detergence et des Produits d'Entretien.
12. (2003) CESIO Recommendation for the Classification and Labelling of Surfactants as 'Dangerous for the Environment', April 2003.
13. CESIO, Avenue E.van Nieuwenhuyse 4, B-1160 Brussels, Belgium.

7.3 The European Commission's New Chemicals Strategy (REACH)

Philip E Clark

This chapter is written during the early days of the legislative process on the European Commission's New Chemicals Strategy and is a snapshot of the current position; the final regulation may be changed from that which is outlined here and the reader must bear this in mind when doing further work on the subject.

7.3.1 Introduction

It has long been thought, both by industry and politicians, that the existing chemicals legislation is ineffective, complicated and, at some times, contradictory. Following a meeting of the European Environment Ministers in Chester, England in June 1999, the EU Commission was instructed to revise its existing European chemicals laws. The Commission published a White Paper entitled *Strategy for Future Chemicals Policy* in February 2001. In May 2003, the first draft of the new chemicals regulation was presented and an 8-week Internet consultation started to allow stakeholders to put forward their concerns on the workability of the proposed regulations. Following considerable criticism of the proposal by industry and the National and European trade associations, a new draft was issued in September 2003 which was further modified taking account of the Commission's Interservice Consultation. On 29 October 2003, the draft of the new Chemicals Regulations was adopted by the European Commission, formally published on 28 November in the Official Journal of the European Communities – COM2003 0644(03) and will go forward for its first vote before the full Parliament. Once adopted, it will apply to all chemicals used within the European Community, including surfactants and will be known as REACH.

7.3.2 History of chemicals legislation

For the last 40 years, the European Community has been constructing chemicals legislation aimed at reducing the risk to human health and the environment from the use of chemicals. Risk is the assessment of the hazard of the chemical taking account of the likelihood of exposure to that chemical, i.e. risk = hazard × exposure.

To understand why the current New European Chemicals Strategy has been proposed, the development of chemicals legislation within Europe must first be considered.

In 1967, the Classification, Packaging and Labelling of Dangerous Substances Directive was implemented and this became known as the CPL Directive [1]. It has been amended many times and some of the significant changes were contained in various amendments. Amongst these were the requirement for a European inventory of chemicals on the market on 18 September 1981 (this is known as the EINECS [2] list) and the New Substances Regulation in 1992 [3] which required a risk assessment for all new notified substances being placed on the market.

In 1993, a regulation on the evaluation and control of risk from existing substances was published: this became known as the Existing Substances Regulation [4] and applied to all substances placed on the market since 18 September 1981. The number of reported substances at that time was 100 106 and it is estimated that 30 000 of these are currently marketed in volumes of over 1 tonne per year. Approximately 140 of these substances have been identified as substances of concern and have been subjected to extensive risk assessments under the Existing Substances Regulation.

Currently, any substance or preparation that gives rise to concern can be reviewed. Risk assessments and adequate analysis of the cost and benefits are required prior to any proposal or adoption of a regulatory measure controlling the chemicals or preparations containing them. If considered necessary, restrictions are applied under the Marketing and Use Directive [5]. This present system works well for the new substances but is weak in regard to the existing substances.

In line with requirement for better regulation, the European Commission issued a draft proposal for consultation on the Registration, Evaluation, Authorisation (and restriction) of CHemicals that has become known as the REACH system.

7.3.3 The principles behind REACH

Manufacturers and importers of substances in quantities over 1 tonne will have to submit a registration dossier to the European Chemicals Agency.

There are exemptions from registration for substances listed in annex II, which contains 69 entries:

- Polymers are exempt for the time being but the intention is to revisit them at a later date.
- Non-isolated intermediates are exempt but there is a reduced registration required for isolated intermediates either for on site use or transported to another site.
- Research chemicals are exempt for a period of up to 5 years but will be monitored by the Central Agency.
- Products registered under the Plant Protection Directive [6] or Biocidal Products Directive [7] are deemed registered under REACH.
- The final principle is that there should be no additional unnecessary testing on animals.

7.3.4 REACH

This system will be co-ordinated by a European Chemicals Agency which will be established under this new legislation: its location is yet to be agreed upon.

The draft legislation separates chemicals into two categories namely new substances that have already been assessed under 92/32 the New Substances Regulation and phase-in substances, that is all substances that are currently on the market but are not designated as new products under 92/32 the New Substances Regulation. This phase-in will be done over 11 years on the basis of the amount of material each manufacturer produces each year. In addition, all existing substances which are classified as CRM (carcinogen, mutagen and toxic to reproduction) categories 1 or 2 and are supplied in quantities of greater than 1 tonne per year will be registered in the first 3 years. The tonnage bands are 10–100 tonnes, 100–1000 tonnes and above 1000 tonnes of active material, not including solvents. The proposed legislation consists of the following stages.

(i) Pre-registration. For a company to make use of the phase-in period, it must pre-register with the European Chemicals Agency. This pre-registration will entail a manufacturer submitting information on every phase-in substance it manufactures or imports.

All information submitted under the pre-registration will be placed on a data base and will become participant in a Substance Information Exchange Forum (SIEF). The SIEF will be set up by the European Chemicals Agency. (The information required for pre-registration must be submitted, at the latest, 18 months prior to the start of the 1000 tonnes or more phased tonnage band or the 1 tonne or more phased tonnage band.) The aim of pre-registration and SIEF is to minimise the duplication of testing and allow data sharing amongst all manufacturers of that substance.

Data sharing is not completely compulsory. If the owner of the data does not want to share his data, then other manufacturers must proceed as if the data did not exist and sanctions may be imposed on the holder of the information. This is also aimed at starting consortia for joint applications.

(ii) Registration. This requires a dossier to be compiled and submitted to the European Chemicals Agency. The dossier should contain all the data available and a proposal for further testing that is considered necessary to fill any gaps in the information.

The information requirements are listed in annexes IV, V, VI, VII and VIII. Annex IV lays down the basic information needed about the manufacturer and the substance. Annex V lists the information needed for substances manufactured or imported more than 1 tonne. Annex VI lists the information needed for substances manufactured or imported more than 10 tonnes. Annex VII lists the information needed for substances manufactured or imported more than 100 tonnes. Annex VIII lists the information needed for substances manufactured or imported more than 1000 tonnes.

A Chemical Safety Report must accompany all dossiers for substances manufactured or imported over 10 tonnes.

Once the European Chemicals Agency has received the dossier, it will assign a registration number and a date for the registration and these will be communicated to the manufacturer or importer for future use. The Agency will conduct a completeness check to ensure all the elements required are present. If the dossier is incomplete the Agency will inform the registrant what further information is needed and set deadlines for its submission.

The Agency shall reject the registration if the registrant fails to complete it within the deadline set.

(iii) Evaluation. A competent authority in the country where the substance is manufactured will carry out the evaluation.

(iv) Dossier evaluation. The competent authority will examine any testing proposals set out in the dossier or in a downstream user report for provision of information specified in annex VII or VIII. On the basis of this examination, the competent authority may make one of the following three decisions:

(a) To carry out the test(s) proposed in the dossier and set deadlines for the study to be completed and the report submitted

(b) To carry out a modified test regime and set deadlines for the study to be completed and a report submitted
(c) To reject the testing proposal

Any information submitted subsequently will be subject to the same decision process. When the competent authority has completed its evaluation, it will notify the Agency.

(v) Substance evaluation. If the competent authority, on examination of any information on a registered substance either from its registration or from other relevant sources, has reason for suspecting the substance may present a risk to health or the environment, it may decide that further information is required for the purpose of clarifying the suspicion.

Examples for suspecting that a substance may need further investigation are:

- Structural similarity of the substance to a known substance of concern which is persistent and liable to bio-accumulate suggesting that the substance or one or more of its transformation products may cause concern
- Aggregated tonnage from the registrations submitted by several registrants

(vi) Authorisation. Substances of very high concern have to be properly controlled or replaced with a suitable alternative. To achieve this, it is intended to produce a list of substances which require authorisation or even restrictions on manufacture or use. This list is known as Annex XIII and is currently a blank sheet of paper. There are however guidelines for the type of substance that will appear on this list and these are:

(a) Substances meeting the criteria for classification as carcinogenic category 1 or 2
(b) Substances meeting the criteria for classification as mutagenic category 1 or 2
(c) Substances meeting the criteria for classification as toxic for reproduction category 1 or 2
(d) Substances which:

(i) Are persistent, bioaccumulative and toxic (PBT) in accordance with the criteria of Annex XII
(ii) Are identified as PBT on the basis of other evidence giving rise to an equivalent level of concern

(e) Substances which:

(i) Are very persistent and very bioaccumulative (vPvB) in accordance with the criteria of Annex XII
(ii) Are identified as vPvB on the basis of other evidence giving rise to an equivalent level of concern

(f) Substances such as those having endocrine disrupting properties which are identified on a case by case basis in accordance with the procedure set out in the Regulation, Article 56 and which are shown to give rise to a level of concern equivalent to the other substances on this list

A manufacturer, importer or downstream user must not place on the market any substance from the Annex XIII list unless its use has been authorised.

(vii) Restriction. If a Member State considers that a substance poses an unacceptable risk to health or the environment, it may propose to the Commission that a restriction be placed on the substance of concern. If it is agreed that there is an unacceptable risk to health or the environment from the manufacture or use of a substance which needs to be addressed on a community wide basis, then a restriction may be placed on that substance. This restriction will not apply to the use of a substance as an on-site intermediate. No substance subject to a restriction shall be placed on the market unless it complies with the conditions of the restriction.

(viii) Dossier content. A dossier consists of various sections of information relating to the registrant, the substance's physicochemical properties, the toxicological and ecotoxicological data and the Chemical Safety Report (if manufactured over 10 tonnes per year).

Annexes IV to XIII contain details of the requirements. The general information is as follows.

Annex IV

1. Information about the registrant.
   - The registrant's name, address telephone number, fax number and e-mail address
   - The name of the contact person
   - Location of the registrant's production and own use site(s), as appropriate

2. Identity of the substance. For each substance, the information given in this section must be sufficient to enable it to be identified. If it is not technically possible or if it does appear scientifically necessary to give information, the reason shall be clearly stated.

   The information required on the identity of the substance will include some of the following:

   - Name or other identifier such as
     - IUPAC name
     - Other names such as trade name
   - EINECS or ELINCS number
   - CAS number
   - Other identity codes if available
   - Information on molecular structure and formula
   - Information on composition including impurities and additives
   - Information on spectral data, HPLC or GC methods
   - Descriptions of other analytical methods

3. Information on manufacture and use(s) of the substance(s). The information on manufacture and use will cover:
   - The tonnages either manufactured or imported each year
   - If manufactured, a brief description of the process used
   - A generic description of the registrant's own use
   - The form of the substance supplied to downstream users
   - A generic description of identified downstream uses

- Composition and quantities of waste resulting from production
- Uses advised against

4. Guidance on safe use. Guidance on the safe use of the substance is required and this information should be consistent with the Safety Data Sheet for the following sections:

   - Section 4 – First Aid Measures
   - Section 5 – Fire Fighting Measures
   - Section 6 – Accidental Release
   - Section 7 – Handling and Storage
   - Section 14 – Transport Information

   If a Chemical Safety Report is not needed (i.e. below 10 tonnes), the following additional sections of information will be required at registration:

   - Section 8 – Exposure controls/personal protective equipment
   - Section 10 – Stability and reactivity
   - Section 13 – Disposal information on recycling and methods of disposal for industry and information on recycling and methods of disposal for the public

   The contents of a Safety Data Sheet may need to be changed from the current format prescribed by the Safety Data Sheet Directive [8], but the list of sections in Article 30 of the Regulation has the same 16 points.

   Annex V (substances manufactured or imported in quantities of 1 tonne or more)

5. Information on the physicochemical properties of the substance. The testing information required on the substance depends on the volumes manufactured or imported. All substances will require a base set of testing information and as the tonnage bands increase the amount of information increases. All substances will need physicochemical testing and the amount of information will change as the tonnages increase:

   - State of substance at 20°C and 101.3 kPa
   - Melting point/freezing point
   - Boiling point
   - Relative density
   - Vapour pressure
   - Surface tension
   - Water solubility (not required for polymers)
   - Water extractivity (only required for polymers)
   - Partition coefficient n-octanol/water
   - Flash point
   - Flammability
   - Explosive properties
   - Self-ignition temperature
   - Oxidising properties
   - Granulometry

6. Toxicological information. All substances will also need the following base set of toxicological information:

- Skin corrosion or irritation
- Eye irritation
- Skin sensitisation
- Mutagenicity

All substances will also need the following ecotoxicological information:

- Aquatic toxicity short-term testing on daphnia
- Ready biodegradation

Annex VI (Substances manufactured or imported in quantities of 10 tonnes or more). If a substance is produced or imported between 10 and 100 tonne the following additional testing is required:

- Acute toxicity either oral, inhalation or dermal
- Repeat dose toxicity, 28-day period
- Reproductive toxicity: the route of administration to be decided

The extra ecotoxicological information is:

- Growth inhibition on algae
- Short-term toxicity testing on fish
- Activated sludge respiration inhibition tests

Degradation testing is

- Biotic: the ready biodegradability testing
- Abiotic: hydrolysis as a function of pH

Also information on its fate and behaviour in the environment will be required.

Annex VII (Substances manufactured or imported in quantities of 100 tonnes or more). If a substance is produced or imported between 100 and 1000 tonnes per year the following additional testing is required:

- Physicochemical information on its stability in organic solvents and identity of relevant degradation products, dissociation constant and viscosity
- Toxicological information on repeat dose toxicity testing, sub-chronic toxicity study (90 day) and reproductive toxicity testing: a two-generation study will be needed
- Ecotoxicological information on long-term testing on daphnia, long-term testing on fish and fish early life stage toxicity testing will be needed

The biodegradation testing will be required:

- Simulation testing on ultimate degradation in surface water
- Soil simulation testing
- Sediment simulation testing

The identification of degradation products will be required.

The effects on terrestrial organisms such as earth worms, soil microorganisms and toxicity to plants is required.

Annex VIII (Substances manufactured or imported in quantities of 1000 tonnes or more). If a substance is produced or imported more than 1000 tonnes per year the following additional testing is required:

Reproductive toxicity

- Two-generation reproductive toxicity study

Ecotoxicological information

- Biotic degradation: further confirmatory testing on rates of biodegradation (aerobic and/or anaerobic) in environmental compartments (water, sediment, soil) with specific emphasis on the identification of the most relevant degradation products.
- Effects on terrestrial organisms:
  - Long-term testing on earthworms
  - Long-term testing on soil invertebrates
  - Long-term toxicity testing on plants
  - Long-term toxicity to sediment organisms
  - Long-term or reproductive toxicity to birds

(ix) Chemical Safety Report. A Chemical Safety Report must accompany all dossiers for substances manufactured or imported more than 10 tonnes per year. This is a written chemical safety assessment that risk assesses the physicochemical, toxic and ecotoxic properties and the potential exposure to the substance and should cover 90% of the known intended uses. Use information must come from the down stream users; an extended Safety Data Sheet will have to be passed on to all users in the supply chain.

(x) Information in the supply chain. Provision has been made in REACH to include the current Safety Data Sheet Directive. The Safety Data Sheet should be consistent with results of the chemical safety assessment. Where a Safety Data Sheet is not required (i.e. where the substance or preparation is not classified as hazardous), a minimal list of information needs to be supplied to the users. This information is:

(a) The registration number assigned to all registrations
(b) Whether the substance is subject to authorisation and any authorisation granted or denied under the authorisation procedure
(c) Any restriction imposed under the authorisation procedure
(d) Any other available and relevant information about the substance that is necessary to enable appropriate risk management measures to be identified and applied

7.3.5 The impact on the surfactant industry

This legislation will have an impact on the surfactant industry. There will be an extra cost of testing, an increase in work load to compile the dossiers which will put up costs and there will be cases where the product will be removed from the market place as it will no longer be commercially viable.

The surfactant industry is well established and the amount of information on these products varies according to their use and target market. It is safe to say that many are made in large tonnages and will fall into the 100–1000 tonne bands for registration; some may be produced in the more-than-1000-tonne band. It is very unlikely that the manufacturer will have all the relevant information for these bands and will need to have testing carried out to fill the gap. The cost of this testing may cause some producers to take the product off the market. This will have a knock on effect down the supply chain, necessitating reformulation. There will be a cost associated with reformulation, which will be particularly in the high tech areas such as aircraft cleaning, electronic industry and wetting agents for the agrochemical industry. Testing can take years for new products to be approved: what will happen in the interim?

Other problems will arise; in some cases the alternatives may be more harmful to the environment or to human health and their substitution would be a retrograde step.

Why is there this information gap? The answer lies in the drivers for test information during the legislative development (outlined above). Testing has been done to keep pace with the latest pieces of legislation and to classify the product accordingly. Classification produces the information thought necessary for the perceived risk at the time; the perception of risk has changed as technology has changed, as pressure from interest groups (usually single issue) increases or when epidemiology has shown previously unknown facts. Customer requirements also generate information on the product but this is quite often 'use specific' and bears no relevance to the information needed for classification. The new strategy sets out a suite of tests to be carried out (based on tonnage bands), many of which would not normally have fallen into the normal sphere of the producer's use or knowledge.

The strategy as currently drafted does not take into account the exposure factors when applying the testing requirements. This will cause extra testing to be carried out which has no relevance to the use of the substance.

The estimated cost of testing as shown below will put a disproportionate burden on the small and medium enterprise surfactant manufacturers where the tonnages may be in the 100–1000 tonne band but where the selling price for some of the products is in the £600–£800 per tonne range and the profit margins on these products as low as £80–120 per tonne.

7.3.6 Testing cost

These figures are based on the current cost using a well-known test house:

- The cost of the basic test information for 1 tonne is estimated at about £15 000
- The cost for testing information for 10–100 tonnes is estimated at about £100 000
- The estimated cost for extra testing for the 100–1000 tonnes band is £332 000
- The total cost of the test is about £447 000

Most producers will have some of the information needed but it is unlikely that they will have all the information for 100–1000 tonne band. To spread £332 000 cost over a single product that only makes £120 per tonne is not economically viable.

7.3.7 Conclusion

The European Chemicals Strategy is still in its early stages and there are many proposals for changes being put forward by governments, trade associations and companies, any one of which has the potential to reduce the burden on the surfactant industry and ensure a future for both large and small companies.

References

1. 67/548/EEC (1967) Classification, packaging and labelling of dangerous substances. Official Journal of the European Communities, L196 (16 August 1967), 1–98.
2. 79/831/EEC (1979/1981) Sixth amendment to Directive 67/548/EEC, Official Journal of the European Communities, L259 (15 October 1979), 10–28 and 81/437/EEC, Commission Decision of 11 May 1981 laying down the criteria in accordance with which information relating to the inventory of chemical substances is supplied by the Member States to the Commission, L167 (24 June 1981), 31–38.
3. 92/32/EEC (1976/1992) Seventh amendment to Directive 67/548/EEC. Official Journal of the European Communities, L154 (27 September 1976, 5 June 1992), 1–29.
4. 93/793/EC (1993) Council Regulation of 23 March 1993 on the evaluation and control of the risks of existing substances, plus corrigendum. Official Journal of the European Communities, L084 (5 April 1993), 1–75.
5. 76/769/EEC (1976) Marketing and Use Directive. Official Journal of the European Communities, L262 (27 September 1976), 201–03.
6. 91/414/EEC (1991) Plant Protection Directive. Official Journal of the European Communities, L230 (19 August 1991), 1–32.
7. 98/8/EC (1998) Biocidal Products Directive. Official Journal of the European Communities, L123 (24 April 1998), 1–63.
8. 2001/58/EC (2001) Safety Data Sheet Directive (second amendment to 91/155/EEC). Official Journal of the European Communities, L212 (7 August 2001), 24–33.

7.4 The Biocidal Products Directive

Mike Bernstein

7.4.1 Introduction

What does an authorisation directive on biocidal products have to do with surfactants? As will be demonstrated later, the scope of this Biocidal Products Directive (BPD) [1] covers wide areas of application. Several of these have not, at least in the United Kingdom, been previously thus regulated. Amongst the latter are the disinfectants. Several surfactants, mostly within the amphoteric and cationic classes, are well known disinfectants. The connection is not, however, limited to these and surfactants have been listed in most of the 23 product types that are covered by the Directive. The Directive itself is probably not the most complex directive or regulation ever issued by the European Commission but it must be high up on the list of such difficult pieces of legislation. Even though it was published in 1998, the EU Member States, Commission and other stakeholders were still, in 2004, discussing various aspects of interpretation. In addition, the transitional measures introduced by this Directive will continue to have an effect beyond 2010. For this reason it will be necessary to have a reasonable understanding of the principles of the Directive and its associated regulations before considering any impact and the reasons for that. Also, this can only be a snapshot, as there will continue to be further impact for the remainder of the transitional period at least.

7.4.2 The Directive

The BPD introduced a European scheme for authorisation of biocidal products. The Directive was closely based on an earlier authorisation directive dealing with plant protection products (agricultural pesticides). The context within which the Commission views biocidal products can be seen from the fact that the Directive was originally, in early drafts, called the 'Non-Agricultural Pesticides Directive'. It is therefore not surprising that the Biocidal Products Directive introduces an onerous authorisation scheme that aligns with a previously stated EC requirement to reduce the number of pesticides on the EU market. The Directive came into force on 14 May 1998 and had to be implemented in all Member States within 2 years. Many Member States, including the UK, were late.

7.4.3 Some definitions

A first issue arises because the definition of 'biocidal product' in the Directive is very general in scope. That definition is: 'Active substances and preparations containing one or more active substances, put up in the form in which they are supplied to the user, intended to destroy, deter, render harmless, prevent the action of or otherwise exert a controlling effect on any harmful organism by chemical or biological means'.

Equally wide ranging is the definition of 'active substance': 'A substance or microorganism, including a virus or a fungus, having general or specific action on or against harmful organisms'.

From these it can be seen that, although the Directive uses the term 'biocidal', there is no need to kill the organisms, to be within scope. Thus amongst others, repellents and attractants are included. It is also worth noting that the definition of products uses the word 'intended', thus it is not strictly necessary to claim an effect on a label in order to be within the scope of this Directive. The authorities do, of course, recognise the difficulties of proving intention and some reference on a label will most likely be necessary.

Two further definitions are needed here only to help understand some items referred to later.

Low-risk biocidal product. A biocidal product, which contains as active substance(s) only one or more of those listed in Annex IA and which does not contain any substance(s) of concern. Under the conditions of use, the biocidal product shall pose only a low risk to humans, animals and the environment.

Basic substance. A substance which is listed in Annex IB, whose major use is non-pesticidal but which has some minor use as a biocide either directly or in a product consisting of the substance and a simple diluent which itself is not a substance of concern and which is not directly marketed for this biocidal use.

7.4.4 Requirements and operation

Notwithstanding the transitional measures (see later), all biocidal products must be authorised before placing them on the EU market. An active substance is authorised by any one EU Member State authority, acting on behalf of the European Commission. In the case of a 'new' biocidal active substance, the applicant may choose the Member State for authorisation. An active substance dossier has to include a large amount of data on the active substance and on at least one example end use biocidal product. On authorisation, the active substance is listed on Annex I, IA or IB of the Directive and the biocidal product is also authorised. Annex I is to contain normal authorised active substances, IA will contain substances authorised for use in 'low-risk biocidal products' and IB will contain 'basic substances'. At the time of writing, these Annexes are empty. Other biocidal products based on such an authorised active have to be authorised by any one Member State, acting on its own behalf. This will also require a dossier with data on the biocidal product. On authorisation, the product should be able to obtain 'mutual recognition' from the other EU Member States. 'Mutual recognition' may only be refused if the use of a product is irrelevant in a particular Member State (e.g. if the target organism does not exist) and this can be justified to the European Commission.

Placing biocidal products on the market for research and development purposes will be permitted without authorisation. However, depending on the nature of the trial and the likely risk, information may have to be notified to or pre-assessed by the Competent Authority.

Annex V of the Directive refers to product types and describes 23 such product types which cover the whole of the EU biocidal product market. Authorisations will be applicable and limited to one or more of these 23 product types and may also carry other conditions. These product types are also of relevance to the transitional measures. As has already been indicated, surfactants have been listed in most of the 23 types.

Main group	Product type
1 Disinfectants and general biocidal products	
	1. Human hygiene biocidal products
	2. Private area and public health area disinfectants and other biocidal products
	3. Veterinary biocidal products
	4. Food and feed area biocidal products
	5. Drinking water disinfectants
2 Preservatives	
	6. In-can preservatives
	7. Film preservatives
	8. Wood preservatives
	9. Fibre, leather, rubber and polymerised material preservatives
	10. Masonry preservatives
	11. Preservatives for liquid-cooling and processing systems
	12. Slimicides
	13. Metalworking-fluid preservatives
3 Pest control	
	14. Rodenticides
	15. Avicides
	16. Molluscicides
	17. Piscicides
	18. Insecticides, acaricides and products to control other arthropods
	19. Repellents and attractants
4 Other biocidal products	
	20. Preservatives for food or feedstocks
	21. Anti-fouling products
	22. Embalming and taxidermist fluids
	23. Control of other vertebrates

7.4.5 Costs

There are several elements of cost that attach to the authorisation of a biocidal active substance or product. These cost elements are:

1. The Directive gives Member States the authority to reclaim from industry all costs in relation to running the authorisation scheme. These will consist of fees for specific tasks (authorisation, mutual recognition, etc.) together with overhead cost. In the UK the latter is provided by an annual charge on all companies involved in the biocidal active/product market and is called the General Industry Charge (GIC). Estimated fees from HSE (the Health and Safety Executive) – the UK Competent Authority for this Directive in the UK – are:

   (a) ~£ 65 000 for an active substance with one supporting product
   (b) ~£ 8000 for additional products based on an authorised active

(c) ~£ 5500 for an experimental authorisation
(d) £650 for notification of process-orientated research and development.

The GIC for the year 2003/2004 was £301.54 per liable company. It is expected that this will increase significantly in future years.

2. The costs of obtaining laboratory data to support an authorisation application dossier. There are no absolute figures for these costs but long agreed estimates are: ~£1–2 million for an active substance with supporting product and ~£50–100 000 for a product. In both cases, these are calculated on the basis of the fact that no suitable data are already available. It is currently not clear if old pre-GLP (good laboratory practice [2]) data will be acceptable. Although the manufacturers of some biocidal active substances will have suitable data to submit, many will have to generate anew the majority of the data for a dossier. Such costs, on their own, make it unlikely that companies will attempt authorisation of products that do not have a significant market presence. This is further exacerbated by problems over data protection, which will be considered later.
3. Many company man-hours will have to be spent in organising and analysing the data, working out waiving arguments, preparing and submitting dossiers, negotiating with authorities and other aspects of the requirements. One company estimated, following the submission of a dossier under the transitional measures, that at least 1 man-year had been spent on the process.

Taking all these costs together, it is clear that marketing a biocidal active substance or a product based on it, in the EU, is a major commitment and one that will not be undertaken lightly. This is certainly true for 'new' active substances and we shall not be seeing many of these, if any, in future. However it is also true of the existing actives and products on the EU market. These are very much in the majority and this brings us to the transitional measures already mentioned several times.

7.4.6 Transitional measures

Article 16 of the Directive refers to the transitional measures. These measures apply to active substances that were on the EU market before 14 May 2000 (existing biocidal active substances) and products based on them. Such actives and products could stay on the EU market, subject to existing individual Member State requirements, for a period up to 10 years from 14 May 2000. During the same 10-year period, the Commission was to commence a programme for the 'review' of these substances. The Directive also refers to a Regulation that was to be published and that would provide a suitable priority system.

In the event, a series of regulations is to be published, the first three of which have been published at the time of writing [3–5]. These Regulations specify requirements in two general areas:

1. All manufacturers or importers of biocidal active substances were to 'identify' or 'notify' their active substances to the European Chemicals Bureau (ECB). The difference between the options was:

   - *Identification.* A simple procedure, based on an Internet-available form, allowing continued marketing until 1 September 2006 only. These actives would be considered 'new' if further marketing would be required. They would have to be taken off the market until authorised as a 'new' active substance. Biocidal products based on these actives would also have to be taken off the market.

- *Notification.* A more complex procedure, which required use of an Oracle-based database package (IUCLID 4 special biocide version). Certain data, a sub-set of the dossier requirements, or waiving arguments were required. In addition, a commitment to eventual dossier submission on 'review' was needed together with timescales for obtaining the remaining dossier data. Notifications were to be submitted for specific product types. Uses under any of the 23 product types for which the 'notification' was not submitted would only be considered 'identified'.

A 'Notification' was checked by the ECB for completeness only and no check was done nor indication given as to suitability of the data for the eventual 'review' dossier. Both options had to be completed by 31 January 2003. This date had originally been 28 March 2002 but owing to the small number of 'identifications' or 'notifications' received, the period for submissions was extended.

The third of the published Regulations provides lists of those biocidal active substances that had been 'identified' and 'notified', with (in the latter case) the product types for which the 'notifications' had been accepted. These lists provide a first measure of the impact of this legislation on the biocide market and this will be considered later. Surfactants of various types have been 'notified' in most, if not all, of the 23 product types.

2. The priority system was specified and is based on product types. Four batches of several product types each were listed with date periods for 'review' dossier submission.

Batch	Product types	Dates for dossier submission
A	8 and 14	No later than 28 March 2004
B	16, 18, 19 and 21	No earlier than 1 November 2005 and no later than 30 April 2006
C	1, 2, 3, 4, 5, 6 and 13	No earlier than 1 February 2007 and no later than 31 July 2007
D	7, 9, 10, 11, 12, 15, 17, 20, 22 and 23	No earlier than 1 May 2008 and no later than 31 October 2008

Each 'review' is to be undertaken by a pre-specified Member State. The third published Regulation lists the Rapporteur Member States for the substances included in the first two batches. A further regulation will be published to identify the Rapporteur Member States for the substances in the third and fourth batches.

At the time of writing, the submission period under the first batch of reviews has ended. The number of submissions will be considered later. Once an active substance has been authorised under the 'review', all biocidal products based on that substance would be called up for dossier submission. The time gap between active dossier submission and product dossier request is not known at this time. It is likely however to be at least 2 or 3 years.

7.4.7 Data protection and 'free-riding'

Data protection has been one of the most contentious aspects of the Directive and is covered under Article 12, 'Use of data held by competent authorities for other applicants'. At the time of writing, the European Commission, the Member State Competent Authorities and industry are still discussing interpretation of Article 12. In its simplest terms, second or subsequent applicants will not be able to make use of data submitted by the first applicant unless the first applicant issues a 'letter of access' to these later applicants. It is the length of time over which these data are protected that causes the first area of difficulty. New active substances and biocidal products based on them are given 15 years of data protection from the date of Annex I, IA entry. This is not in contention. However, the situation is more complex for existing active substances. If data have been newly generated and not submitted to an authority anywhere within the EU prior to the BPD authorisation, then 10 years of data protection for those data, from the date of Annex I, IA entry, is given. If data have previously been submitted to a Member State anywhere within the EU, those data will get any relevant protection under any national scheme up to a maximum of 10 years from 14 May 2000 or, in the absence of national protection, until 10 years from 14 May 2000. This means that some existing biocidal active substances, especially those in the third and fourth 'review' batches will have very little data protection left by the time they achieve listing on Annex I, IA. This is making it very difficult for biocidal active substance manufacturers to justify the costs associated with the 'review'. This issue has been recognised by the Member State Competent Authorities and the European Commission and some modification may be forthcoming for data that were submitted to an authority after 14 May 2000.

A further problem, commonly known as the 'free-rider' issue, is related to the above. When an active substance is added to Annex I, IA, it will not be linked to the original applicant. A formulator will need a 'letter of access' from the original applicant in order to gain authorisation on a biocidal product based on this active substance. However, once the product authorisation is granted, the formulator may purchase this active substance from a company not associated with the original active authorisation. This is a further disincentive for active manufacturers on submission of 'review' dossiers. The European Commission has not been receptive to suggestions for closing this apparent loophole. It argues that it is for industry to sort out.

7.4.8 Impact

The main elements responsible for impact have already been discussed: costs in obtaining an authorisation; lack of data protection and the free-rider issue once an authorisation is granted. The ultimate measure of this impact is the number of biocidal active substances and biocidal products on the EU market. This can be broken down into stages and these are discussed below.

7.4.8.1 First stage impact

From an original exercise undertaken by CEFIC (the European trade association for the chemical industry) and contributed to by the European Commission, some 1500 biocidal substances were thought to be on the European market before 2000. At a first level of visible impact, this can be compared with the number of biocidal active substances 'notified' and 'identified' under the first 'review' Regulation.

Source	Number
CEFIC biocidal substances	~1500
Identified substances (lost from 2006)	~600
Notified substances	~365
Overall loss after 2006	~1135

In other words, some 75% of all biocidal active substances on the EU market before 2000 will have disappeared from the marketplace, by the end of 2006, for this reason alone.

7.4.8.2 Second stage impact

As already covered, the first batch of review dossiers had to be submitted by 28 March 2004. In theory, all 'notifiers' within that batch should have submitted a dossier. However, several 'notifiers' had advised the European Commission of withdrawal from one or more product types and at the time of writing, the Commission had published three such withdrawal lists. In addition, the experience has been that further 'notifiers' failed to submit dossiers. By the end of the submission period, the received figures compared with the expectations were as follows.

	PT 8 wood preservatives	PT 14 rodenticides
Expected (notified)	81	17
Actual dossiers received	38	13
Loss %	53%	23%

This represents a further significant loss in addition to that experienced from the first level impact. Several of the missing dossiers in product type 8 would have been for quaternary ammonium type surfactants. However, dossiers for other quaternary ammonium compounds were submitted in this group and this may simply represent product rationalisation. In general, at this time, it is not clear how many substance losses are due to long overdue product rationalisations. Nevertheless, as already suggested, owing to the data protection cut-off in 2010 for many substances, this loss of actives at the dossier stage could be even greater for the later 'review' batches.

7.4.8.3 Third stage impact

This relates to the fact that not all dossiers that have been submitted will be accepted, leading to the authorisation of the actives and their addition to Annex I. Indeed it has been noted that the majority of the 51 dossiers submitted in this first batch were not entirely satisfactory and would require further work. Some of these were accepted as complete but having issues for addressing during evaluation. Others were not even considered complete. Just how successful these dossiers will be is impossible to say at this time. One hopes that most dossiers will ultimately succeed and that the substances referred to will be added to Annex I. It is not possible, at this time, to assess the level of this impact.

7.4.8.4 Fourth stage impact

Most of the discussion on impact so far has dealt with active substances. In addition to this, of course, formulators of biocidal products will be impacted by the loss of active substances. In this case, the formulating companies will be forced to look for alternatives or drop their products. Once an alternative is found, price, stability, efficacy and user acceptability have to be considered. After this, the products themselves will have to be authorised. Thus, in this case, it is not just the authorisation costs that will have to be found.

7.4.8.5 Wider impact

Clearly, the Directive is designed to remove or control risks to humans, animals and the environment. In doing so, the Directive acknowledges that biocidal products are necessary for the control of organisms that are harmful to human or animal health and for the control of organisms that cause damage to natural or manufactured products. However, much of the loss of substances witnessed to date and expected in the near future, as described above, is or will be due to purely commercial considerations. It is not and will not be based on any consideration of the risks that the Directive is aimed to control. This in turn raises concerns that insufficient substances may be left in some areas to achieve adequate control of the intended targets. This is especially important where the target is known to be able to build resistance to specific control agents. The problems with resistant bacteria in hospitals are already public knowledge.

7.4.9 Final comment

The basis and likely impact of this Directive have been described. Apart from some very brief comments, there has been no specific reference to surfactants. This is because the Directive covers any chemical that falls within the scope as described and whilst this will include many surfactants, it also includes many other types. Having said this, there is no reason to suppose that surfactants will fare better than any other class of chemicals. Also, as indicated at the beginning, this chapter describes the situation in autumn 2004 when much of the impact due to the transitional measures is still to be realised. This Directive is considered by many to be 'over the top' control for this group of products. However, this is an inevitable consequence of biocidal products also being known as pesticides. Only time will tell whether this Directive will actually improve or increase the levels of risk to humans, animals or the environment.

References

1. (1998) Directive 98/8/EC of the European Parliament and of the Council of 16 February 1998 concerning the placing of biocidal products on the market. Official Journal L123 (24 September 1998), 1–63.
2. (2004) Directive 2004/10/EC of the European Parliament and of the Council of 11 February 2004 on the harmonisation of laws, regulations and administrative provisions relating to the application of the principles of good laboratory practice and the verification of their applications for tests on chemical substances. Official Journal L 050 (20 February 2004), 44–59.
3. (2000) COMMISSION REGULATION (EC) No 1896/2000 of 7 September 2000 on the first phase of the programme referred to in Article 16(2) of Directive 98/8/EC of the European Parliament and of the Council on biocidal products. Official Journal L 228 (8 September 2000), 6–17.
4. (2002) COMMISSION REGULATION (EC) No 1687/2002 of 25 September 2002 on an additional period for notification of certain active substances already on the market for biocidal use as established in Article 4(1) of Regulation (EC) No 1896/2000. Official Journal L 258 (26 September 2002), 15–16.
5. (2003) COMMISSION REGULATION (EC) No 2032/2003 of 4 November 2003 on the second phase of the 10-year work programme referred to in Article 16(2) of Directive 98/8/EC of the European Parliament and of the Council concerning the placing of biocidal products on the market, and amending Regulation (EC) No 1896/2000. Official Journal L 307 (24 November 2003), 1–96.

Chapter 8
Relevant Legislation – Australia, Japan and USA

8.1 Relevant Legislation – Australia

John Issa

8.1.1 Introduction

Australia is a country which is a federation of six states and two territories. Each of these States and Territories has its own government and laws. In addition there is an overarching Federal Government which governs the whole country and has a set of federal laws. Consequently, Australia has both Federal and State Agencies with responsibilities for the regulation of chemicals. Furthermore, in each jurisdiction chemicals are regulated by several different agencies depending on the use of the chemical. As can be seen from previous chapters, surfactants have many different applications including use as emulsifiers, water treatment chemicals, cosmetics, detergents, in agricultural/pesticide formulations, fire fighting foams, as sanitisers and biocides, etc. It is not surprising, therefore, that surfactants come under the scope of several different regulatory agencies and regulatory instruments in Australia.

Table 8.1 contains a list of the chemical types and the Agency responsible for regulating that type of chemical.

The role of these agencies and their empowering legislation will be discussed below.

8.1.2 National Industrial Chemicals Notification and Assessment Scheme

The importation and manufacture of new chemicals in Australia is regulated under Industrial Chemicals (Notification and Assessment) Act 1989 [1] which is administered by National Industrial Notification and Assessment Scheme (NICNAS) [2]. NICNAS was operated within the National Occupational Health and Safety Commission until November 2001, when it was transferred to the Therapeutic Goods Administration (TGA).

The Australian scheme is mostly based on the European chemicals notification scheme (EINECS), except that polymers are also covered. The requirements for notification of polymers are similar to those in the United States and Canada. The scheme commenced operations on 18 July 1990.

Table 8.1 Chemical regulatory agencies in Australia

Type of chemical	Federal agency	State and territory agency
Industrial chemicals	• National Industrial Chemicals Notification and Assessment Scheme (NICNAS)	Approvals at Federal level. Labelling, transport, storage, handling and compliance under State and Territory OH&S Authorities and Environment Departments
Cosmetic ingredients	• National Industrial Chemicals Notification and Assessment Scheme (NICNAS) • National Drugs and Poisons Scheduling Committee (NDPC) • Australian Consumer and Competition Commission (ACCC) • Therapeutic Goods Administration (TGA)	State Health Departments
Agricultural chemicals	• Australian Pesticides and Veterinary Medicines Authority (APVMA)	State Health Departments and Agriculture Departments
Consumer chemicals	• National Drugs and Poisons Scheduling Committee (NDPSC)	Approvals at Federal level
Food ingredients	• Food Standards Authority for Australia and New Zealand (FSANZ)	State Health Departments
Therapeutic goods (drugs and devices)	• Therapeutic Goods Administration (TGA)	Approvals at Federal level
Dangerous goods	• The Department of Transport& Regional Services (DOTARS)	State and Territory OH&S Authorities and Environmental Departments
Hazardous substances	• National Occupational Health and Safety Commission (NOHSC)	State and Territory OH&S Authorities

The objectives of the NICNAS scheme are:

- Protection of people and the environment by determining the risks to occupational health and safety, to the public and to the environment which are associated with the importation, manufacture or use of a chemical
- Providing information and making recommendations to State authorities which have responsibility for regulating industrial chemicals
- Maintaining Australia's obligations under international agreements
- Collecting statistic in relation to chemicals

The Act excludes pharmaceutical chemicals and their intermediates, agricultural and food chemicals. Thus, any chemical which does not fall under these categories is, by definition, an industrial chemical. Thus, ingredients in cosmetic products come under the scope of this Act since they are not excluded.

There are two parts to NICNAS; the New Chemicals Assessment and the Existing Chemicals Assessment programmes. A new industrial chemical is defined as one which is not listed on the Australian Inventory of Chemical Substances (AICS). Conversely, an existing chemical is one which is listed on the inventory. A new chemical must be notified and assessed prior to its introduction (through import or local manufacture) into Australia. New chemicals which have undergone assessment by NICNAS are added to the AICS, but this does not occur until 5 years after assessment. This provides the applicant company some degree of exclusivity to cover its costs. Once the chemical is entered on the inventory, other companies can import the chemical without notification. However, under a recent amendment of the Act, notifiers can opt to have the chemical listed on the AICS immediately after the NICNAS assessment has been completed.

The assessment which is undertaken by NICNAS covers the assessment of the health and aquatic toxicity hazards of the chemical, occupational exposure, public exposure and environmental exposure and fate. A risk assessment is performed and recommendations are made to control and minimise the risks. The results of the assessment are published in a report which is made available to the public via the NICNAS Web site [3].

The AICS was collected over a 13-year period commencing in 1977 in preparation for the introduction of an industrial chemical assessment scheme. Companies were requested to have their chemicals, which were in commerce in Australia, nominated and listed on the AICS. The inventory was closed off when the NICNAS commenced operations in 1990. There are approximately 40 000 chemicals on the AICS. The AICS also has a confidential section, which at present has less than 1000 chemicals. The confidential status of chemicals on this section of the AICS is reviewed once every 3 years. Chemicals must meet strict criteria before they can be relisted on the confidential section. If the confidentiality claim is rejected by NICNAS then the chemical is transferred to the non-confidential AICS. Similarly, for assessed new chemicals, there is an opportunity for a company to elect to have its chemical listed on the confidential section of the AICS. However, they must meet the criteria which are designed to balance commercial interests for secrecy against the public interest for disclosure (public right to know).

The non-confidential AICS can be accessed on-line via the NICNAS Web site. Alternatively, a request to search the AICS can be made to NICNAS. Furthermore, a search of the confidential section of the AICS can be requested but only through this mechanism. The AICS is also available on the Chemical Abstract Services Chemlist database or National Chemical Inventories (NCI) [4].

There are several different notification categories which can be used by applicants depending on the type of chemical (polymer or discrete chemical), annual import volume and other exposure or intrinsic hazard criteria. The level of notification and the extent of data required to support the application are matched to the *a priori* level of risk posed by the chemical.

More recently, several new notification categories have been introduced under the Low Regulatory Concern Chemicals (LRCC) initiative [5]. These changes to the scheme have been introduced in order to streamline the notification process, reduce the regulatory burden on and costs to industry while maintaining adequate safeguard to assess the risk to human health and to the environment from the introduction of a new industrial chemical. One major innovation amongst the LRCC changes is the introduction of a set of self-assessment categories. These are for chemicals and polymers which are of low hazard and risk. Notifiers conduct their own risk assessment by completing a template assessment report. This report is submitted to NICNAS for checking and once it is of an acceptable standard it is published as one of the NICNAS assessment reports.

The available new chemical notification categories and processing times are presented in Table 8.2.

The following exemptions from notification are available:

- Articles
- Incidentally produced chemicals/impurities
- Reaction intermediates
- Research and development <100 kg/year
- Chemicals <100 kg/year
- Cosmetic ingredient introduced at concentration of <1% in end-use product (no annual volume limit)
- Cosmetic ingredient introduced at 10–100 kg/year
- Cosmetic ingredient introduced at <10 kg/year

Several chemicals have been reviewed by the existing chemicals review programme over the past 14 years. Table 8.3 contains a list of existing chemicals which have been reviewed and for which an assessment report has been published.

The outcomes of the review are similar to those for new chemicals. However, existing chemicals may have health and environmental hazard classification being assigned and recommendation for either setting or revising an existing occupational exposure standard.

NICNAS is involved in several international cooperative projects relating to chemical assessment and management. Over the years, NICNAS has been a keen contributor to the concept of exchanging assessment information between countries which have comparable chemical assessment schemes. NICNAS has membership on the OECD New Chemicals Task Force which reports each year to the OECD Joint Meeting of the Chemicals Committee and the Working Party on Chemicals, Pesticides and Biotechnology. The principal aim of the Task Force is to reduce the burden on industry by aligning notification requirements between the countries and by standardising assessment reports. The harmonisation will facilitate work sharing and the exchange of information on new chemicals between member countries. The Task Force has developed a work plan consisting of the following seven work elements, namely:

(i) Learning from experience with multilaterals in sharing and comparing assessments
(ii) Standardised notification form
(iii) Standardised formats for assessment reports
(iv) Hazard assessment – promoting the exchange of common elements
(v) Minimal and no-notification requirements for low concern and exempt chemicals (exclusions and exemptions)
(vi) Confidential business information (CBI) – proprietary information
(vii) Mutual acceptance of notifications (MAN)

Table 8.2 NICNAS new chemical notification categories

Notification category	Description	NICNAS time frames (days)
Standard notification	For chemicals introduced at >1000 kg/year or for polymers with number average molecular weight (Mn) <1000 daltons. The full set of physicochemical, mammalian toxicity (acute oral, acute dermal, skin and eye irritation, skin sensitisation, mutagenicity, chromosomal aberration, repeat dose toxicity) and aquatic toxicity (acute fish, Daphnia immobilisation, algal toxicity, biodegradation) studies are required	90
Limited notification	Chemical introduced at <1000 kg/year or for polymers with Mn >1000. Physicochemical data are required	90
Polymer of low concern (PLC)	For polymers which meet specific criteria. The criteria are the same as those under the US TSCA. These are polymers with Mn >1000 and have low levels of low molecular weight species, do not contain reactive functional groups, are not polycationic, are stable and not intended to undergo further reaction. Biopolymers which meet the criteria can also be treated as polymers of low concern. However, those with high BOD are excluded	90
LRCC polymer of low concern (self-assessment)	Most polymers which meet the PLC criteria can be submitted as self-assessed notifications	28
Low risk chemical (self-assessment)	These are chemicals which are not classified as hazardous and are used in applications where exposure to workers, the public and the environment is minimal	28
Certificate extension	Allows a secondary notifier to obtain an extension to the assessment certificate which has already been issued to the primary notifier. The approval of the primary notifier is required	50
Commercial evaluation permit	This is a permit which can be obtained for a maximum of 2 years and for a maximum of 4000 kg of the chemical or polymer. This permit is usually obtained as a precursor to one of the main notification categories above. The intention is to evaluate the market feasibility of the chemical before committing to a full notification. Chemicals which are used in consumer end-use products do not qualify for this permit	14

(*Continued*)

Table 8.2 (*Continued*)

Notification category	Description	NICNAS time frames (days)
Low volume chemical permit	For chemicals introduced at <100 kg per year. This permit can be used for chemicals which do not qualify for the <100 kg exemption	21
Early introduction permit	This is a permit which can be obtained for a non-hazardous and non-dangerous goods chemical or polymer. It is submitted in conjunction with one of the full notification categories and allows the notifier to introduce and use the chemical ahead of the assessment of the full notification being completed	28
Controlled use permit	Some specific uses where the risks are minimal can be introduced under this permit. Examples include chemicals which are intended for export with no internal marker use	

8.1.2.1 Australian Pesticides and Veterinary Medicines Authority

The Australian Pesticides and Veterinary Medicines Authority (APVMA)[6] is the federal authority which is responsible for reviewing and registering agricultural and veterinary products. Some surfactants for which a pesticidal claim is made come under the scope of the APVMA. Thus surfactants which are used as biocides, as adjuvants (which assist in the distribution and penetration of the active into the plant and enhance the activity), to kill pests in post-harvest treatments and in antifouling or water treatment processes are regulated by the APVMA.

Prior to the establishment of this federal, central Authority, the registration of all agricultural and veterinary chemical products was an individual State/Territory government responsibility. From the early 1960s the Federal government coordinated and provided a national approach to the assessment and clearance of selected classes of agricultural and veterinary chemical products. In July 1991, the Commonwealth, State and Territory Ministers responsible for agricultural issues decided to establish the National Registration Scheme. This national scheme would be responsible for the registration of agricultural and veterinary chemicals up to the point of retail sale, with States and Territories authorities being responsible for control of use.

In August 1992, the Commonwealth announced that it would establish a National Registration Authority for Agricultural and Veterinary Chemicals (NRA) to undertake registration activities, with associated policy issues being the responsibility of the Department of Primary Industries and Energy.

Legislation to establish the NRA received Royal Assent on 24 December 1992 and came into effect on 15 June 1993. On 30 July 2004, the name of the Authority was changed to The Australian Pesticides and Veterinary Medicines Authority (APVMA).

Table 8.3 Existing chemicals assessed under NICNAS

Chemical	Report no.	Report date
TGIC – triglycidylisocyanurate	PEC/1	April 1994
TGIC – triglycidylisocyanurate – secondary	PEC/1s	Feburary 2001
Savinase – proteolytic enzymes in detergents	PEC/2	Feburary 1993
Glutaraldehyde	PEC/3	July 1994
HCFC – 123	PCE/4	March 1996
HCFC – 123 secondary notification	PEC/4s	July 1999
Sodium ethyl xanthate	PEC/5	May 1995
Sodium ethyl xanthate secondary notification	PEC/5s	February 2000
2-butoxyethanol	PEC/6	October 1996
1,4-dioxane	PEC/7	June 1998
Trichloroethylene	PEC/8	March 2000
Chrysotile asbestos	PEC/9	February 1999
Arcylonitrile	PEC/10	February 2000
N-vinyl-2-pyrrolidine (NVP)	PEC/11	April 2000
Glycolic acid	PEC/12	April 2000
para-dichlorobenezene	PEC/13	December 2000
ortho-dichlorobenzene	PEC/14	February 2001
Tetrachloroethylene	PEC/15	June 2001
Short chain chlorinated paraffins (SCCPs)	PEC/16	June 2001
Trisphosphates	PEC/17	June 2001
Ammonium, potassium and sodium persulphate	PEC/18	June 2001
Hydrofluoric acid (HF)	PEC/19	June 2001
Polybrominated flame retardants (PBFRs)	PEC/20	June 2001
Benzene	PEC/21	September 2001
Limonene	PEC/22	May 2002
Acrylamide	PEC/23	May 2002
Methylcyclopentadienyl manganese tricarbonyl (MMT)	PEC/24	June 2003
Alkyl phosphate anti-valve seat recession additive	PEC/25	July 2003
Sodium alkylbenzene sulfonate anti-valve seat recession additive	PEC/26	February 2004

All new agricultural and veterinary chemical products must be registered by the APVMA before they can be supplied, distributed or sold anywhere in Australia. In addition, active constituents within these products must be approved by the APVMA either before or at the same time that the product is being registered.

Agricultural chemical products include any substance or organism used to:

- Destroy, stupefy, repel, inhibit the feeding of or prevent pests in plants or other things
- Destroy a plant or modify its physiology
- Modify the effect of another agricultural chemical product or
- Attract a pest for the purpose of destroying it

This encompasses all herbicides, insecticides and fungicides. Dairy cleansers for on-farm use, crop markers, insect repellents for use on humans, swimming pool disinfectants and algaecides, rodenticides, antifouling paints, timber preservatives and household and home garden products for pest and weed control come under the scope of the APVMA.

The data requirements for an APVMA submission are contained in the Requirements Manual [7]. The extent of data required to support an application is dependent on the registration category. Where a product contains a new active ingredient, the full data set is required. For new products which are based on existing actives, less data are required. Furthermore, the closer a new product resembles an existing product (i.e. in formulation, in pest/crop claims, use situations, application rates), the less data are required. The general categories of data required are:

- Chemistry and manufacture of product
- Toxicology of the product
- Metabolism and kinetics
- Residues
- Overseas trade aspects of residues in food commodities
- Occupational Health and Safety
- Environment
- Efficacy and crop safety

8.1.3 Food Standards Australia New Zealand

Food Standards Australia New Zealand (FSANZ) is a joint agency of Australia and New Zealand. It is a statutory authority that develops food standards for composition, labelling and contaminants, including microbiological limits, that apply to all foods produced or imported for sale in Australia and New Zealand.

In Australia, the development and enforcement of laws relating to the manufacturer and sale of food are the responsibility of the individual States and Territories. The responsibilities of the Federal Government with respect to food are largely restricted to import/export controls, through the Australian Quarantine Inspection Service (AQIS) [8], and the management of coordinated food standards development through Food Standards Australia New Zealand [9]. Under this system, each jurisdiction has its own individual food laws although significant steps have been taken to harmonise the requirements in each jurisdiction by developing a uniform model food act which the states and territories have adopted into their own legislation.

The Food Standards Code [10] is the main regulatory instrument which controls the quality of food, contaminant levels, approved additives, processing aids, sanitisers and disinfectants and these standards are performance based. If a chemical or a group of chemicals is covered by a food standard then they must only be used in food in accordance with the standard. However, if a chemical is not mentioned in a standard, then this does not preclude its use in food. For a new chemical not previously used in food production, it would be necessary for the supplier to undertake a detailed risk analysis of the product to demonstrate its safety and suitability. The assessment would need to consider both the toxicological profile of the chemical and the levels of human exposure that are likely to arise from residues in food.

Surfactants, as cleaners and sanitisers, play an important role in controlling microbial and chemical contamination in food. AQIS maintains a list of chemical compounds approved for use at establishments registered to prepare goods prescribed for the purpose of the Export Control Act 1982 [11]. Whilst specifically targeted at export meat processing, the list provides a useful tool to identify cleaners and sanitisers that are considered suitable for food contact applications in Australia.

8.1.4 National Drugs and Poisons Scheduling Committee

The National Drugs and Poisons Schedule Committee (NDPSC) [12] was established under the *Therapeutic Goods Act 1989*. The Committee comprises State and Territory government members and other persons appointed by the Minister such as technical experts and representatives of various stakeholders, such as chemical industry representatives. The NDPSC is responsible for the scheduling of chemical substances which are used in consumer products. The NDPSC publishes the Standard for the Scheduling of Drugs and Poisons (SUSDP). This standard establishes labelling and packaging requirements for consumer products which contain scheduled substances. The standard also specifies any restrictions or conditions which apply. The legal effect is provided by State and Territory laws which call up the SUSDP.

As well as the name of the product and its manufacturer, a product's label also may be required, by legislation, to contain one or more of the following elements:

- Signal word(s) – to warn the consumer of a potential hazard
- Cautionary statements – concise general precautions to be observed
- Safety directions – precautions to be taken for safe use
- Warning statements – advice about specific hazards to avoid
- First aid instructions – advice on what do if poisoned
- Dangerous goods classification symbols – class labels for transport purposes
- Name and quantity, proportion or strength of its constituents
- Directions for use

The selection of the signal word is determined by the schedule into which the product falls. The list of signal words is shown in Table 8.4.

Table 8.4 Signal word statements used for chemicals in consumer products

Schedule no.	Signal word
2	PHARMACY MEDICINE
3	PHARMACIST ONLY MEDICINE
4	PRESCRIPTION ONLY MEDICINE or PRESCRIPTION ANIMAL REMEDY
5	CAUTION
6	POISON
7	POISON DANGEROUS POISON
8	CONTROLLED DRUG

Packaging requirements can include child resistant closures, the word POISON being embossed along the side of the packaging so that the container can be identified by visually impaired people or in the dark.

8.1.5 Therapeutic Goods Administration

The Therapeutic Goods Administration (TGA) [13] has the prime responsibility for the evaluation and approval of new therapeutic goods in Australia. Therapeutic goods are divided into two general areas: drugs and medical devices. Drugs are further divided into ethical prescription medicines, which require a high level of assessment, and over-the-counter (OTC) medicines (which do not require a prescription from a medical practitioner), which require a lower level of evaluation. Apart from registration, there is also a streamlined 'listing' process for low risk products. Thus, there are both listable medicines and listable device categories. The TGA has recently introduced an innovative, Web-based interface for electronic submissions of listing applications for therapeutic goods. This Web-based interface is called SIME [14].

Some biocidal surfactants, which are used as disinfectants or sterilants, come under the control of the TGA. These types of products are regulated as medical devices (Therapeutics Goods Order No, 54 (Standard for disinfectants and sterilants) (TGO54) [15].

Under TGO 54, disinfectants are categorised ranging from those used to clean critical medical devices to hospital grade disinfectants and to household/commercial grade disinfectants. This standard specifies the efficacy testing which is required to support the claims made on the product label. The standard also controls the packaging and labelling requirements for these types of products. Sterilants, instrument grade disinfectants and hospital or household/commercial grade disinfectants with therapeutic claims require registration. Hospital grade disinfectants without claims require only listing with the TGA, while household/commercial grade disinfectants, which do not make any claims on the label, are exempt from any listing or registration with the TGA. However, the efficacy, packaging and labelling required still apply to household/ commercial disinfectants.

The functions of the TGA will be transferred to a new Australia/New Zealand joint agency called the Trans-Tasman Joint Agency for Therapeutic Goods in the near future.

8.1.6 Hazardous substances

The Hazardous Substances Model Regulations have been developed by the National Occupational Health and Safety Commission[16] which is a federal agency. These regulations are mostly based on the EU Dangerous Preparations Directive 67/548/EEC (1967). The Hazardous Substances regulatory package consists of the following documents:

- National Model Regulations 1994
- Code of Practice for Labelling 1994
- Code of Practice for MSDS 1994, updated April 2003
- List of Designated Hazardous Substances 1994, updated 1999
- Approved Criteria for Classifying Hazardous Substances 1994, updated 1999, and September 2004

- Scheduled Carcinogenic Substances 1994
- Guidance Note for Assessment of Risks 1994
- Guidance Note for Retail Section 1994

In Australia, OH&S laws are developed by individual States and Territories. Therefore, each State and Territory government has adopted the Hazardous Substances National Model Regulations into its own OH&S regulations.

The Hazardous Substances regulations define a hazardous substance as one which:

- Is listed on the List of Designated Hazardous Substances (based on EU Annex I of Directive 67/548/EEC)
- Has been classified as a hazardous substance by the manufacturer or importer in accordance with the Approved Criteria for Classifying Hazardous Substances.
- Has an occupational exposure standard (different from EU)

The Approved Criteria cover toxicity end points such as acute toxicity by oral, dermal or inhalation routes, skin and eye irritation, corrosivity, skin sensitisation, genotoxicity, repeat dose toxicity, reproductive toxicity and carcinogenicity studies. One notable exception is that while the classification system is based on the EU Dangerous Preparation Directive, in Australia it does not utilise the risk phrases pertaining to physical hazards (i.e. R1-R19). Furthermore, the hazard pictograms which are required to appear on packages containing chemical substances are not used in Australia. For physical hazards, the Dangerous Goods transport labels are used (see below).

An Australian supplier of a hazardous substance must:

- Classify products
- Prepare labels
- Prepare MSDS
- Send MSDS to customers
- Review MSDS and labels at least every 5 years
- Send notification to NOHSC for a hazardous substance
- Provide the chemical identity of any ingredient in a hazardous substance when requested by a person

Similarly the regulations place the following responsibilities on employers with respect to hazardous substances:

- Identify hazardous substances on their premises
- Obtain MSDS from supplier
- Ensure products are properly labelled
- Set up a Register of Hazardous Substances
- Conduct workplace risk assessment
- Minimise risk to workers
- Provide training
- Provide atmospheric or biological monitoring if required
- Provide medical surveillance programme
- Maintain certain records for 30 years

Australia intends to adopt the Globally Harmonised System (GHS), and the NOHSC is responsible for coordinating efforts in Australia in implementing the GHS. Thus the elements of the GHS will be incorporated into the Hazardous Substances regulations progressively over the next few years with a target completion date of 2007.

8.1.7 Dangerous goods

Dangerous goods are chemical substances which present a hazard to people, property or to the environment by having certain physical properties such as being an explosive, a compressed gas (cryogenic, toxic, flammable), a flammable liquid or solid, an oxidising substance, toxic liquid or solid, a corrosive or an environmentally hazardous substance. Dangerous goods are assigned to the following classes:

- Class 1, explosives (1.1, 1.2, 1.3, 1.4, 1.5, 1.6)
- Class 2, gases
    - Class 2.1, flammable gas
    - Class 2.1, non-flammable, non-toxic gases
    - Class 2.3, toxic gas
- Class 3, flammable liquids (packing groups I, II, III)
- Class 4, flammable solids
    - Class 4.1, flammable solid (packing groups II, III)
    - Class 4.2, flammable (can undergo spontaneous combustion) (packing groups II, III)
    - Class 4.3, when in contact with water can emit flammable gases (packing groups I, II, III)
- Class 5, oxidising agents
- Class 5.1, oxidising substances which can supply oxygen to a fire (packing groups I, II, III)
    - Class 5.2, organic peroxides
- Class 6, toxic and infectious substances
    - Class 6.1, toxic (packing groups I, II, III)
    - Class 6.2, infectious
- Class 7, radioactive materials
- Class 8, corrosive substances (packing groups I, II, III)
- Class 9, miscellaneous DG (environmentally hazardous)

Substances which fall into these classes are labelled and placarded using the class labels which are shown in Figure 8.1.

Dangerous goods are regulated at the State/Territory level in Australia and each State/Territory has its own Dangerous Goods Act. A federal body, The Department of Transport & Regional Services (DOTARS) [17], has a coordinating function. As well as the Dangerous Goods Acts and accompanying Regulations, the other key document is the Australian Dangerous Goods (ADG) Code for Road and Rail (6th edition). The current edition of the ADG Code reflects the 9th edition of the UN Recommendations on the Transport of Dangerous Goods ('Orange Book') which is currently up to the 13th edition [18]. Thus, the Australian Code is lagging behind the latest version of the UN document. The 7th edition of the ADG Code is expected to be available in early 2006. As well as the UN Recommendations,

Figure 8.1 Class labels for stated dangerous goods.

Australia recognises other international transport organisation documents such as the International Maritime Dangerous Goods (IMDG) Code [19], International Air Transport Association (IATA) Dangerous Goods Regulations [20] and International Civil Aviation Organisation (ICAO) guidelines [21]. Similarly, The European Agreement concerning the International Carriage of Dangerous Goods by Road (ADR) [22], while not applicable to Australia, is used at times to assist with testing and classification criteria.

In addition to these documents, there are several Australian Standards which pertain to the storage, packaging, labelling and handling of particular classes of dangerous goods. Some examples of Dangerous Goods Standards include:

- AS 1940 The storage and handling of flammable and combustible liquids
- AS 4332 The storage and handling of gases in cylinders
- AS 4326 The storage and handling of oxidising agents
- AS 3780 The storage and handling of corrosive substances

8.1.8 Eco labelling in Australia

8.1.8.1 Standards Australia

Standards Australia [23] has established AS/NZS ISO 14021:2000, Environmental Labels and Declarations – Self-Declared Environmental Claims (type II environmental labelling). This Standard is an adoption with national modifications of ISO 14021:1999 and it specifies requirements applying to the development of environmental claims, expressed in either words or symbols, about products and services. The objective of the standard is to harmonise the various national guidelines on environmental claims used on product labels and in marketing generally, in order to facilitate trade in the global marketplace and to improve consumer confidence in environmental claims.

8.1.8.2 Australian Greenhouse Office

The Australian Greenhouse Office (AGO) [24] has a number of initiatives to facilitate practices that reduce greenhouse impact:

- 'Greenhouse Challenge Program'. Members may display the Greenhouse Challenge logo on products and corporate information to publicly market greenhouse efforts.
- 'Greenhouse Friendly'. This is a product certification programme where the company needs to demonstrate how all emissions from the production and use of the product have been offset by other greenhouse gas reduction programmes being conducted by the company. A full life cycle assessment of all greenhouse gas emissions associated with the product needs to be conducted.

8.1.8.3 Australian Competition and Consumer Commission (ACCC)

The ACCC administers the Trade Practices Act 1974 (Act). The consumer protection provisions of the Act cover advertising and selling and aim at any commercial conduct that could be misleading, deceptive or untruthful. This includes statements made on labelling and packaging. It is the responsibility of companies to ensure that consumers form an accurate impression of a product and that environmental terms such as 'green', 'environmentally safe' or 'fully recycled' should not be used to mislead consumers.

8.1.8.4 Australian Environmental Labelling Association Inc.

The Australian Environmental Labelling Association (AELA) [25] is a non-government, non-profit association. The AELA has developed an independent national environmental labelling programme for Australian consumers in conformance to the international standard, ISO 14 024 for third party environmental labelling. The intention is that the environmental product label will work with current labelling programs in Australia that address single environmental issues. It awards products which meet or exceed voluntary standards of environmental performance the Environmental Choice logo.

References

1. http://scaleplus.law.gov.au/cgi-bin/download.pl?/scale/data/pasteact/0/440
2. http://www.nicnas.gov.au
3. http://www.nicnas.gov.au/
4. http://www.cas.org/ONLINE/CD/NATCHEM/sources.html
5. http://www.nicnas.gov.au/publications/gazette/pdf/2004-specialgazette.pdf
6. http://www.apvma.gov.au/
7. http://www.apvma.gov.au/guidelines/subpage_guidelines.shtml
8. http://www.affa.gov.au/content/output.cfm?ObjectID=3E48F86-AA1A-11A1-B6300060B0AA00014
9. http://www.foodstandards.gov.au/
10. http://www.foodstandards.gov.au/foodstandardscode
11. http://www.affa.gov.au/content/publications.cfm?Category=Australian%20Quarantine%20fixand%20Inspection%20Service&ObjectID=60037EF4-ED35-4055-8081C9727693C3D
12. http://www.tga.gov.au/ndpsc/index.htm
13. http://www.tga.gov.au/
14. http://www.tgasime.health.gov.au/SIME/home.nsf
15. http://www.tga.gov.au/docs/html/tgo/tgo54.htm
16. http://www.nohsc.gov.au/
17. http://www.dotars.gov.au/transreg/str_dgoodsum.htm
18. http://www.unece.org/trans/danger/danger.htm
19. http://www.imo.org/home.asp
20. http://www.iata.org/whatwedo/dangerous_goods
21. http://www.icao.org/
22. http://www.unece.org/trans/danger/publi/adr/ADRagree_e.pdf
23. http://www.standards.com.au/catalogue/script/Search.as
24. http://www.greenhouse.gov.au/greenhousefriendly/consumers/index.html
25. http://www.aela.org.au/

8.2 Japanese Legislation Relating to the Manufacture and Use of Surfactants

Yasuyuki Hattori

8.2.1 Chemical substances control law and industrial safety and health law

The manufacturer and/or importer of a new surfactant in Japan is required to register it beforehand under the Chemical Substances Control Law, a law to regulate chemical substances, and the Industrial Safety and Health Law, a law to protect the health of workers. A chemical substance that is not registered beforehand or is not exempt from the registration under the laws cannot be handled as a chemical substance in Japan. If you use such a substance without registration, you must not only pay a penalty but also recall your products in the worst case. This legislation is applicable to all chemical substances including surfactants.

The Chemical Substances Control Law was established in 1973 in order to introduce a prior examination system of chemical substances in Japan ahead of any other countries in the world. Subsequently, the United States established the Toxic Substances Control Act (TSCA) in 1977 to introduce a prior examination system similar to that of Japan. The EEC also introduced in 1979 a notification system for new chemical substances by the Directive 79/831/EEC, the 6th amendment to the 'Directive 67/548/EEC on Dangerous Substances', and the Directive came into force in all Member States on 18 September 1981.

A major amendment was made to the Japanese Chemical Substances Control Law in April 2004 on the recommendations made by the Organization for Economic Co-operation and Development (OECD) in the Environmental Conservation Review – *Further Expansion of the Regulation Scope for Conservation of Ecosystem* in line with the Agenda 21 – 'Human Action Program for Sustainable Development' of the United Nations Conference on Environment and Development (UNCED). In order to conform to international harmonisation, the amended Chemical Substances Control Law has introduced a new examination and regulation system to prevent not only hazards to human health but also damages to animals and plants.

The Chemical Substances Control Law regulates chemical substances produced by chemical reaction excluding natural products. However, articles and substances regulated by other laws, such as the Food Sanitation Law in Japan and the Pharmaceutical Affairs Law, and exclusively used for the purposes covered by such laws, together with substances used for research and development, are also exempt from the Law.

The total number of chemical substances registered under the Chemical Substances Control Law is approximately 28 000 including approximately 20 000 substances as existing substances as of 1973 and approximately 8000 substances as new substances registered thereafter. Until the amendment, they were classified into non-regulated substances, class I specified chemical substances, class II specified chemical substances and designated chemical substances according to their hazard levels on humans caused by their environmental pollution. In the Chemical Substances Control Law amended in 2004 there have been newly established:-

- Type I monitoring chemical substances that are not biodegradable, highly accumulative and not clear in their toxicity to humans

- Type II monitoring chemical substances that are not biodegradable and not so accumulative, but suspected of being toxic to humans and, therefore, placed under government monitoring until their toxicity is identified (Substance suspected of being toxic to humans at screening level: same as the conventional Designated Chemical Substance)
- Type III monitoring chemical substances that are not biodegradable and not so accumulative, but suspected of being toxic to the ecosystem and, therefore, placed under government monitoring until their toxicity is identified (substance suspected of being toxic to the environment)

Table 8.5 shows the criteria of the chemical substances regulated by the amended Chemical Substances Control Law and the legal regulations on them. The substances newly regulated by the amended Chemical Substances Control Law are shaded in grey in the table.

Figure 8.2 is a simplified review flow chart of a new chemical substance under the amended Chemical Substances Control Law. The following is a brief explanation of the practical registration flow based on Figure 8.2.

Firstly, a chemical substance is checked to determine whether or not it is a new substance under the Chemical Substances Control Law. Whether or not the substance has already been registered can be confirmed using 'Total Search System for Chemical Substances' on the homepage of the National Institute of Technology and Evaluation (http://www.safe.nite.go.jp/english/db.html). In the Chemical Substances Control Law, CAS-registry numbers have not been assigned to all of the listed chemical substances. Therefore, a 'no hit' in searching for a substance using CAS-RN does not always mean that it is not listed. In such a case, partial match search should be run by chemical substance name.

If the substance is a new substance, it should be checked whether or not it is manufactured and/or imported in an amount of 1 tonne/year or less or whether it is assumed to be little released into the environment (e.g. intermediate, chemical substance used in the closed system). If the substance falls under either of such categories, simple notification can be made for it. In simple notification, a chemical substance is reviewed only by information such as chemical name, structural formula and manufacturing flow and, if there is no problem, manufacture and/or import of the substance is permitted. After registration, however, the actual quantities manufactured/imported and uses of the substance which is assumed to be little released into the environment must be reported to the competent authority (the Ministry of Economy, Trade and Industry). A new substance that does not fall under these categories must normally be notified.

Such a new substance is subjected to a biodegradation test before notification. If the substance is biodegradable, it is notified to the competent authority (the Ministry of Economy, Trade and Industry) with the test results and necessary documents without carrying out additional safety evaluation. If there is no problem, the substance is registered as a new chemical substance approximately 3 months after notification. There is neither a duty to report the manufactured and/or imported amount of the substance nor regulation on the use, etc. of the substance and, therefore, it can be freely manufactured, imported or used as a non-regulated substance.

If a substance has been found to be persistent in a biodegradability test, it should be tested for bioaccumulation. If it is not highly bioaccumulative and manufactured and/or imported in an amount of 10 tonnes/year or less, the substance can be notified with the biodegradability and bioaccumulation test data obtained without additional safety evaluation within the framework newly set by the recent amendment.

Table 8.5 Regulatory criteria of chemical substance under 'The amended Chemical Substances Control Law' in Japan

Classification	Criteria	Regulation
Class I Specified chemical substance	Persistent Highly bioaccumulative Has long-term toxicity to humans, or to primates	Prior permission is required for manufacture and/or import (virtually prohibited) Any use other than specified use is prohibited Import of certain products specified by the Japanese Cabinet order is prohibited Other restrictions
Class II Specified chemical substance	Persistent Not highly bioaccumulative Has long-term toxicity to humans or has toxicity to flora and fauna in the human life environment Confirmed to have the potential to cause environmental pollution	Mandatory reporting of planned and actual amounts of manufacture and/or import If deemed necessary, the government issues an order to change the planned manufacture and import amounts Government provides technical guidelines and recommendations Mandatory labelling, related recommendations from government, etc.
Type I monitoring chemical substance	Persistent Highly bioaccumulative Long-term toxicity to humans, or to primates is not clear	Mandatory reporting of actual amounts of manufacture and/or import Obedience to guidance, advice, etc. Government can direct manufacturers and importers to investigate long-term toxicity to humans and/or toxicity to primates (when necessary)
Type II monitoring chemical substance ('designated chemical substance' before revision of the law)	Persistent Not highly bioaccumulative Suspected long-term toxicity to humans (screening level)	Mandatory reporting of actual amounts of manufacture and/or import Obedience to guidance, advice, etc. Government can direct manufacturers and importers to investigate long-term toxicity to humans (when necessary)
Type III monitoring chemical substance	Persistent Not highly bioaccumulative Has ecotoxicity	Mandatory reporting of actual amounts manufactured and/or imported Obedience to guidance, advice, etc. Government can direct manufacturers and importers to investigate long-term toxicity to flora and fauna in the human-life environment (when necessary)

(newly regulated substances in grey)

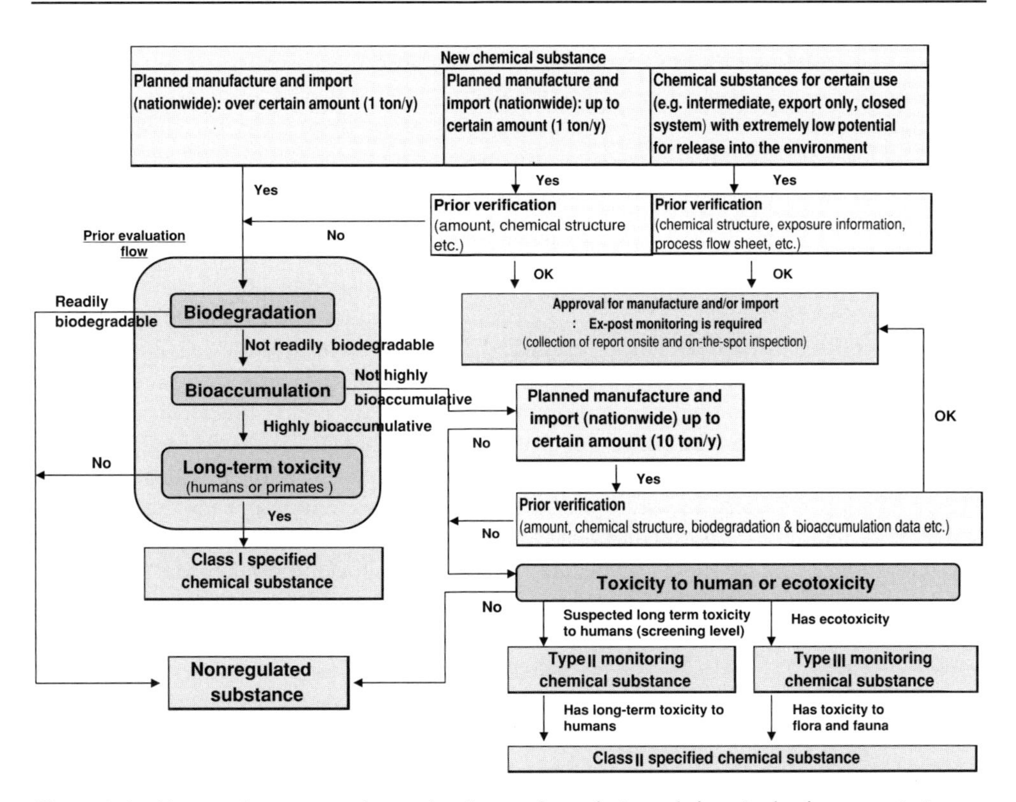

Figure 8.2 New review system for evaluation and regulation of chemical substances in Japan.

A chemical substance that is neither biodegradable nor highly bioaccumulative and intended to be manufactured and/or imported in an amount more than 10 tonnes/year should be tested for toxicity to humans and the ecosystem. If the substance is toxic to either of them, it is assigned to class II or III monitoring substance. If it is toxic to neither of them, it is listed as a non-regulated substance.

A substance that is not biodegradable and highly bioaccumulative should be evaluated for potential long-term toxicity to humans or primates. If the substance is toxic, it is registered as class I specified chemical substance requiring prior approval for manufacture and/or import (virtually prohibited from manufacture and/or import). If the substance is not toxic, it is registered as a non-regulated substance. It takes hundreds of millions of yen to carry out a long-term toxicity study in humans or primates and, therefore, a company will give up industrialisation of such a substance upon finding that it is not biodegradable and highly accumulative unless an extraordinarily large economic effect can be expected from the substance application.

A new chemical substance registered under the Chemical Substances Control Law through the above process is not publicly disclosed for its registration contents for 5 years from the registration date. The company which registered the substance, therefore, can exclusively, in effect, manufacture and/or import it for 5 years after the registration.

As stated above, a new chemical substance must also be registered prior to manufacture and/or import not only under the Chemical Substances Control Law but also under the Industrial Safety and Health Law, which is a law to protect workers. If not registered under the latter law, the chemical substance cannot be handled by workers in Japan. An examination as to novelty of a chemical substance can be done using the chemical substance database of 'the Japan Advanced Information Center of Safety and Health' of 'the Japan Industrial Safety and Health Association' (no English version is available at present) http://www.jisha.or.jp/frame/index org jaish.html.

Under the Industrial Safety and Health Law, unlike the Chemical Substances Control Law, all chemical substances including a chemical substance regulated by other laws, such as the Food Sanitation Law in Japan and the Pharmaceutical Affairs Law, and used exclusively for the purpose permitted by such laws must be notified. Notification of a substance is accepted if it shows a negative result for the test and the notification document is complete. On acceptance, the substance can be manufactured and/or imported. If the substance shows a positive result for the test, the competent authority requires that additional tests be performed, such as a chromosomal aberration test, and the provision to workers who handle the substance with the information on safety, handling methods and emergency measures for the substance described in a Material Safety Data Sheet (MSDS).

Table 8.6 shows the Japanese laws concerning chemical substances (Chemical Substances Control Law and Industrial Safety and Health Law) in comparison with the corresponding laws in the United States (TSCA) and the EU (79/831/EEC). In the present amendment to the Chemical Substances Control Law, the concept of ecotoxicity was adopted as one of the prior examination items, which finally harmonised the Japanese chemical examination system with the corresponding U.S. and European systems. The substances subject to registration are compared in the grey part of the table. There are no large differences in the concept of a registration substance (e.g. articles and substances used for R & D are exempt from all the laws because their amounts released into the environment are small) although there are some minor differences among the laws in Japan, the United States and EU.

8.2.2 Pollutant release and transfer register system

The pollutant release and transfer register (PRTR) system is one in which enterprises voluntarily calculate the amounts of specific chemical substances released into the environment that are potentially toxic to human health and/or the ecosystem, and the amounts of such substances contained in waste and transferred out of the firm's area, and report these amounts to their local governments. The national government adds up the released or transferred amounts of such a substance based on the reports from the enterprises, statistical data, etc. and publicly announces the results.

The Law Concerning Reporting, etc. of Releases to the Environment of Specific Chemical Substances and Promoting Improvements in their Management which focused on the introduction of the PRTR system was promulgated on 13 July 1999 and enforced on 6 January 2001. The law obligates enterprises to report the amounts of specific chemical substances released or transferred by them but does not obligate them to reduce the identified substances. However, anyone can obtain the data on the amounts of an identified substance released or transferred by an enterprise upon request. Therefore, enterprises handling general consumer products and environment-friendly enterprises cannot but try to limit the use of these substances resulting in lower environmental impact.

Table 8.6 Comparison of laws to control chemical substances in Japan, USA and EU

	Japan		USA	EU
	Chemical substance control law	Industrial safety and health law	TSCA	Directive 79/831/EEC (6th amendment to directive 67/548/EEC)
Enforcement date	16 Apr. 1974 1 Apr. 2004 (amended)	1 Oct. 1972	1 Jan. 1977	(18 Sep. 1979)→ 18 Sep. 1981 (30 Apr. 1992 (amended))
Purpose	Prevention of damage to humans and living organisms caused by environmental pollution from chemical substances	Prevention of damage to health of workers caused by chemical substances	Prevention of damage to humans and living organisms caused by environmental pollution from chemical substances	Prevention of damage to humans and living organisms caused by environmental pollution from chemical substances
Definition of substance	Chemical element Compound generated by chemical reaction	Chemical element Chemical compound	Chemical element Isolated radical Compound generated by chemical reaction or under natural circumstances	Chemical element Isolated radical Compound generated by chemical reaction or under natural circumstances
Number of existing substances	Approx. 28 000	Approx. 30 000	Approx. 77 000	Approx. 100 000 (EINECS)

(*Continued*)

Table 8.6 (*Continued*)

○ Registraation is required

	Japan		USA	EU
	Chemical substance control law	Industrial safety and health law	TSCA	Directive 79/831/EEC (6th amendment to directive 67/548/EEC)
Natural substance	Not applicable	○	○	○
Article	Not applicable	Not applicable	Not applicable	Not applicable
R&D use	Not applicable	Not applicable	Not applicable	Not applicable (≤100 kg only)
Medical & cosmetics use	Not applicable	○	Not applicable	○
Food use	Not applicable	○	Not applicable	Not applicable
New polymer	○ (≥1%)	○	○ (>2%)	○ (≥2% as each new monomer)
Impurity and by-product	○(≥1%)	○(≥10%)	Not applicable	○ (>10%)

Manufacturers with 21 or more employees fall under the enterprises obligated to report the release and transfer amounts. Substances subject to reporting are specified class I designated chemical substances and class I designated chemical substances. In the case of a specified class I designated chemical substance, the substance is subject to reporting if it is contained in an amount of 0.1% mass or more in a raw material used and if it is handled by an enterprise in an amount of 0.5 tonne per year or more. In the case of a class I designated chemical substance, the substance is subject to reporting if it is contained in an amount of 1.0% mass or more in a raw material used and if it is handled by an enterprise in an amount of 1 tonne per year or more. An enterprise that has failed to make the notification or has made a false notification is fined a maximum of ¥200 000.

The statistics notified to date are as follows:

- Specified class I designated chemical substances: 12 substances
  (carcinogenic substances among the class I designated chemical substances)
  The notified amount of all specified class I designated chemical substances released and transferred totaled 20 000 tonnes during the period from April 2002 to March 2003.
- Class I designated chemical substances: 354 substances
  Substances that have been proved to be toxic to humans and/or the ecosystem (including ozone-depleting substances) and have spread widely in the environment (substances manufactured and imported in an amount of 100 tonnes or more per year and detected in two or more places in the environment)
  The notified amounts of all class I designated chemical substances released and transferred totalled 290 000 tonnes and 217 000 tonnes, respectively, during the period from April 2002 to March 2003. In addition, 589 000 tonnes of such substances not subject to notification (e.g. substances released from enterprises each with less than 21 employees and from households) were released in a year according to an estimate made by the government.
  In addition, there are chemical substances designated as class II designated chemical substances under the PRTR system.
- Class II designated chemical substances: 81 substances
  Substances which are as toxic as class I designated chemical substances, but to which humans and the environment are not so much exposed as class I designated chemical substances (substance manufactured and imported in an amount of 1 tonne/year or more and detected in only one place in the environment).

The amount of a class II designated chemical substance released and transferred needs not be reported to the government. When an enterprise tries to sell or provide a product containing such a substance to another enterprise, the seller must prepare an MSDS containing information on the name, content, properties and warnings in handling of the substance and provide it to the buyer before delivery of the product (the system was enforced in January 2001).

Several of the surfactants generally used as daily commodity materials (the four listed below) fall under class I designated chemical substances in the PRTR system and, therefore, their release and transfer amounts must be reported to the government. These surfactants were designated because they are produced and imported in large amounts, have ecotoxicity (LC_{50} <10 mg/L), and were detected in two or more places in the environment (having no noteworthy toxicity to humans).

Table 8.7 Release amount (tonnes/year) from PRTR data and the comparison between actual concentration in environment and maximum allowable toxic concentration (MATC) concerning LAS, AE, DHTDMAC and AO

LAS: n-alkylbenzenesulphonic acid and its salts (alkyl C = 12–14)
AE: Polyoxyethylene alkyl ether (alkyl C = 12–15)
DHTDMAC: Bis(hydrogenated tallow) dimethylammonium chloride
AO: N,N-dimethyldodecylamine N-oxide

		Release amount (tonnes/year) from PRTR[a]						
			Government estimation					
Surfactant	Period	Notification	Minor enterprise	Other business category	Household use	Total	Surfactant concentration detected at several river points in JAPAN: maximum (average)[b]	Predicted no effect concentration (PNEC)[c]
LAS	'01.4–'02.3	47	5914	2923	24 216	33 099	81 $\mu g\ l^{-1}$ (11 $\mu g\ l^{-1}$)	250 $\mu g\ l^{-1}$
	'02.4–'03.3	41	2322	1824	16 014	20 201		
AE	'01.4–'02.3	231	1521	1632	15 301	18 684	12 $\mu g\ l^{-1}$ (1.2 μg^{-1})	110 $\mu g\ l^{-1}$
	'02.4–'03.3	227	1970	1911	17 289	21 396		
DHTDMAC	'01.4–'02.3	1	1	37	149	188	3.8 $\mu g\ l^{-1}$ (0.9 $\mu g\ l^{-1}$)	94 $\mu g\ l^{-1}$
	'02.4–'03.3	1	0	102	134	237		
AO	'01.4–'02.3	0.5	0	292	1 544	1 836	0.34 $\mu g\ l^{-1}$(0.05 $\mu g\ l^{-1}$)	26 $\mu g\ l^{-1}$
	'02.4–'03.3	0.1	0	393	1 152	1 545		

[a] The data were summarised or estimated by the Japanese government using the submitted PRTR data.
[b] The data were obtained at seven points of four urban rivers in Japan, normally in March, June, September and December from June 1998 to September 2003.
[c] Concentration at which no effect on aquatic lives is predicted.

No. 24: n-alkylbenzenesulfonic acid and its salts (alkyl C = 12–14) (hereinafter referred to as LAS)
No. 166: N, N-dimethyldodecylamine N-oxide
CAS-RN=1643-20-5 (hereinafter referred to as AO)
No. 251: Bis (hydrogenated tallow) dimethylammonium chloride
CAS-RN=61789-80-8 (hereinafter referred to as DHTDMAC)
No. 307: Polyoxyethylene alkyl ether (alkyl C = 12–15) (hereinafter referred to as AE)

Table 8.7 gives the amounts of these four surfactants released into the environment in 2002 (April 2001 to March 2002) and 2003 (April 2002 to March 2003) including the total released amount notified to the government and the released amount estimated by the government except the notified amount[1]. As is evident from the table, the amounts of surfactants released into the environment during production and handling are much smaller than those released from the general households.

The Japan Soap and Detergent Association has voluntarily implemented monitoring surveys since 1994 on the concentration of surfactants in the surface layer water of rivers in Japan in order to monitor the situation of persistence of surfactants in public waters and evaluate the effects of surfactants on aquatic organisms [2]. Table 8.7 shows the measurement results of the concentrations of four surfactants regulated by the PRTR system in the major Japanese rivers during the period from June 1998 to September 2003 (mean levels and highest levels) [2]. The highest levels in the table were all lower than the predicted no-effect concentrations (PNEC) [3, 4] and therefore it was considered that the surfactants cause almost no risks to the ecosystem in the environmental waters.

These results are not a cause for concern, but it is considered that the Association should continue the monitoring of surfactant levels in the environment to check the levels periodically.

References

1. http:/www.meti.go.jp/policy/chemical management/law/kouhyo.htm
2. http://jsda.org/3kankyo 10.htm
3. Nishiyama, N., Yamamoto, A. and Takei, T. (2004) Ecological risk assessment of surfactants. The 38th Annual Meeting of Japan Society on Water Environment (Sapporo, Japan).
4. van de Plassche, Erik J., de Bruijn, Jack H.M., Stephenson, Richard R., Marshall, Stuart J., Feijtel, Tom C. and Belanger, Scott E. (1999) Predicted no-effect concentrations and risk characterization of four surfactants: linear alkyl benzene sulfonate, alcohol ethoxylates, alcohol ethoxylated sulfates, and soap. *Environ. Toxicol. Chem.* **18**(11), 2653–663.

8.3 Relevant US Legislation

Arno Driedger

8.3.1 General

In the United States of America (US), surfactants are regulated by a number of federal agencies and by some individual states as well. Applicable regulations are usually dependent on the use of a surfactant or the efficacy claim made for a particular product that may contain a given surfactant as part of its composition. Surfactants are usually marketed to end users as part of formulated products. These products may be household or industrial cleaning solutions, cosmetics, drugs, food, agricultural adjuvants, oil field chemicals, etc. There are no regulations that specifically ban or limit the use of surfactants in consumer or industrial products.

The primary regulatory vehicle for surfactants in the US is the Toxic Substances Control Act (TSCA) which is administered by the US Environmental Protection Agency (EPA). The provisions of TSCA contain a number of exemptions because of other, existing legislations. Thus, surfactants used in drugs, foods or cosmetics are exempt from the provisions of TSCA because these products are regulated under the Food, Drug and Cosmetics Act (FDCA) which is administered by the Food and Drug Administration (FDA). Likewise, pesticides are exempted from the provisions of TSCA because they are regulated under the Federal Insecticide Fungicide and Rodenticide Act (FIFRA) which is also administered by the EPA. Thus, a cationic surfactant, such as benzalkonium chloride, could be used as a cleaning component in a household cleanser, as a preservative in a cosmetic product or as an active ingredient in a disinfectant product that would be regulated under TSCA, FDCA and FIFRA for the three said applications, respectively.

Surfactant suppliers and vendors should take great care that a surfactant intended exclusively for application in cosmetics, and therefore exempt from TSCA, does not suddenly find application in non-exempt TSCA uses because unauthorised manufacture or import of chemicals could lead to serious enforcement actions by EPA. Manufacturers or importers of surfactants also should not assume that the requirements for registration under TSCA, FDC or FIFRA are similar or harmonised between or within the regulatory agencies. The enforcement and penalty programmes are also completely different. Thus, a new surfactant intended solely for cosmetics could require only minimal resources to meet FDCA requirements, one intended for cleaning agents would require an 'intermediate' amount for registration while a new active FIFRA chemical would have 'high' registration resource demands.

8.3.2 TSCA

The US Congress enacted the TSCA in 1976 and gave the Environmental Protection Agency (EPA) broad authority to control chemicals not regulated by other statutes. The Act was prompted by the commercial introduction of new chemicals in large quantities and their subsequent release into the environment. Some such releases elicited dire health and environmental consequences as evidenced by asbestos and PCBs. Surfactants were part of the justification of the need for governmental chemical control because the introduction of branched alkylbenzene sulphonates (ABS) into laundry detergents resulted in massive foaming of effluent streams from sewage treatment out-falls because ABS did not biodegrade. While this problem was solved with the switch to biodegradable linear sulfonates, pictures of foaming rivers provided graphic evidence of the need for control of new and existing chemicals.

This review is not intended to be a comprehensive text of TSCA. Detailed information is found on the EPA Web site – epa.gov – and in the TSCA handbook by Conner et al. [1]. The regulations of the Act are codified under Title 40 of the Code of Federal Regulations, Parts 700–799.

TSCA chemicals are classified as "existing" or "new". The former consists of chemicals that are listed on the TSCA Inventory of Chemicals. These are substances manufactured, imported or processed for commercial purpose in the US between 1 January, 1975 and 1 June, 1979, and chemicals that were subsequently added to the inventory via the premanufacture notification (PMN) process. "New" chemical substances are chemicals not listed on the TSCA inventory. The obligations of chemical manufacturer, processor or importer and the ability of the EPA to regulate a chemical depend on whether a chemical is new or an existing one. Therefore, it is critical to understand the contents of the inventory, the regulatory changes that occur over time and the EPA enforcement procedures when manufacturing or importing chemicals.

In general, chemicals that are listed on the inventory are "approved" for manufacture or import into the US subject to any record keeping, reporting or testing requirements that may be in effect. "New" chemicals must be notified prior to their manufacture or import. To notify, a manufacturer must first obtain a Chemical Abstract Registry Number for the chemical and then file a PMN form with the Agency. The notification describes the chemical, intended use, method of manufacture, exposure to workers and amounts released to the environment during manufacture. A notification submitter must also include any test data in its possession or control that are related to environmental or health effects. But there is no requirement to conduct specific toxicological testing. The Agency has 90 days to review the submission; however, the review clock can be interrupted if the PMN document is deficient. The EPA then approves the PMN as written or may place restrictions on the manufacture or use of the PMN substance. The submitter can then manufacture or import the chemical subject to any restrictions and must also submit a Notice of Commencement 90 days after the first day of manufacture. It is only then that a "new" chemical is considered "listed" on the TSCA chemical inventory.

Manufacture or import of "new" chemicals is illegal and EPA has an aggressive program to prosecute violators. Fines can easily be millions of dollars because of the formula used by the Agency to calculate penalty. Chemical companies are usually most vigilant with TSCA compliance because of the high price of non-conformance.

8.3.3 FDCA

Separate bureaux within the FDA regulate surfactants used in cosmetics, foods or drugs. The specific regulations of the Act for each of the three product categories are codified in different sections of Title 21, Code of Federal Regulations. The intended product category determines how a surfactant is regulated under FDCA.

Cosmetics constitute a major commercial end use for surfactants. When used as components of products such as decorative cosmetics, skin care, oral hygiene, hair care or other such products intended to beautify the body, surfactants are regulated as cosmetic ingredients by the FDA [2]. A practical manual for the selling of cosmetics has been issued by The Cosmetics, Toiletries and Fragrances Association (CTFA) [3]. Estrin and Akerson [4] edited an extensive review on cosmetic products. However, while cosmetic products have specific regulations, cosmetic ingredients such as surfactants are not specifically regulated. There is no official list of approved ingredients, i.e. surfactants that are officially approved for use in cosmetics. Manufacturers of cosmetics are not required to register manufacturing sites or formulation composition with the FDA. However, each ingredient and finished product must be adequately substantiated for safety prior to marketing. Otherwise, the product must be labelled "warning: the safety of this product has not been determined" [5]. The determination of "adequate" is left to the discretion of the supplier, formulator or marketer of the surfactant. Thus, the burden of proof falls on the FDA to show that a product or ingredient is not safe and remove or ban it from the marketplace. This has led to a form of self-regulation within the US cosmetic industry. Consequently, the CTFA published the International Cosmetic Ingredient Dictionary and Handbook [6], a compilation of ingredients used in cosmetics. The CTFA also convened the Cosmetic Ingredient Review (CIR) composed of a panel of toxicology experts to review commonly used ingredients [7]. The CIR review monographs are published in the peer reviewed *International Journal of Toxicology.* These monographs have become the cornerstone of "adequate" safety substantiation of ingredients for the cosmetic industry. However, the CIR reviews are not legally binding and cosmetic manufacturers have a wide window of safety substantiation options. While the process of self-regulation allows for rapid entry into the cosmetic market without costly testing protocols or lengthy registration procedures, it also can be an opportunity for misuse on introducing unsafe cosmetics into the market because FDA can only act after adverse human reactions have occurred and then initiate action against an inadequately tested cosmetic product or ingredient.

Surfactants also find application in the food industry as direct or indirect food additives. "Direct" food additives are substances that are incorporated into foods for a specific intended purpose. "Indirect" food additives are chemicals that enter the food chain in an indirect manner. For example, components that leach from a container into the food are considered to be indirect additives. Since 1958, substances were required to be "generally recognised as safe" (GRAS). Surfactants are currently among thousands of food additives and have been subject to GRAS and food additive petitions to the FDA. These petitions contain all information pertaining to the safety and intended use and they are subject to FDA approval. The conditions of approval for these food additives and limitations of use are spelled out in specific regulations codified in Parts 170–199, Title 21 of the Code of Federal Regulations. The submission and approval process of GRAS and food additive positions was lengthy and could take over a decade for completion. FDA has since instituted a GRAS notification procedure [8] whereby manufacturers of a new surfactant submit a self-determined GRAS petition. The FDA then has a limited amount of time to raise an issue with the petition and the proposed use of the substance. Otherwise, it can be used after a limited time period (75 days) has expired. However, there will be no official Federal Register notice to inform industry or other interested parties that a new surfactant can be used in foods. FDA has also recently revised the procedures for pre-market notification of food contact substances (indirect food additives) [9]. These regulations define the requirements as of 2002 for food contact use of surfactants.

Sometimes the FDA will determine that a particular claim made by a chemical supplier is actually a drug claim rather than a food related claim. Claims pertaining to drug effects trigger US federal regulations that govern testing, manufacture and sale of pharmaceutical agents. The general regulations that are applied to drugs are covered under the Code of Federal Regulations Title 21, Subsection C (Parts 200–299). These regulations deal with drug active compounds and finished formulated products. Surfactants used in the manufacture of drugs are usually not "active" chemicals. Rather, they are classified as "excipients", chemicals that impart desired formulation (handling, storage, preservation, etc.) characteristics to the final dosage form of a pharmaceutical preparation. Most commonly used excipient surfactants are described in monographs published by the US Pharmacopoeia (USP), a quasi-agency organisation. These monographs have become the standard of identity for commonly used drug excipient chemicals.

8.3.4 FIFRA

The FIFRA, administered by the EPA, was enacted by Congress in 1947 and has been amended several times. The regulations are described in Title 40, Code of Federal Regulations (Parts 150–189). These regulations control pesticides which are defined as "any substance or mixture of substances intended for preventing destroying, repelling or mitigating any pest, or for use as a plant regulator, defoliant or desiccant" [10]. Pesticide formulations have two major components – "active" and "inert" ingredients. The former are chemicals that elicit the pesticidal effect that is claimed on the label of a product. The latter are those chemicals added to a pesticidal product to stabilise the formulation or to promote physical attributes such as stickiness, dispersion, odour, etc. of a formulation. The regulations deal primarily with the active components of a pesticide. The EPA is currently in the process of promulgating regulations for the registration of inert ingredients.

Surfactants are used primarily as inert ingredients of pesticides, especially in agricultural applications. These surfactants usually have TSCA as well as inert pesticidal regulatory status in the US. Some surfactants are active components in pesticides. For example, the cationic surfactants benzalkonium chloride and didecyl ammonium chloride and their various derivatives are registered as active ingredients in disinfectant products with many household and industrial applications.

FIFRA regulations control not only the safety of a chemical but also the labelling (application dosage, safety precautions, pest control, etc.). Registration of pesticides includes approval of label uses, setting of tolerances in foods and environmental and human safety. Registration of new pesticidal surfactants is very expensive and time consuming.

8.3.5 Other pertinent regulations

Manufacturers and importers of surfactants must also be aware of a number of other regulations that impact the commerce of chemicals in the US. This section will briefly describe the requirements of the Occupational Safety and Health Act (OSHA), administered by the

US Department of Labor and the Hazardous Materials Transportation Act (HMTA) administered by the US Department of Transportation (DOT).

The manufacture or import of surfactants into the US results in worker exposure to these chemicals. OSHA hazard communication requirements were promulgated to protect workers and are specified in Title 29 of the Code of Federal Regulations, Part 1900. They define the testing conditions under which a surfactant is deemed to be hazardous which then trigger the requirement for a Material Safety Data Sheet (MSDS). It must be freely available for inspection by workers. Although OSHA requires MSDS documents only for "hazardous" materials, practically all chemicals in commerce have MSDS papers prepared by chemical producers.

The transport of surfactants from manufacturer to formulated product producer is regulated under HMTA. The regulations are codified in Title 49 of the Code of Federal Regulations, Parts 171–180. Although the transport of chemicals originates at chemical plants where OSHA regulates worker exposure, the criteria for hazard classification of chemicals are not the same. There is no harmonisation between the two regulatory agencies. The OSHA health hazard definitions [11] are different from those defined under the HMTA transportation classifications [12]. Thus hazard descriptions of chemicals in a MSDS, which is an OSHA required document, are sometimes mistakenly applied to HMTA situations. This could result in a misclassification of DOT packaging groups and the erroneous placarding of chemical shipments.

The regulations described above constitute the major regulatory statutes in the US that affect surfactants. Readers should also be aware of environmental regulations for the manufacture of chemicals such as the Clean Air Act, Clean Water Act, Safe Drinking Water Act, Resource Conservation and Recovery Act, Emergency Planning and Community Right-to-Know Act and the Comprehensive Environmental Response, Compensation and Liability Act. Finished household products that contain surfactants are regulated by the Federal Hazardous Substances Act. The Fair Packaging and Labeling Act regulates advertising claims such as biodegradability of surfactants.

In general, regulations continue to be promulgated to control the import, manufacture and use of surfactant chemicals. Chemical suppliers must be vigilant in their compliance programs to avoid fines or litigation due to non-compliance as part of their product stewardship initiatives. New surfactants introduced into the market will remain free of regulatory constraints provided they have the following characteristics: biodegradable; not carcinogenic, reproductive toxins or mutagens; low mammalian and aquatic toxicity.

References

1. Conner, J.D. Jr., Ebner, L.S., Landfair, S.W., O' Connor, C.A. III, Weinstein, K.W., Boucher, M., Brophy, R.C., Brown, E.C., Neilson, M.R., Wax, D.E., Jablon, C.S. and Johnston, T.B. (1997) *TSCA Handbook*, 3rd ed., Government Institutes, Inc., Rockville, MD.
2. Food and Drug Administration (1981) 21CFR 700.3(e).
3. Beckley, C.C. and Gregory, C.H. (eds) (2001) *CTFA Labeling Manual. A Guide to Cosmetic and OTC Drug Labeling and Advertising.* 7th ed. The Cosmetic Toiletry and Fragrance Association, Inc., Washington, DC.
4. Estrin, N.F. and Akerson, J.M. (eds) (2000) *Cosmetic Regulations in a Competitive Environment.* Dekker, New York, NY.

5. Food and Drug Administration (1975) 21CFR 740.10(a).
6. Gottschalck, T.G. and McEwan, G.N. (eds) (2004) *International Cosmetic Ingredient Dictionary and Handbook.* Cosmetic Toiletry and Fragrance Association, Washinton, DC.
7. Bergfeld, W.F. and Anderson, F.A. (2000) The cosmetic ingredient review. In N.F. Estrin and J.M Akerson, (eds). *Cosmetic Regulation in a Competitive Environment.* Dekker, New York, NY.
8. Food and Drug Administration (1997) Substances generally recognized as safe; proposal rule. federal register Vol. 62, No. 74, p. 18937.
9. Food and Drug Administration (2002) 21CFR 170.100.
10. Environmental Protection Agency (2001) 40 CFR 152.3 (5).
11. Occupational Safety and Health Administration (1994) 29 CFR 1910.1200.
12. Department of Transportation (1992) 49 CFR 173.132 and 173.133.

Chapter 9
Surfactant Manufacturers

Richard J Farn

This chapter contains a list of many of the major manufacturers of surfactants throughout the world. It is neither exhaustive nor the list of surfactants which each company makes but the main series are given. The location of each company is that of its headquarters and full addresses together with those of regional offices can be found on the company's Web site.

Manufacturer	Trade name	Composition
3M Speciality Chemicals St Paul, MN, USA www.3m.com/paintsandcoatings	Fluorad FC Series	Nonionoic polymeric fluorochemicals
Air Products & Chemicals Inc. Allentown, PA, USA www.airproducts.com	Dynol 604	Ultra wetting agent
	Envirogem Series	Gemini surfactants
	Surfynol CT Series	Gemini surfactants
	Surfynol DF Series	Gemini surfactants
Akzo Nobel Surface Chemistry AB Stenungsund, Sweden www.surface.akzonobel.com	AG Series	Alkylglucosides
	Ampholac Series	Amphoterics
	Arneel Series	Nitriles
	Armeen Series	Fatty amines
	Aromox Series	Amine oxides
	Arquad & Ethoquad Series	Quaternary ammonium compounds
	Berol Series	Many different chemical types
	Duomeen Series	Fatty diamines and salts
	Ethylan Series	Alcohol ethoxylates and alkoxylates
	Lankropol K Series	Sulphosuccinates
	Perlankrol Series	Ether sulphates
	Phospholan Series	Phosphate esters
	Triameen Series	Polyamines
BASF Aktiengesellschaft Ludwigshafen, Germany www.basf.com/detergents	Degressal Series	Foam suppressors
	Emulan Series	Emulsifiers
	Lupasol Series	Polyethyleneimines

Manufacturer	Trade name	Composition
	Lutensit Series	Anionic surfactants and blends
	Lutensol A, AO, AT,ON & TO Series	Fatty alcohol ethoxylates
	Lutensol AP Series	Alkyl phenol ethoxylates
	Lutensol FA Series	Fatty amine ethoxylates
	Lutensol XP & XL Series	Guerbet alcohol ethoxylates
	Nekal BX Series	Alkylnaphthalene Sulphonates
	Plurafac LF Series	Fatty alcohol alkoxylates
	Pluriol A Series	Polyalkylene glycols
	Pluriol E Series	Polyethylene glycols
	Pluronic PE & RPE Series	EO-PO block copolymers
	Tamol Series	Naphthalene sulphonic acid condensates
Chemax Performance Products (Ruetgers Organics Corp.)	Chemal Series	Block copolymers Ethoxylated and alkoxylated alcohols
Piedmont, SC, USA www.ruetgers-organics-corp.com	Chemax Series	Defoamers, PEG and PPG Fatty acid mono- and diesters Ethoxylated castor oils
	Chemeen Series	Fatty amine ethoxylates
	Chemfac Series	Phosphate esters
	Chemid Series	Amides
	Chemquat Series	Specialty quaternaries
	Chemsulf Series	Sulphates
	Sorbax Series	Sorbitan esters and ethoxylates
Clariant Functional Chemicals Div.	Antimussol Series	Defoamers
Frankfurt, Germany www.ipc.clariant.com	Arkopal N Series	Nonyl phenol ethoxylates
	Dispersogen Series	Dispersing agents
	Dodigen Series	Quaternary ammonium compounds
	Emulsogen CLA & COL Series	Alkyl PEG ether carboxylic acids
	Emulsogen EL & HCO Series	Castor oil ethoxylates
	Emulsogen EPN Series	Alkyl polyglycol ethers
	Emulsogen TS Series	Tristyrylphenol phenol ethoxylates
	Genagen C Series	Fatty acid polyethylene glycol esters
	Genamin C Series	Fatty amine ethoxylates
	Genamin OX Series	Amine oxides

(*Continued*)

Manufacturer	Trade name	Composition
	Genapol BE Series	End capped alkyl ethoxylates
	Genapol C Series	Coconut alcohol ethoxylates
	Genapol EP Series	Alcohol EO-PO adducts
	Genapol LA Series	C12-C14 alcohol ethoxylates
	Genapol O Series	Oleyl alcohol ethoxylates
	Genapol OA & OX Series	OXO alcohol ethoxylates
	Genapol PF Series	EO-PO block copolymers
	Genapol T Series	Tallow alcohol ethoxylates
	Genapol UD Series	C11 OXO alcohol ethoxylates
	Genapol X Series	C13 OXO alcohol ethoxylates
	Hostaphat Series	Phosphoric acid mono- and diesters
	Hostapur OS Series	Olefin sulphonates
	Hostapur SAS Series	Secondary alkyl sulphonates
	Praepagen Series	Quaternary ammonium compounds
	Sapogenat T Series	Tributyl phenol ethoxylates
Cognis Deutschland GmbH Dusseldorf, Germany www.cognis.com	Arlypon F Series	Ethoxylated fatty alcohols
	Comperlan Series	Fatty acid alkylolamides
	Dehydol Series	Fatty alcohol ethoxylates
	Dehymuls Series	Sorbitan esters
	Dehypon Series	Fatty alcohol alkoxylates
	Dehyquart Series	Quaternary ammonium compounds
	Dehyton Series	Amphoteric surfactants
	Glucopon Series	Alkyl polyglucosides
	Sulphopon Series	Fatty alcohol sulphates
	Texapon Series	Fatty alcohol sulphates and ether sulphates
Cytec Industries Inc. West Paterson, NJ, USA www.cytec.com	Aerosol 18 & 22	Sulphosuccinamates
	Aerosol A, OT & AY Series	Mono and diester sulphosuccinates
	Aerosol NPES Series	Nonyl phenol ether sulphates
DeForest Enterprises Inc. Boca Raton, FL, USA www.deforest.net	DeIONIC Series	Alkyl phenol ethoxylates Alcohol ethoxylates and alkoxylates EO/PO block CoPolymers
	DeMIDE Series	Alkanolamides
	DeMOX Series	Amine oxides
	DeMULS Series	Sorbitan esters and ethoxylates
	DePEG Series	PEG and castor oil ethoxylates
	DePHOS Series	Phosphate esters
	DeSULF Series	Sulphates and ether sulphates Alkyl polyglucosides and sulphonates

Manufacturer	Trade name	Composition
	DeTAINE/DeTERIC Series	Betaines and amphoterics
	DeTHOX Series	Ethoxylated alcohols and glycerin Ethoxylated fatty acids and amines
Degussa AG Essen, Germany www.degussa.com	ABIL Series	Silicone surfactants
	REWOLAN Series	Ethoxylated lanolins
	REWOMID Series	Fatty alkylolamides
	REWOMINOX/VAROX Series	Amine oxides
	REWOPAL Series	Ethoxylated alcohols
	REWOPOL SB Series	Sulphosuccinates
	REWOPON Series	Imidazolines
	REWOQUAT/ VARI-QUAT/VARISOFT Series	Quaternary ammonium compounds
	REWOTERIC Series	Amphoteric surfactants
	TAGAT/TEGOSOFT/ TETOGEN Series	Polyethylene glycol fatty acid esters
	TEGIN Series	Glycerol fatty acid esters
	TEGO Betaine Series	Amphoteric surfactants
	TEGOPREN/ TEGOSTAB Series	Silicone surfactants
	VARONIC Series & Esters	Ethoxylated amines
Dow Corning Corp Midland, MI, USA www.dowcorning.com	DC Series	Silicone surfactants, alkoxylates Based mainly on ABA and graft polymer
	Sylgard 309	Ethoxylated trisiloxane
Huntsman Performance Products Houston, Texas, USA www.huntsman.com	Dehscofix Series	Naphthalene sulphonic acid, salts, etc.
	Eltesol Series	Aromatic Sulphonic acids and salts
	Empicol Series	Fatty alcohol sulphates, ethoxysulphates, etc.
	Empigen Series	Amphoterics, amine oxides and quats.
	Empiphos Series	Phosphate esters
	Empilan Series	Alkylolamides and ethoxylates of alcohols, alkyl phenols and amines

(*Continued*)

Manufacturer	Trade name	Composition
	Empimin Series	Sulphosuccinates, Alcohol sulphates and ether sulphates
	Nansa Series	Alkyl benzene sulphonic acids and salts Olefin sulphonates
Kao Corporation Tokyo' Japan chemical@kao.co.jp	Aminon Series	Fatty acid diethanolamides
	Amphitol Series	Amphoterics and amine oxides
	Demol Series	Arylsulphonate formaldehyde condensates
	Emal Series	Alkyl sulphates and ether sulphates
	Emanon Series	Polyethylene glycol fatty acid esters
	Emulgen Series	Polyoxyethylene alkyl ethers
	Neopelex Series	Alkyl benzene sulphonates
	Quartamin Series	Quaternary ammonium compounds
	Rheodol Series	Sorbitan and polyoxyethylene sorbitan esters
Kolb Distribution Ltd (Dr W Kolb) Hedingen, Switzerland www.kolb.ch	Hedipin CFA, PO, PS & PT Series	Fatty acid ethoxylates
	Hedipin ED Series	Polyethylene glycol esters
	Hedipin P, R & SO Series	Triglyceride ethoxylates
	Imbentin AG & POA Series	Linear fatty alcohol ethoxylates
	Imbentin C, U & T Series	Branched chain fatty alcohol ethoxylates
	Imbentin N & O Series	Alkyl phenol ethoxylates
	Imbentin PAP Series	EO/PO block copolymers
	Imbentin SG Series	Low foaming alcohol ethoxylates
	Imbentin CAM, SAM & TAM Series	Fatty amine ethoxylates
	Imbentin PEG Series	Polyethylene glycols
	Kosteran & Kotilen Series	Sorbitan esters and ethoxylates
Libra Chemicals Ltd Manchester, England www.librachem.co.uk	Libraphos Series	Phosphate and esters
	Libratex Series	Alkyl benzene sulphonic acid and salts
	Librateric Series	Amphoteric dipropionates
Nikko Chemicals Co. Ltd. Tokyo, Japan www.nikkol.co.jp	Nikkol Series:-	
	AMCA,CA & Amidoamine	Amphoterics
	BB,BC,BD,BL,BO,BS & BT	Polyoxyethylene alkyl ethers

Manufacturer	Trade name	Composition
	CMT,LMT,MMT,PMT & SMT	N-acyl taurates
	CO & HCO	Castor and hydrogenated castor oil ethoxylates
	DDP & TDP	Polyoxyethylene alkyl ether phosphates
	ECT	Alkyl ether carboxylates
	GL, GO & GS	Polyoxyethylene sorbitol fatty acid esters
	MG	Glyceryl fatty acid esters
	MYL, MYO & MYS	Polyethylene glycol fatty acid esters
	PBC & PEN	Alcohol ethoxy/propoxy ethers
	Phosten	Alkyl phosphates and ether phosphates
	RW, BWA & GBW	Lanolin/beeswax derivatives
	Sarcosinate	Sarcosinates
	SBL & NES	Alkyl ether sulphates
	SI, SL, SO, SP & SS	Sorbitan fatty acid esters
	SLS,KLS,TEALS,ALS	Alkyl sulphates
	Tetra-,Hexa- & Decaglyn	Polyglyceryl fatty acid esters
	TI,TO,TP & TS	Polyoxyethylene sorbitan fatty acid esters
	TMG	Polyoxyethylene glyceryl fatty acid esters
Omnichem s.a. Louvain-la-Neuve, Belgium www.omnichem.be	Tensiofix Series	Comprehensive series of surfactants for agricultural formulations
OMNOVA Solutions Inc. Akron,Ohio, USA www.omnova.com	PolyFox Series	Fluorinated surfactants
Pilot Chemical Company Santa Fe Springs, CA, USA www.pilotchemical.com	Calamide Series	Alkanolamides
	Calfax Series	Diphenyl oxide disulphonates
	Calfoam Series	Alcohol sulphates and ether sulphates
	Caloxylate N-9	Nonyl phenol ethoxylate
	Calsoft Series	Alkyl benzene and alpha olefin sulphonates
	Pilot SXS Series	Sodium xylene sulphonate
Raschig GmbH Ludwigshafen, germany www.raschig.de	Ralufon A,D,MDS & OH Series	Sulphobetaines
	Ralufon EA 15-90	Polyethyleneglycol-(2-ethylhexyl)-(3-sulphopropyl)-diether, K-salt

(*Continued*)

Manufacturer	Trade name	Composition
	Ralufon EN 16-80	Ethylhexyl ethoxylate
	Ralufon F Series	Polyethyleneglycol-alkyl-(3-sulphopropyl)-diether, K-salt
	Ralufon NAPE 14-90	Sulphopropylated polyalkoxylated b-naphthol, K-salt
Rhodia SA Boulogne, France www.rhodia-hpcii.com	Alkamide Series	Fatty alkanolamides
	Alkamuls Series	Ethoxylated fatty acids and oils
	Amphionic Series	Amphoteric surfactants
	Antarox Series	EO.PO block copolymers
	Igepal BE Series	Nonyl phenol ethoxylates
	Miranol Series	Amphoteric surfactants
	Mirataine Series	Amphoteric surfactants
	Rhodacal Series	Alkyl aryl sulphonic acids and salts
	Rhodamox LO	Amine oxide
	Rhodaquat Series	Quaternary ammonium compounds
	Rhodasurf Series	Fatty alcohol ethoxylates
Sasol Germany GmbH Marl, Germany www.sasol.de	Alfonic, Biodac, Slovapol, Safol,	Linear and branched alcohol ethoxylates
	Amphodac LB & Ampholyt JB	Amidopropyl betaines
	Emulgante, Lialet & Marlipal Series	
	Cosmacol AES, Marlinat 242 & Daclor Series	Alcohol ether sulphates
	Dacamid Series	Fatty acid alkylolamides
	Dacpon Series	Alcohol sulphates
	Marlon B24 Series	Terminally blocked linear alcohol ethoxylates
	Marlox & Biodac Series	Alcohol EO/PO adducts
	Marlon A Series	Alkyl benzene sulphonic acid + salts
	Marlon PS Series	Paraffin sulphonate, sodium salts
	Marlowet 45 Series	Alkyl-(aryl)-alkoxylated carboxylic acids
	Marlazin & Diammin Series	Ethoxylated amines
	Otix, NPE, Marlophen NP,	Alkyl phenol ethoxylates
	Nonfix & Slovafol Series	

Manufacturer	Trade name	Composition
	Slovacid,Coster K & Emulgante A & EL Series	Ethoxylated fatty acid esters and amides
Seppic SA Paris, France www.seppic.com	Amonyl Series	Amphoteric surfactants
	Montane Series	Sorbitan esters
	Montanox Series	Ethoxylated sorbitan esters
	Octaron Series	Nonyl phenol ether sulphates
	Oramide Series	Fatty alkylolamides and ethoxylates
	Simulsol Series	Ethoxylated fatty alcohols and acids Alkyl ether sulphates
Shell Chemicals Houston, Texas, USA www.shell.com/chemicals/neodol	Neodol1 Series	C11 fatty alcohol ethoxylates
	Neodol 91 Series	C9–C11 fatty alcohol ethoxylates
	Neodol 23 Series	C12/C13 fatty alcohol ethoxylates
	Neodol 25 Series	C12–C15 fatty alcohol ethoxylates
	Neodol 45 Series	C14/C15 fatty alcohol ethoxylates
Stepan Company Northfield, Illinois, USA www.stepan.com	Accosoft Series	Amido amine cationics
	Actosoft Series	Imidazolines
	Ammonyx Series	Quaternary ammonium compounds Amine oxides
	Amphosol Series	Betaines
	Biosoft Series	Alkyl benzene sulphonic acids and salts Fatty alcohol ethoxylates
	Drewpol Series	Polyglycerol esters
	Igepal Series	Alkyl phenol ethoxylates
	Ninol Series	Amides and ethoxylated amides
	Polystep Series	Sulphates, ether sulphates and phosphates
	Steol Series	Alcohol and alkyl phenol ethoxy sulphates
	Stepanate Series	Sodium aryl sulphonates and salts
	Stepanol Series	Fatty alcohol sulphates
	Stepanquat Series	Quaternary ammonium compounds
	Stepantex Series	Ester quat cationics
	Toximul Series	Agrochemical surfactants

(Continued)

Manufacturer	Trade name	Composition
The Dow Chemical Company Midland, MI, USA www.dow.com/surfactants	Dowfax Series	Alkyldiphenyloxide disulphonates
	Tergitol 15-S series	Secondary alcohol ethoxylates
	Tergitol L & X Series	EO/PO copolymers
	Tergitol NP Series	Nonyl phenol ethoxylates
	Triton GR Series	Dioctyl sulphosuccinates
	Triton X Series	Octylphenol ethoxylates
Tomah Products Inc. Milton, WI, USA www.tomah3.com	Tomadol Series	Alcohol ethoxylates
	Tomah Amphoteric Series	Amphoteric surfactants
	Tomah AO Series	Ether amine oxides
	Tomah DA Series	Synthetic ether diamines
	Tomah E Series	Ethoxylated ether amines and fatty amines
	Tomah PA Series	Synthetic ether amines
	Tomah Q Series	Quaternary ammonium compounds
Unger Fabrikker AS Fredrikstad, Norway www.unger.no	Emulgator F Series	Fatty alcohol ethoxylates
	Ufacid Series	Alkyl benzene sulphonic acids
	Ufanon Series	Alkylolamides and amphoterics
	Ufarol Series	Alkyl sulphates (liquids and powders)
	Ufaryl Series	Alkyl benzene sulphonates (powders)
	Ufasan Series	Alkyl benzene sulphonates (liquid/paste)
	Ungerol Series	Alkyl ether sulphates
Uniqema Gouda, The Netherlands www.uniqema.com	Arlacel Series	Sorbitan esters
	Arlasolve Series	Polyoxyethylene alcohols
	Atlox Series	Ethoxylated sorbitol esters
	Atphos Series	Phosphate esters
	Atpol Series	Ether carboxylates
	Atsurf Series	Sorbitan fatty acid esters
	Brij Series	Linear alcohol ethoxylates
	Cirrasol Series	Fatty acid or alcohol ethoxylates
	Lubrol Series	Linear alcohol ethoxylates
	Monafax Series	Phosphate esters
	Monalux Series	Amine oxides
	Monamate Series	Sulphosuccinates
	Monamid & Monamine Series	1:1 and 2:1 alkanolamides
	Monaquat Series	Quaternary ammonium compounds

Manufacturer	Trade name	Composition
	Monateric Series	Amphoteric surfactants
	Monawet Series	Sulphosuccinates
	Monazoline Series	Imidazolines
	Myrj Series	Polyoxyethylene fatty acid esters
	Phosphoteric Series	Phosphorylated amphoterics
	Promidium Series	Propoxylated amides
	Span Series	C12–C14 sorbitan esters
	Synperonic Series	Alkoxylates
	Tween Series	Ethoxylated sorbitan esters
Zohar Dalia Kibbutz Dalia, Israel www.zohardalia.com	Emulgit Series	Calcium linear alkyl benzene sulphonates
	LABS Series	Linear alkyl benzene sulphonic acids
	Lauramide Series	Fatty acid alkylolamides
	Zoharphos Series	Alkyl ether phosphates
	Zoharpol Series	Nonyl phenol ether sulphates
	Zoharpon Series	Alkyl sulphates and ether sulphates
	Zoharquat Series	Quaternary ammonium compounds
	Zoharsoft Series	Quaternary imidazoline derivatives
	Zoharteric Series	Amphoteric surfactants
	Zoramox Series	Amine oxides

Index

1:4 Dioxane, 93, 119, 120

Acyl sarcosinates, 127
  Applications, 129
  Chemistry & general properties, 127
  Structure vs. properties, 128
Adsorption at surfaces, 38
  At liquid–gas & liquid–liquid interfaces, 38, 46
  At liquid–solid interface, 39, 46, 48
  Models, 51
Aggregate structures & shapes (*see also* Micelles), 35
Alkoxylation, 133
  Batch/continuous production units, 134
  General reactions, 133
Alkyl benzene, 3, 94
  High & low, 2-phenyl content, 95
Alkylbenzene sulphonates, 93
  Linear alkyl benzene sulphonates, 93
Alkyl diphenyloxide disulphonates, 100
  Applications, 101
  Chemistry & general properties, 100
  Composition vs. performance, 101
Alkyl ether sulphates, 118
  Applications, 121
  Chemistry & general properties, 118
  Composition vs. performance, 120
  Raw materials, 120
Alkyl phenol ethoxylates, 135
  Biodegradability, 236
  Manufacture, 136
  Oestrogenic activity, 241
Alkyl phthalamates, 128
  Applications, 129
  Chemistry & general properties, 128
  Structure vs. properties, 128
Alkyl polyamine polycarboxylates, 183
Alkyl poly glucosides, 149
  Manufacture, 149
  Uses, 150
Alkyl sulphates, 113
  Applications, 117
  Chemistry & general properties, 113
  Composition vs. performance, 116
  Raw materials, 114
Alkyl sulphonates – *see* Paraffin sulphonates, 104
Alpha olefin sulphonates, 102
  Applications, 104
  Chemistry & general properties, 102
  Chlorsultone formation, 103
  Composition vs. performance, 103
  Raw materials, 103
Amine ethoxylates, 142
  Manufacture, 142
Amine oxides, 144
  Applications, 144
  Manufacture, 144
Aminopropionates & iminodiptopionates, 170
  Structures & manufacture, 171
  Properties & uses, 172
Amphoacetates, 173
  Applications, 173
  Manufacture, 174
  Structures & manufacture, 174, 175
Amphohydroxypropylsulphonates, 178
  Manufacture, 178
  Structures & manufacture, 179
Amphopropionates, 176
  Applications, 177
  Manufacture, 176
  Structures & manufacture, 177, 178
Amphoteric surfactants, 170
Anhydrohexitol esters, 147
  Manufacture & applications, 147
Anionic surfactants, 91
Arlacel series, 148
Aromox series, 145

Australian legislation, 269
  Australian Pesticides & Veterinary Medicines Authority, 274
  Chemical notification categories, 273
  Chemical regulatory agencies, 270
  Dangerous goods & symbols, 280, 281
  Eco labelling, 282
  Existing chemicals assessed under NICNAS, 275
  Food Standards Australia New Zealand, 276
  Hazardous substances, 278
  National Drugs & Poisons Scheduling Committee, 277
  National Industrial Chemicals Notification & Assessment Scheme (NICNAS), 269
  Therapeutic Goods Administration, 278

Baby shampoos, 174, 182
Betaines, 180
  Applications, 182
  Manufacture, 180
  Structures, 180, 181
  Toxicology, 181
Biocidal Products Directive, 260
  Costs, 262
  Data protection, 265
  Definitions, 260
  Impact, 265
  Product types, 262, 264
  Requirements & operation, 261
  Transitional measures, 263
Biocides, 167
Biodegradability, 135, 157, 172, 236
  Biodegradation of surfactants, 135
  Detergents Regulation, 243
  Legislation, 239
  Measurement, 238
  Primary biodegradation, 238
  Sewage treatment plants, 237
  Test procedures, 240, 241
  Ultimate biodegradation (mineralisation), 238

Carboxylates, 124
Cationic surfactants, 153
  Applications, 156
  Biodegradability, 157
  Composition & structure vs. properties, 157
  Fabric softeners, 156
  Manufacturing processes, 153
  Quaternisation, 155
Chemicals legislation – history, 250
Chlorsulphonic acid, 92
Classification & labelling of surfactants, 248
Cloud point, 65, 136, 186, 194
Critical micelle concentration, 1, 30, 33, 54, 149, 163, 191

Defoaming, 76
Detergency, 47
  Fundamental processes, 47
  Impact of phase behaviour, 66
Detergents, 18, 137
  Basic formulae, 48, 49
  Detergents Regulation, 243
  Effect of ingredients other than surfactants, 60
  Foam & foam control, 76
  Ingredients declaration, 245
  Labelling, 245
  Softergents, 161
  Use of cationics in detergents, 163
Detergents Regulation, 243
Dialkyl benzene, 96
Dowfax, 101
Drag reduction with cationics, 165
Draves wetting test, 112, 193

Eco labelling, 242, 282
Electrophoretic mobility, 53
Emulsion polymerisation, 221
  Applications, 222
Emulsions, 69
  Breakdown, 74
  Coalescence, 75
  Creaming & sedimentation, 75
  Definitions, 69
  Effect of temperature, 73
  Flocculation, 76
  Types, 70
  Use of silicone surfactants, 197
Esters of polyhydric alcohols & fatty acids, 145
Ether carboxylates, 126
  Applications, 127
  Chemistry & general properties, 126
  Composition vs. performance, 127
  Raw materials, 127
Ethomeens, 143
Ethoxylation, 133
European legislation, 236
  Biocidal Products Directive, 260
  Biodegradability, 236
  Classification & labelling, 248
  History, 250
  New chemicals strategy (REACH), 250
Eutrophication, 17

Fabric softeners, 156
  Dryer sheets, 160
Fatty alcohol ethoxylates, 136
  Raw materials, 137
  Use in detergents, 137
Fatty alkanolamides, 142
  Manufacture, 143

Mono vs. diethanolamides, 143
Ninol type, 144
Fluorad, 232
Fluorinated surfactants, 227
Applied theory & properties, 228
Environmental considerations, 231
Latest developments, 231
Structures, 232, 234
Surface tension, 229, 233
Uses, 227
Foaming, 76, 236
Foam control, 81, 199
Foam stability, 7, 77
Foaming & interfacial parameters, 78
Fuel applications of cationics, 166
Functionalised monomers, 204
Examples, applications and structures, 206, 207

Gas hydrate inhibitors, 165
Gemini surfactants, 24, 150
Glucopons, 150
Glycerol esters, 146
Manufacture & uses, 146
Glycol esters, 146
Guerbet alcohols and ethoxylates, 111, 139

Hair conditioning, 162
Hard surface cleaners, 164
Hostapur SAS, 105
Hydraulic fracturing fluids, 166
Hydrophilic groups, 2, 24,
Hydrophilic–lipophilic balance (HLB), 70
Calculation, 71
Values for sorbitan esters & polysorbate derivatives, 148
Hydrophobic groups, 3, 24
Hydroxysultaines, 183

Imidazoline-based amphoteric surfactants, 172
Structures, 173
Inisurfs, 208
Structures, 209
Interfacial tension, 28
Anionic/cationic surfactant mixtures, 163
Measurement, 31
Reduction, 28
Ionic Surfmers, 214
Structures, 215, 216
Isethionates, 21, 129
Applications, 130
Chemistry & general properties, 129
Composition vs. properties, 130
Raw materials, 129

Japanese legislation, 284
Amended Chemical Substances Control Law, 286
Chemicals released to environment, 292
Comparison of laws, Japan, USA & EU, 289, 290
Pollutant release & transfer system, 288

Labelling of surfactants, 248
Legislation, 236
Australia, 269
Europe, 236
Japan, 284
USA, 294
Linear alkyl benzene sulphonates, 93
Applications, 97
Chemistry & general properties, 94
Composition vs. performance and properties, 96
Raw materials, 94

Maleate Surfmers, 216
Performance enhancement, 219
Properties & applications, 216, 217
Structures, 218
Maxemul, 221
Micelles, 1, 33, 63, 83
Microemulsions, 195
Methyl ester ethoxylates, 140
Myrj, 140

Neodols, 138
New chemicals strategy (REACH), 250
Non-ionic surfactants, 133
Nonionic Surfmers, 219
Properties & applications, 220, 221
Nonyl phenol ethoxylates, 135
Biodegradability, 136, 242, 245
Cloud points, 136
Relevant legislation, 136
Water & PEG content, 135
Novec, 232

Oestrogenic activity, 241
Organo clays, 164
Oxo Process, 115

Paraffin sulphonates, 104
Applications, 105
Chemistry & general properties, 104
Composition vs. performance, 105
Raw materials, 105
Peaked (narrow range) ethoxylates, 138
Perfluoro octanoic acid & sulphonate, 231
Personal care, 197
Use of silicone surfactants, 197

Petroleum sulphonates, 98
  Applications, 99
  Composition vs. performance, 99
  "Green" & "Mahogany", 99
  Overbased sulphonates, 100
  Raw materials, 98
Phase behaviour, 62
  Impact on detergency, 66
Phosphate esters, 122
  Applications, 123
  Chemistry & general properties, 123
  Composition vs. performance, 123
  Mono ester vs. diester, 123
  Raw materials, 123
Phosphoamphoterics, 184
Phosphobetaines, 184
Pluronic grid, 141
Polyalkylene oxide block copolymers, 141
  Manufacture, 141
  Pluronic grid, 141
  Properties & uses, 142
PolyFox, 232
Polyglycerol esters, 146
Polymerisable surfactants, 204
  Functionalised monomers, 206
  Surfmers, 212
Polyoxyalkylene polyol esters, 148
  HLB values, 148
  Manufacture, 148
Polyoxyethylene esters of fatty acids, 139
  Properties & uses, 140
  Raw materials & manufacture, 139
Polyurethane foam manufacture, 196
Properties & surfactant choice, 3

Reactive surfactants, 204
Registration, Evaluation, Authorisation of Chemicals (REACH), 250
  Costs of testing, 258
  Impact on the surfactants industry, 257
  Principles of REACH, 251
  Requirements, 251
Reverse micelles, 37
Rheology, 82
  Bilayer phases, 86
  Definitions, 83
  Shear stress, 84
  Yield stress, 85

Secondary alcohol ethoxylates, 139
Sewage treatment plants, 237
Shell Higher Olefin Process (SHOP), 103
Silicone surfactants, 186, 205
  Applications, 196
  HLB values, 186
  Hydrolytic stability, 191
  Microemulsions, 195
  Personal care, 197
  Phase behaviour, 194
  Spreading, 192
  Structures, 187, 188, 189
  Surface activity, 191
  Synthesis, 189
  Wetting, 192
Soaps, 124
  Applications, 125
  Chemistry & general properties, 124
  Composition vs. performance, 125
  Raw materials, 124
Softergents, 161
Span, 147
Sulphates, 112
Sulphonated amphoterics, 183
Sulphonated fatty acids, 108
  Applications, 109
  Chemistry & general properties, 108
  Composition vs. performance, 109
  Raw materials, 109
Sulphonated methyl esters, 105
  Applications, 107
  Chemistry & general properties, 106
  Composition vs. performance, 107
  Raw materials, 124
Sulphonates, 92
Sulphosuccinates, 109
  Applications, 112
  Chemistry & general properties, 110
  Composition vs. performance, 111
  Raw materials, 111
Sulphur trioxide, 92
Sultaines, 183
Surface active initiators (Inisurfs), 208
  Applications, 208
  Structures & properties, 209, 210
Surface active transfer agents (Transurfs), 21
Surfactant aggregates, 32, 63
  Structures & shapes, 35
Surfactant applications, 5, 15
  Agricultural formulations, 8
  Civil engineering, 9
  Cosmetics & toiletries, 9
  Detergents, 9
  Food industry, 10
  Household products, 10
  Industrial, 21
  Leather, 11
  Metal & engineering, 11
  Miscellaneous industrial, 10
  Paints, inks, coatings, adhesives, 12, 198
  Paper & pulp, 12

Petroleum & oil, 12
Plastics, rubber & resins, 13
Textiles & fibres, 13, 199
Surface activity, 26
Surface tension, 26, 54, 229, 233
Measurement, 31
Reduction, 28
Typical values, 27
Surfactant choice, 3
Surfactant manufacturers, 300
Surfactant market, 14
Consumption by application area, 16, 17, 19, 21, 22
Household products, 14
Industrial products, 21
Personal care, 20, 21
Surfactant phases, 62
Cloud point, 65
Isotropic, hexagonal, lamellar & reverse, 64
Liquid crystalline, 65
Temperature effect, 71, 82
Surfactant structure, 24
Molecular structure, 24
Surfmers, 212
Amphoteric Surfmers, 219
Cationic Surfmers, 219
In emulsion polymerisation, 223
Maleate Surfmers, 216
Nonionic Surfmers, 219
Properties & applications, 212, 214
Structures, 213
Ionic Surfmers, 214

Taurates, 130
Applications, 131
Chemistry & general properties, 130
Composition vs. properties, 131
Raw materials, 130
Tergitols, 139
Tetralins, 95
Thickeners, 163
Transurfs, 211
Typical hydrophilic groups, 2, 24
Typical hydrophobic groups, 3, 24

USA legislation, 294
Federal Insecticide, Fungicide & Rodenticide Act (FIFRA), 297
Food, Drug & Cosmetics Act (FDCA), 295
Toxic Substances Control Act (TSCA), 294

Wetting, 54, 57
Draves test, 193
Superwetting, 193

Ziegler process, 103, 115